MODERN HUMANITIES RESEARCH ASSOCIATION
TEXTS AND DISSERTATIONS
VOLUME 89

TIME IN THE PHILOSOPHY OF GABRIEL MARCEL

MODERN HUMANITIES RESEARCH ASSOCIATION
TEXTS AND DISSERTATIONS

Established in 1970, the series promotes important work by younger scholars by making the most accomplished doctoral research available to a wider readership. Titles are selected and edited by a Board of distinguished experts from across the modern Humanities.

Editorial Board

English: Professor Catherine Maxwell, Queen Mary, University of London
French: Professor William Brooks, University of Bath
Germanic: Professor Ritchie Robertson, University of Oxford
Hispanic: Professor Derek Flitter, University of Exeter
Italian: Professor Brian Richardson, University of Leeds
Portuguese: Professor Thomas Earle, University of Oxford
Slavonic: Professor David Gillespie, University of Bath

Managing Editor: Dr Graham Nelson

Time in the Philosophy of Gabriel Marcel

by
Helen Tattam

Modern Humanities Research Association
2013

Published by

The Modern Humanities Research Association
1 Carlton House Terrace
London SW1Y 5AF
United Kingdom

© *Modern Humanities Research Association 2013*

Helen Tattam has asserted her right under the Copyright, Designs and Patents Act 1988 to be identified as the author of this work. Parts of this work may be reproduced as permitted under legal provisions for fair dealing (or fair use) for the purposes of research, private study, criticism, or review, or when a relevant collective licensing agreement is in place. All other reproduction requires the written permission of the copyright holder who may be contacted at rights@mhra.org.uk.

Copy-Editor: Richard Correll

First published 2013

ISBN 978-1-907322-83-9 (hardback)
ISBN 978-1-907322-84-6 (paperback)
ISSN (MHRA Texts and Dissertations) 0957–0322

CONTENTS

	Acknowledgements	vii
	Chronology of Significant Events and Works	viii
	Abbreviations of Frequently Cited Works	xi
	Introduction: An Unplaced French Philosopher	1
PART I: TIME AND SUBJECTIVITY		
1	Being and Time	14
2	Phenomenological Time	47
	Conclusion: Metaphysics and Presence	79
PART II: TIME AND THE PROBLEM OF HERMENEUTICS		
3	Narrative Time	86
4	Marcel's Theatre: An-Other Time	119
	Conclusion: Between Ricœur and Lévinas	154
PART III: TIME AND ETERNITY		
5	Time and God	160
	General Conclusion: Toward what Metaphysics?	193
	Bibliography	197
	Index	212

ACKNOWLEDGEMENTS

This book emerged from doctoral research that would not have been possible without financial support from the Arts and Humanities Research Council and the Graduate School at the University of Nottingham. For this I am very grateful. I must also thank the Modern Humanities Research Association, with special thanks going to William Brooks for his editorial guidance and critical engagement, Richard Correll for his care and attention to detail, and Graham Nelson for his professionalism and support. I am grateful to Camille de Villeneuve and Eric Severson for inspiring and encouraging me; and additional thanks must go to my former colleagues and research associates in the French Department at the University of Nottingham, who supported me continually and offered valuable advice. I am especially appreciative of the discussions I had with John Marks and James Helgeson, who helped me to think creatively and see the broader implications of my ideas; and I must equally thank Mike Kelly from the University of Southampton, whose insightful comments have helped me to focus and clarify my arguments. Above all, I am indebted to William Grainger and Rosemary Chapman — to William, my husband, for his constancy and understanding; and to Rosemary, my primary supervisor, for thinking both with and against me and for her exemplary rigour, expert guidance, and unparalleled support.

Finally, I must thank the editors of *Romance Studies* and *American Catholic Philosophical Quarterly* for permission to reprint some of the material in Chapter 3, as reworked versions of some of the chapter's content have previously been published as the following journal articles: 'Storytelling as Philosophy: The Case of Gabriel Marcel', *Romance Studies*, 28 (2010), 223-34; 'Philosophers' Stories: Gabriel Marcel and Narrative', *American Catholic Philosophical Quarterly*, 84 (2010), 711-27.

H. T., September 2012

CHRONOLOGY OF SIGNIFICANT EVENTS AND WORKS

1889	Gabriel Marcel is born on 7 December in Paris, to Henry Marcel and Laure Marcel (née Meyer).
1893	Death of Marcel's mother (15 November).
1898	Marcel's father marries his mother's sister, Marguerite, and becomes French Minister to Sweden. The family moves to Stockholm.
1899	Marcel's family returns to Paris, where his father becomes *Directeur des Beaux-Arts*.
1908–09	Marcel studies at the Sorbonne and attends Bergson's lectures at the *Collège de France*.
1910	Marcel completes his *agrégation de philosophie*.
1911–12	Marcel teaches in Vendôme.
1914	Marcel begins writing his *Journal métaphysique*.
1914–18	Marcel works for the French Red Cross during World War I.
1915–18	Marcel teaches at Lycée Condorcet, Paris.
1916–17	Marcel has metapsychical experiences.
1919	Marcel marries Jacqueline Boegner.
1919–22	Marcel teaches in Sens.
1921	Marcel works for the *Nouvelle Revue française* and meets Charles Du Bos.
1922	Marcel and his wife adopt their son, Jean. Marcel is the theatre critic for *L'Europe nouvelle* and a member of the review committee at both Plon and Grasset. He also writes for *Sept*, *Temps présent*, and *La Vie intellectuelle*.
1926	Marcel founds the 'Feux Croisés' series at Plon. Also, his father dies this year (6 March).
1927	Publication of *Journal métaphysique*.
1929	Marcel converts to Catholicism and is baptized on 23 March, with François Mauriac as his sponsor.
1933	Publication of the lecture 'Position et approches concrètes du mystère ontologique' (also published in *L'Homme probématique*; see below) in the same volume as Marcel's play *Le Monde cassé* (also published in *Cinq pièces majeures*; see below).
1935	Publication of *Être et avoir* (2 vols).
1936	Marcel hosts informal philosophical gatherings ('les vendredis chez

Gabriel Marcel') for his students and fellow philosophers (21 rue de Tournon, Paris).

1939–40 Marcel teaches at Louis-le-Grand, Paris.
1940 Publication of *Du refus à l'invocation*. Marcel's step-mother also dies this year (1 January).
1940–41 Marcel teaches in Montpellier.
1944 Publication of *Homo viator: prolégomènes à une métaphysique de l'espérance*.
1945–65 Marcel is the theatre critic for *Nouvelles littéraires*.
1946 Marcel meets Heidegger in Freiburg.
1947 Death of Marcel's wife. Marcel writes his autobiographical essay 'Regard en arrière'.
1948 Marcel directs a UNESCO conference in Beirut.
1949 Marcel receives the *Grand Prix littéraire* from the Académie française.
1949–50 Marcel delivers the Gifford Lectures at the University of Aberdeen, Scotland.
1951 The Gifford Lectures are published under the title *Le Mystère de l'être* (2 vols). *Les Hommes contre l'Humain* is also published, and Marcel travels to North Africa and South America.
1952 Marcel is elected to the Institute of France (Académie des sciences morales et politiques).
1954 Publication of *L'Homme problématique: position et approches concrètes du mystère ontologique*.
1956 Marcel is awarded the Goethe Prize of the City of Hamburg.
1956–66 Marcel makes a number of visits to Canada, the United States of America, and Japan.
1958 Marcel is awarded the *Grand Prix national des lettres*.
1959 Publication of *Présence et immortalité: journal métaphysique (1938–1943) et autres textes*.
1961 Marcel delivers the William James Lectures at Harvard University, USA.
1962 Publication of *Fragments philosophiques: 1909–1914*.
1963 Marcel is awarded the Osiris Prize.
1964 Marcel is presented with the Peace Prize of the German Book Trade in Frankfurt. The William James Lectures are also published this year, under the title *La Dignité humaine et ses assises existentielles*.
1967 Marcel presents his conference paper 'Mon temps et moi (Temps et valeur)'.
1968 Publication of *Pour une sagesse tragique et son au-delà*.
1969 Marcel is awarded the *Prix Érasme*.

1971 Publication of *En chemin, vers quel éveil?* (autobiographical work).
1972 The *Dignité de Grand Croix de l'Ordre national du mérite* is conferred on Marcel.
1973 Death of Marcel (Paris, 8 October), and publication of *Cinq pièces majeures*.

Only frequently cited works have been listed above. For a complete chronological list of Marcel's works, see the Bibliography.

ABBREVIATIONS OF FREQUENTLY CITED WORKS

Works by Marcel

CPM	Cinq pièces majeures
DH	La Dignité humaine et ses assises existentielles
EA	Être et avoir
EC	En chemin, vers quel éveil?
FP	Fragments philosophiques: 1909–1914
HH	Les Hommes contre l'Humain
HP	L'Homme problématique: position et approches concrètes du mystère ontologique
HV	Homo viator: prolégomènes à une métaphysique de l'espérance
JM	Journal métaphysique
ME I	Le Mystère de l'être I
ME II	Le Mystère de l'être II
MT	'Mon temps et moi (Temps et valeur)'
PI	Présence et immortalité: journal métaphysique (1938–1943) et autres textes
PST	Pour une sagesse tragique et son au-delà
RA	'Regard en arrière'
RI	Du refus à l'invocation

Other Works

AE	Lévinas, Autrement qu'être: ou au-delà de l'essence
DMT	Lévinas, Dieu, la mort et le temps
DVI	Lévinas, De Dieu qui vient à l'idée
EAGM	Belay and others (eds), Entretiens autour de Gabriel Marcel
EM	Parain-Vial (ed.), L'Esthétique musicale de Gabriel Marcel
EN	Lévinas, Entre nous: essais sur le penser-à-l'autre
EPR	Ricœur, Entretiens Paul Ricœur Gabriel Marcel
GM	Bouëssée (ed.), Gabriel Marcel interrogé par Pierre Boutang
HS	Lévinas, Hors sujet
Œ	Bergson, Œuvres
SA	Ricœur, Soi-même comme un autre
TA	Lévinas, Le Temps et l'Autre
TI	Lévinas, Totalité et infini: essai sur l'extériorité
TR I	Ricœur, Temps et récit I: l'intrigue et le récit historique
TR III	Ricœur, Temps et récit III: le temps raconté

Full references for the above can be found in the Bibliography.

INTRODUCTION

An Unplaced French Philosopher

> Le temps philosophique est [...] un temps grandiose de coexistence, qui n'exclut pas l'avant et l'après, mais les *superpose* dans un ordre stratigraphique. C'est un devenir infini de la philosophie, qui recoupe mais ne se confond pas avec son histoire. La vie des philosophes [...] obéit à des lois de succession ordinaire; mais leurs noms propres coexistent et brillent [...]. La philosophie est devenir, non pas histoire; elle est coexistence de plans, non pas succession de systèmes.
> DELEUZE and GUATTARI 1991: 58–59

The aim of this book is to shed light on the philosophy of a twentieth-century French thinker who is notoriously difficult to situate, namely Gabriel Marcel (1889–1973). Marcel stands somewhat outside the canon of twentieth-century French philosophers. As J. J. Benefield rightly notes: 'Among the names of thinkers who are commonly listed as belonging to the French school of existentialism[,] that of Gabriel Marcel is often relegated to one side and promptly overlooked[,] if not forgotten' (1973: 6). Furthermore, virtually every scholar who has attempted to engage with Marcel's thought has commented on how difficult his work is to follow or to categorize. Brendan Sweetman, for instance, speaks of Marcel's 'inchoate and often scattered thoughts' (2002: 269); Kenneth Gallagher describes Marcel as a 'relentlessly unsystematic thinker' (1962: ix); Benefield describes Marcel's methodology as simply 'unmethodological' (1973: 12); and Seymour Cain observes how Marcel 'does not fit the usual picture of the philosopher — we associate no university chair, no philosophical system, nor ponderous treatise with his name' (1963: 12). This raises the question as to how one responds to an *œuvre* that fits no identifiable mould, an apparently inconsistent or even incoherent *œuvre* which fails to be recognizable as 'philosophy' in any standard sense. One possible approach, I will argue, requires a willingness to critique the standards we take for granted as philosophical convention. As a result, this book will not simply be a reflection on Marcel's theory, but also a broader reflection on the relationship between philosophy and systems, on balancing the desire to think rigorously with a desire to enquire openly, without imposing conditions on what might constitute 'truth' from the start.

For this reason, I have striven not to work from any preconceived understanding of Marcel's project, nor indeed an overly rigid notion of what 'philosophy' is, in itself. I therefore do not follow in the footsteps of other critics and interpret Marcel's writings in relation to the wider context of existentialism.[1] That is not to suggest that it is never justifiable or helpful to do so. On the contrary, it is no accident that those who do 'place' Marcel in the French intellectual tradition receive his work as a form of 'Christian existentialism'.[2] Such a label has been particularly encouraged by Sartre, who, in his 1945 lecture 'L'Existentialisme est-il un humanisme?', divided existentialism into two varieties — the Christian and the atheistic — and aligned Marcel with the former, drawing comparisons between his work and that of Jaspers.[3] Similarities can certainly be found between Marcel and Jaspers, most notably with respect to their conception of authentic subjectivity, and in particular the importance both place on faith and its grounding in concrete situations, and on a form of intersubjective communication that transcends the subject–object distinction. However, common themes do not automatically define common projects (indeed, Marcel came to feel that his conception of transcendence was fundamentally different from that of Jaspers, since he did not want to point beyond experience to something ineffable and inaccessible),[4] and hence to interpret Marcel from the starting point of 'Christian existentialism', or even simply 'existentialism', tends only to beg the questions one wishes to answer. Critics may explain their difficulties otherwise, as for example Philip Stratton-Lake does, by claiming that it is Marcel's 'fragmented and evocative style of writing [which] makes it difficult to get a clear picture of his [...] brand of existentialism' (1998: 345). Nevertheless, the real barrier to elucidating Marcel's project, I believe, has been the way in which commentators have framed his philosophy from the start. Assuming Marcel's philosophy to be akin to that of Sartre and his contemporaries, I contend, has prevented us from considering his writings on their own terms.

At this stage — and just to contextualize Marcel and his philosophical situation a little more — it is worth noting that Marcel rejected the existentialist label.[5] The label is also retrospective (for Marcel especially, since he was born nearly sixteen years before Sartre, in December 1889), and, being extremely difficult to define itself, can pick out only a 'family resemblance' that is of little help in pinpointing the individuality of his project. Marcel actually introduced the term, employing it in 1943 when making reference to Sartre's developing ideas.[6] It then appeared in the autumn of 1945, in the title of Sartre's lecture; and as David Cooper informs us:

the label was soon to be stuck on many other writers. To begin with, it was attached to the two German philosophers of *Existenz*, Martin Heidegger and Karl Jaspers, whose influence upon Sartre had been considerable. [...] Then the label was fixed [...] on a number of Sartre's French contemporaries and friends, notably Albert Camus and Maurice Merleau-Ponty, eventually returning like a boomerang upon the neologizing Marcel. (1991: 1)

Philosophers such as Heidegger refused the label immediately, not wishing their ideas to be considered 'equivalent' to those outlined in Sartre's lecture. Marcel tolerated the association for a while, but he too came to renounce the description. Even Sartre came to look back on his lecture with contempt, and, disliking what he himself had encouraged the existentialist label to represent, disowned the title.[7] So if, to begin with, there was any uncertainty as to what 'existentialism' meant, the term becomes all the more vague once the main philosopher to have appropriated the label abandons it. At best the categorization is superficial; but more often than not it is misleading, and as Gallagher notes with respect to Marcel in particular: '[Marcel] does not derive from the line of descent to which so many of the "existentialist" thinkers owe their origin, the line that is vaguely drawn from Kierkegaard and Nietzsche; the influence of these thinkers on his formation was next to nil' (1962: ix–x). We need, therefore, to consider a different interpretative approach.

It must also be noted that Marcel's allegedly 'Christian' or 'Catholic' philosophy has rather unreligious origins.[8] His earliest writings date from 1909 (his *mémoire de diplôme d'études supérieures*, on the metaphysical ideas of Coleridge and their relation to the philosophy of Schelling).[9] His conversion to Catholicism, on the other hand, was not until 23 March 1929, when Marcel was thirty-nine — and even then, Marcel did not become a resolute defender of the Church; quite the opposite, he is often very critical of its dogmatism. In addition, the motivation for his conversion is highly unclear.[10] A wave of French writers and intellectuals were converting to Catholicism at the time (others included Claudel, Maritain, and Péguy), rendering this religious movement an equally literary and artistic phenomenon. As Frédéric Gugelot observes: 'La conversion [à cette époque est souvent] un événement partagé, l'expression sociale d'amitiés littéraires'; 'de l'aveu même des convertis, des livres, des lieux, des hommes ont influencé leur cheminement spirituel' (1998: 361). Indeed, for Marcel the conversion of his close friend Charles Du Bos was a huge influence: 'son exemple suffisait à me persuader' (*EC*: 139), he writes in his autobiographical work *En chemin, vers quel éveil?* (1971); and in 1970 he even confessed, in an interview with Pierre Boutang, that 's'il n'y avait pas eu près de moi la présence encourageante de Du Bos, je ne serais peut-être pas devenu catholique parce que, jusque là, au fond, le catholicisme m'avait très peu attiré' (*GM*: 20).[11] Gugelot also remarks how Catholic conversion was a way of responding to '[l']anxiété du moment qu'est l'angoisse de la décadence', observing that, 'au moment où l'anticléricalisme et la sécularisation de la société semblent l'emporter[, ...] ces écrivains et ces artistes apparaissent comme une des voies de l'irruption du sacré dans un temps et un monde désenchantés' — as is testified to by 'l'accent moral de la plupart des conversions' (1998: 361). Thus, while Marcel may indeed fit into a certain Christian, or more specifically Catholic, context, it does not follow that straightforward implications concerning his position can be drawn from this. Indeed, both the notion of 'le sacré' and the notion of 'le moral' are problematized in this book, for their significance emerges to be much more complex than any tidy, contextually based interpretation would suggest. If Marcel's philosophy is in any way religious, what that actually means remains to be determined. To avoid making unfounded assumptions from the outset, therefore, I have postponed the question of religion to the final chapter. There, in light of preceding discussions, I then explore the extent to which it is helpful to understand Marcel's philosophy as 'Christian'.

Marcel had many other allegiances, but these only continue to manifest his 'unplaceability'. He may be associated with Mounier's personalism for example — both generally, because Marcel viewed the human person as the starting point for all philosophical analysis,[12] and with specific respect to the *Esprit* movement, since Marcel played a key role in the foundation of Mounier's journal (1932) by helping to persuade Desclée de Brouwer to offer administrative services, inexpensive printing, and funding for *Esprit*.[13] Marcel was also an acquaintance of Maritain (their philosophical relations extended from 1928 to the end of World War II, approximately)[14] and dabbled with his Thomism (c. 1929–1933), before rejecting it as too rationalistic, too far removed from existence and its experience;[15] and Marcel had some involvement with the Oxford Group, or the Moral Re-Armament (*Réarmement moral*) as it became known in 1938. Led by Frank Buchman, this international Christian movement reacted against the social focus of mainstream churches, by emphasizing the importance of individual spirituality. Marcel, who first encountered the group in 1934,[16] appreciated its nondenominational focus and later attempts to appeal to all faiths, and sympathized with its argument that the crisis facing the world (and arguably Europe in particular) in the late 1930s was, in the first instance, moral. As a testament to this, in 1958 he edited and introduced a volume of articles dedicated to the movement (*Un changement d'espérance: à la rencontre du Réarmement moral*), as well as attending some of its events.[17]

The above facts undoubtedly provide useful contextualization for the reader, and all begin to paint a picture of Marcel's intellectual interests, formative experiences, and concerns. However it is difficult — and moreover, misleading — to define Marcel's position in relation to any of these contexts in particular, for all these associations were relatively fleeting. Instead, what emerges is the fluidity of his stance, and the general independence of his position. Thus, as

Neil Gillman commented in 1980, what is needed (indeed, still needed) is 'a serious attempt to take Marcel's thought on its own terms, uncover its primary impulses and internal stresses, and evaluate it according to its coherence and faithfulness to its own stated purposes' (1980: x). This is what I aim to do.

The way to achieve this, I will argue, is not to attempt a general exposition of his philosophy's key ideas and themes.[18] This has been done on numerous occasions — most comprehensively in Roger Troisfontaines's *De l'existence à l'être* (1953), which draws on material in Marcel's unpublished manuscripts as well as all the works he had published to date, but also (for example) by Gallagher (1962), Cain (1963), Jeanne Parain-Vial (1966; 1989), and the collections of essays compiled by Paul Schilpp and Lewis Hahn (1984), and by William Cooney (1989). The analyses in these works are too neat, however. None really confronts the complexities and contradictions in Marcel's work, as I intend to do. Although other commentators may see Marcel's unsystematic mode of presentation as a barrier to understanding or taking seriously his thought,[19] I suggest that such a fragmented style of philosophizing is precisely what can enable us to grasp his distinct contribution, despite its apparent incoherence. Furthermore, I contend, in order truly to engage with Marcel's ideas it is necessary to bring him into contact with a range of critical partners, so that he can be (re-)considered in the broader context of twentieth-century French thought. I will not be tracing the development of his work chronologically, therefore. Rather, focusing my questioning on specific problematics relating to the theme of time, I seek to engage with his work in a philosophical manner and to consider his ideas in conjunction with others'. After all, thinking does not develop in a vacuum. As Deleuze and Guattari write (and as I quote at the beginning of this introduction): 'La philosophie est devenir, non pas histoire; elle est coexistence de plans, non pas succession de systèmes' (1991: 59).

To approach Marcel's philosophy in this way is not to present a systematic reading of his work as a 'whole', which ignores chronological developments and shifts by treating all aspects with (atemporal) indifference. There is more than one way to read a philosophical *œuvre* atemporally, I would contend. I do not claim to present a definitive reading of Marcel's philosophy, which rigidifies it for all time. On the contrary, I do pay attention to the chronology of his thought; but I do not wish for this to dominate the structure of my own thinking about Marcel's work because this would also render the book's reading undesirably rigid, by forcing its analysis into a 'cause and effect' mould. My approach, therefore, will be to present multiple readings, to propose a series of departures or explorations that offer insights into the motivations that drove various shifts in philosophical approach during the twentieth century, in response to a general dissatisfaction with Western philosophy's traditional conception of metaphysics.[20] The book thus concentrates less on finding a 'place' for Marcel's philosophy and more on exploring its time — that is, both Marcel's understanding of time, as a philosophical notion, and his understanding of the philosophical time at which he was writing and to which he was responding. In this sense, the book is about ideas above all else: it is a philosophical discussion about the nature of philosophy, which starts with Marcel's reflections concerning its task and status and thinks both with and against these, introducing other philosophers as critical partners where it is illuminating to do so — both with respect to our understanding of Marcel and the wider twentieth-century context.

Although the theme of time is not explicitly discussed by Marcel at length, nor in general by other commentators of his work,[21] it is a particularly effective focus for this book because Marcel's presentation of time lends itself to a range of different but — crucially — equally possible interpretations: (i) ontological; (ii) phenomenological; (iii) narrative; (iv) Lévinassian; (v) Augustinian. In one sense, this simply manifests inconsistency; but this is not what I wish to emphasize. Rather, I want to draw attention to how such indeterminacy allows Marcel's writings to be (productively) read with a range of other philosophers — a 'community' in which Marcel's work finds meaning. Each chapter will therefore draw on a number of sources, from different periods, and read these with and against a number of other thinkers, the most significant being (in order of appearance): Bergson (1859–1941), Ricœur (1913–2005), and Lévinas (1906–1995).

It is worth saying something about the diversity of Marcel's philosophical *œuvre* at this stage, since it comprises an unusual range of genres. Rather than writing formal treatises, for instance, Marcel preferred to philosophize in a diary. As Étienne Gilson notes: 'L'œuvre philosophique publié de Gabriel Marcel commence par son *Journal métaphysique* et l'on peut dire que tous ses écrits postérieurs ne sont, en un certain sens, que d'autres fragments du même journal' (1947: 1). Cain (1963: 14) and James Collins (1968: 129) also remark upon the diary form of Marcel's work, but neither they, nor Gilson, comment in any detail on its further significance. I, on the other hand, will argue that Marcel's metaphysical journal is more than a 'distinctive literary form', as Collins suggests. It is the key to discovering what is distinctive to his philosophy as a whole. I must concede, here, that Marcel's diary entries are not typical, in that they are not primarily accounts of day-to-day events; they are much more cerebral. When I make reference to the 'diary form' of his work, therefore, what I wish to draw attention to is the fluid and fragmented structure of diary writing that Marcel embraces, as well as the possibility of making reference to everyday situations, which the diary context helps Marcel to legitimize. As will be seen, both of these aspects — a less rigid mode of presentation and a more intimate connection to the concrete and everyday — are crucial to Marcel's philosophical project.

Marcel's different forms of philosophizing extend further, however, and I will thus equally be engaging with essays, lectures, journal articles, conference papers, and theatrical works, as well as a number of interviews — with Ricœur (conducted *c.* 1967), Boutang (conducted 1970), Marianne Monestier (also conducted 1970), and Pierre Lhoste (conducted 1973).[22] Two major questions are raised as a result of this varied corpus. The first, more general question concerns how one separates Marcel's philosophical voice from his political or personal voice, or from his voice as a literary critic. As will be seen, Marcel has a tendency to gloss over the distinctions between these different modes, making it difficult to delineate his philosophical position. The second question is more specific, and relates to the relationship between Marcel's philosophy and his theatre. For many years, Marcel considered his theatrical creations to be a separate enterprise from his philosophical writing. He began writing dramatic works years before he even began to study philosophy (this he started as a *lycéen*, in 1906),[23] with his first (documented) play dating from 1902.[24] Thus, as Marcel tells Lhoste in 1973, his plays were 'tout à fait détachées d'un programme [philosophique]' (Lhoste 1999: I, interview 4).[25] Later on, however ('aux alentours de 1930' (*RA*: 297), he estimates in his 1947 autobiographical essay 'Regard en arrière'), Marcel identifies a *rapprochement* between his theatrical and philosophical writings,[26] and henceforth comments increasingly on the intimate relation between his theatre and philosophy,[27] stressing how his philosophy was often an articulation of themes he had explored earlier in his plays and using dramatic scenes as illustrations of his theoretical ideas. This trend begins when Marcel publishes one of his plays (*Le Monde cassé*) alongside the landmark lecture he gives in 1933 ('Position et approches concrètes du mystère ontologique'), and arguably culminates in the William James Lectures he delivers at Harvard University in 1961, in which he announces 'l'intime solidarité entre [le versant philosophique et le versant théâtral] de mon œuvre' (*DH*: 8). What, however, does it mean for Marcel to decide that his dramatic writing is not such a separate enterprise from his theory, and what implications does this have for interpreting his philosophical position? I will engage with these questions in some detail in Chapter 4.[28]

As will become evident, considerable structural importance is given to Derrida's critique of the 'metaphysics of presence' in this book, since this challenge is precisely what was at stake for philosophy at the time Marcel was writing — and thus what was at stake for Marcel himself, because he was seeking a new approach to philosophy. Fiercely opposed to deductive, analytic approaches to investigating the nature of human existence, Marcel sought a style of philosophizing that was less impersonal and dogmatic, and more in tune with life. Although he was initially drawn to idealist works by philosophers such as Schelling (1775–1854),[29] Bradley (1846–1924),[30] Royce

(1855–1916),[31] and Hocking (1873–1966),[32] he became critical of their approach and tried to dissociate his philosophy from their totalizing methods. In 'Regard en arrière' he writes:

> je me suis insurgé [...] contre la façon dont un certain idéalisme majore la part de la construction dans la perception sensible, au point de paraître juger insignifiant [...] tout le détail concret et imprévisible qui ne constitue pas seulement la parure de l'expérience, mais lui confère sa saveur de réalité. (*RA*: 308)

Marcel recollects the First World War as the major turning point for his philosophical position — specifically his experiences when working for the French Red Cross. Responsible for relaying information about missing soldiers to their family members, Marcel became closely involved in the complex interpersonal and circumstantial networks of a great many individuals — something that contrasted sharply with the military documentation with which he also dealt. As was only too evident when confronted with these soldiers' grief-stricken relatives, the people he was concerned with were more than just names with ranks and reference numbers. They were whole existences, intimately connected to the lives of other existences, the reality of which no (abstract) military file could ever begin to convey. Thus, as Marcel explains in the introduction to his first major work, *Journal métaphysique* (1927), he rejected his original philosophical intention to write something rigorous and dogmatic, vowing instead to 'rendre à l'existence cette priorité métaphysique dont l'idéalisme a prétendu la priver' (*JM*: xi).[33]

As a consequence, Marcel's understanding of 'existence' changes in his philosophy. Whereas the first part of the *Journal*, written in the months prior to the outbreak of war in 1914, defined it objectively in relation to the physical and the spatial,[34] the introduction to the work as a whole (which also includes a second part composed between 1915 and 1923, which aims to take account of Marcel's revised approach) re-presents his philosophy, insisting that 'l'existence ne peut être à proprement parler ni posée, ni conçue, ni même peut-être connue, mais seulement reconnue à la façon d'un terrain qu'on explore' (*JM*: xi). It is 'un mystère', as Marcel will later describe it, not simply an objectifiable 'problème' that may be philosophized about at a distance, in isolation from our own existential situation; and in keeping with the above emphasis on exploration, the foreword to *Présence et immortalité* (1959) therefore declares Marcel's philosophy to be 'un voyage de découvertes' (*PI*: 10), where his philosophical explorations (as opposed to explanations) represent 'un cheminement parfois hasardeux comportant des tâtonnements, des arrêts, des reprises, des remises en question' (*PI*: 10). Philosophy itself thus becomes an experience,[35] and as a result, the form of Marcel's philosophy also becomes significant — to the extent, I will argue, that it cannot be dissociated from the content of his arguments.

More specifically, what proves of interest is Marcel's first-person narrative[36] style — which shaped his philosophy from the very beginning, for, as mentioned above, his theory was largely developed through the writing of a personal diary.[37] Although Marcel did not initially intend his original 'metaphysical journal' to be made public, he later decided to present it as a philosophical work in itself.[38] He then continued to philosophize in a diary until 1943, its entries forming the first volume of Marcel's second work, *Être et avoir* (1935), and the main content of *Présence et immortalité*. This narrative style and use of 'je' are also a feature in Marcel's essays and lectures, which are similarly self-conscious, meandering, and strewn with autobiographical anecdote. Whereas many thinkers tend to put the 'messiness' of their own subjectivity in parentheses when writing their philosophy, the complexities with which Marcel struggles remain on the page. As a result, not only is (temporal) existence explored in the content of Marcel's philosophy; the time of Marcel's own lived experience is equally visible in the first-person form of his philosophy.

Whether or not Marcel successfully reconceives philosophy and escapes idealist totalization is another question — and a question that is especially prompted by the importance he accords to eternity and a separate, allegedly more authentic notion of 'être', over and above his discussions of time and *existence*. Does this not compromise his philosophy? This is the question Chapter 1 will ask, and the rest of the book will continue to debate. The book is divided into three parts, and because it experiments with a number of different interpretative approaches conclusions will be drawn at the end of all three, with the final conclusion reflecting more broadly on Marcel's philosophy, in light of preceding discussions. Chapter 1 provides an introduction to Marcel's philosophy, presenting his initial preoccupations and the direction in which they developed in relation to Bergson, who had a profound influence on the foundation of Marcel's philosophical project — especially with regard to time. Chapter 2 then examines, in more detail, Marcel's position as he attempted to move away from Bergson, in order to determine the specific preoccupations underlying his different metaphysical discourse on time — a discourse that starts to appear problematic, for its indeterminate status becomes all the more apparent and the importance of (a more authentic notion of?) eternity only seems to be reinforced. Part II (Chapters 3 and 4) then considers the implications that the varying forms of Marcel's philosophy have for an understanding of time in his work, examining his thought in relation to Ricœur and Lévinas. In addition to Marcel's first-person style, this requires attention to be paid to the other significant mode in which Marcel's philosophy is expressed: his theatre. While this has not been neglected to the same extent as the narrative form of Marcel's writings, the unity of his theory and theatre is taken for granted, to the exclusion of each genre's individual style. This book's study, however, enables a reconsideration of the relation between the two, for as will be seen, the presentation of time in his plays is rather different from the discourse one finds in his philosophy. Part III (Chapter 5), finally, will reassess the relation between Marcel's philosophy and religion. It is at this point that I ask whether Marcel's discussions of time and eternity need to be understood in the context of his Christian belief.

Notes to the Introduction

1. Blackham (1951), Wahl (1954), Mounier (1962), Collins (1968), Macquarrie (1972), Lescoe (1974), and Cooper (1991) all interpret Marcel in specific relation to this context.
2. Gilson's 1947 collection of essays on Marcel, for example, is entitled *Existentialisme chrétien: Gabriel Marcel*.
3. Sartre (1996: 26).
4. Marcel comments on the specifics of Jaspers's philosophy in *RI*: 284–326.
5. *EAGM*: 10; *EC*: 228–31; *EPR*: 73–75; *GM*: 72; Marcel (1967b: 9; 1969: 254–55); *ME I*: 5.
6. Benefield (1973: 11); Cooper (1991: 1); Daigle (2006: 5).
7. '[L]'existentialisme, je ne sais pas ce que c'est', Beauvoir (1963: 50) quotes Sartre as saying in the summer of 1945.
8. *EPR*: 77; *RA*: 300.
9. Published in 1971 under the title *Coleridge et Schelling*.
10. See for example Fouilloux (1989).
11. See also *GM*: 27.
12. See Sayre (1997).
13. Hellman (1981: 39). See also Hellman (1981: 45), and Loubet-del-Bayle (1969: 131, 135).
14. Bars (1991: 231).
15. Useful discussions of Marcel, Maritain, and Thomism can be found in Alexander (1948: 201–19), Bars (1991: 243), Benefield (1973), Devaux (1974: 91–92), Hughes (2002: chap. 3, part IV), and Sweetman (2008: chap. 8).
16. Sacquin (1989: 326).
17. For general discussion of the Oxford Group/*Réarmement moral*, see *EC*: 163–66, Marcel (1969: 261), and Monestier (1999: I, interview 2).
18. This is in fact extremely difficult in Marcel's case, the difficulty arising — as Benefield explains — 'from the diffusive range of his interests and the interrelation of his major themes. His philosophy may be likened to an intricate tapestry of ideas and insights in which his central themes are so intimately interwoven that to extricate one thread would draw out several others with it' (1973: 11–12).
19. Cf. also Cooper, who treats existentialism as 'a relatively systematic philosophy', and therefore has 'rather little to say about those, like Camus [or Marcel, one might add], who make a virtue out of being neither a philosopher nor systematic' (1999: 9).
20. 'Metaphysics' is difficult to define, but on a broad level it can be said to be concerned with the nature of reality, seeking, more specifically, to explain aspects that are not immediately discoverable (e.g. the nature of time, space, identity, or the mind). It is therefore one step removed from empirical analysis (meta-physics), investigating how the world needs to be in order for our (direct) experience of the world to be the way it is. What, more precisely, it means to do metaphysics is what the history of philosophy debates. 'Ontology' (the study of Being) has traditionally been treated as a synonym of 'metaphysics' — that is, it has often been assumed that the study of Being just is the

study of reality. However, this is contested by various philosophers discussed in this book, raising (again) the question as to what metaphysics entails.
21. Alexander (1948), Parain-Vial (1976; 1985b; 1989: 145–57), and Vigorito (1984) are exceptions, and it will be important to engage with their analyses.
22. Interviews with Marcel are listed in the 'Other Works' section of the Bibliography, under their respective editors.
23. *EC*: 60–61.
24. *La Duchesse de Modène*, held at the Harry Ransom Humanities Research Center, University of Texas at Austin.
25. See also *EC*: 85, *EPR*: 53, and Marcel (1914: 2).
26. In addition to *RA*: 297, see Marcel (1969: 257).
27. E.g. *PI*: 11, 17; Marcel (1952; 1966: 78–79).
28. There is not sufficient space in Chapter 4 to analyse Marcel's theatrical practice in itself; the chapter engages with Marcel's plays only insofar as philosophical status is attributed to them. Tattam (2007) discusses Marcel's drama in more detail, however, exploring ethical discourses within a selection of Marcel's plays and analysing these in relation to a range of plays by Camus and Sartre. Particular attention is given to the ways in which each thinker adapts and experiments with the dramatic genre to address ethical issues. The interplay of philosophy and drama in their respective *œuvres* is then examined, in order to consider the influence that theatre has on each philosopher's ethical voice.
29. Marcel felt an affinity with the post-Kantian notion 'intellectual intuition' (see for example *FP*: 66–67), embraced by Schelling and Hegel in defence of the metaphysics Kant had declared impossible (Beiser 2002: chap. 8). The style and spirit of Schelling's philosophy were especially inspiring for Marcel: 'pour celui qui regarde la philosophie comme une aventure héroïque comportant des risques et côtoyant des abîmes, [Schelling] demeurera toujours un compagnon exaltant, et même [...] un inspirateur' (1957: 87).
30. 'La conviction que le réel ne saurait précisément se laisser appréhender sous les espèces d'une somme, qu'à vrai dire il n'est aucunement "sommable", est sans doute une de celles qui se sont imposées à moi le plus tôt, avant tout à la lecture de Bradley' (*RA*: 316).
31. Along with Schelling, Royce was one of the first philosophers that Marcel worked on in depth, writing four articles on him for the *Revue de métaphysique et de morale* between 1918 and 1919, republished in 1945 as *La Métaphysique de Royce*. In the foreword to the American edition of this work (1956; reprinted in Marcel 2005: 241–44), Marcel states: 'Royce's philosophy — and this is its great value — marks a kind of transition between absolute idealism and existentialist thought' (2005: 244). He also declares that 'Royce helped to orient my own thought toward what I will call [...] the discovery of the *"thou"*' (2005: 241); and Royce's thoughts on loyalty helped to inspire Marcel's ethics, paving the way for Marcel's philosophy of fidelity (2005: 244; *DH*: 96–97).
32. Marcel's *Journal métaphysique* was dedicated to Hocking as well as Bergson. In his foreword to the American edition of *La Métaphysique de Royce*, Marcel writes: 'Hocking's book [*The Meaning of God in Human Experience* (1912)] was an advance on Royce's thought, an advance in the direction of that metaphysical realism toward which I resolutely tended' (2005: 242; see also 223–24, 238).
33. For more detail concerning the impact of World War I on Marcel's thought, see *EC*: 89–96; *EPR*: 12–21; *GM*: 14–17; *RA*: 311–12.
34. '[R]ien ne peut être dit exister que ce qui peut entrer en relations de contact, en relations spatiales avec mon corps. Ce sera là, si on veut, une définition' (*JM*: 26; 20 January 1914).

35. '[La philosophie] est d'une certaine manière une expérience, je dirais presque une aventure à l'intérieur d'une aventure beaucoup plus vaste, qui est celle de la pensée humaine dans son ensemble' (*PST*: 28).
36. By 'narrative' I understand the semiotic, mediated mode of enunciation where a storyteller, or 'narrator' (explicit or implicit), uses the form of a story to both represent (spatio-temporally; in a possible world), and meaningfully relate, a sequence of events to an audience.
37. The significance of the narrative form of Marcel's philosophy has been overlooked. Admittedly, unlike other philosophers of existence such as Sartre, Beauvoir, or Camus, who all explored their ideas through the writing of novels as well as theory, Marcel failed to find a narrative voice in fictional prose (*EAGM*: 105; *EC*: 11, 16, 127), abandoning his only attempted novel, *L'Invocation à la nuit* (1921; the unfinished manuscript is held at the Harry Ransom Humanities Research Center in Austin, Texas), in favour of theatre. Narrative thus appears extraneous to Marcel's work. However, to accept it as irrelevant is to succumb to the common assumption that philosophy is somehow 'above' narrative, thought to concern itself directly with universal truths. Indeed, narratologists have, in general, neglected the role of narrative in philosophical texts, even though narrative theory's move away from structuralism has resulted in a shift toward more applied, multidisciplinary concerns (Rée (1987; 1991) and Trainor (1988) are exceptions).
38. *EC*: 130.

PART I

Time and Subjectivity

CHAPTER 1

~

Being and Time

It is in reaction to Bergson, perhaps more than any other philosopher, that Marcel came to establish his own philosophical position: 'sans l'aventure bergsonienne et l'admirable courage dont elle témoigne, il est probable que je n'aurais jamais eu ni la vaillance, ni même simplement le pouvoir de m'engager dans ma propre recherche' (*EM*: 79), he confesses in the 1952 article 'Méditation sur la musique'. Marcel's encounter with Bergson's thought can therefore be understood as the catalyst behind his entire philosophical project.[1] '[Bergson] a joué pour moi un rôle de libérateur même s'il n'est pas extrêmement facile de dire ce que je lui dois. J'ai eu l'impression tout de même, qu'en un certain sens, je lui dois l'essentiel' (*GM*: 29), he declared in a 1970 interview with Boutang. If Marcel cannot pinpoint what, precisely, his philosophy owes to Bergson, this is likely to be because he went on to develop his own independent position. Bergson was influential with respect to the early formation and individuation of his thought, this chapter argues; and this, I will show, can be seen to hinge particularly on the question of time. The chapter is divided into three sections. The first describes Marcel's encounter with Bergson and details the similar ways in which both approach human reality — especially their insistence on the importance of recognizing temporality for an authentic understanding of Being. The second then observes the rather different metaphysics underlying their apparently homologous descriptions of the human immediate, exposing remarkably dissimilar philosophies of time. Finally, the third section relates time in Marcel and Bergson to their conceptions of eternity, and uses a comparison of the differing dialectics that emerge to reflect on the specificity of the position Marcel established against Bergson, as well as its philosophical consistency.

Marcel's Encounter with Bergson

In 1908 and 1909, while completing his *diplôme d'études supérieures* at the Sorbonne, Marcel attended Bergson's lectures at the *Collège de France*;[2] and as Marcel readily admits, these were an inspiration to him.[3] In conversation with Ricœur (*c.* 1967) he recounts:

> J'eus le bonheur de suivre [les cours de Bergson] pendant deux ans au Collège de France et je n'y penserai jamais sans émotion; chaque fois qu'on s'y rendait, c'était en quelque sorte le cœur battant et comme avec l'espoir d'entendre une révélation. [...] oui, on avait bien le sentiment que Bergson était en train de découvrir quelque chose. (*EPR*: 15)

Particularly influential for Marcel was Bergson's mode of philosophizing, which prompted him to reorient his philosophical investigations in favour of concrete reality: 'c'est à Bergson que je dois de m'avoir libéré d'un esprit d'abstraction dont je devais beaucoup plus tard dénoncer les méfaits' (*EC*: 81), he acknowledges in *En chemin, vers quel éveil?* (1971). Indeed, Bergson launched a critique against philosophy's lack of precision,[4] which, he contended, derived precisely from its attempts to offer universality of scope, for this rendered philosophy indifferent to the detail and movement of reality. In the introduction to *La Pensée et le Mouvant* (1934) he writes:

> Les systèmes philosophiques ne sont pas taillés à la mesure de la réalité où nous vivons. Ils sont trop larges pour elle. Examinez tel d'entre eux, convenablement choisi: vous verrez qu'il s'appliquerait aussi bien à un monde où il n'y aurait pas de plantes ni d'animaux, rien que des hommes; où les hommes se passeraient de boire et de manger; où ils ne dormiraient, ne rêveraient ni ne divagueraient; où ils naîtraient décrépits pour finir nourrissons; où l'énergie remonterait la pente de la dégradation; où tout irait à rebours et se tiendrait à l'envers. C'est qu'un vrai système est un ensemble de conceptions si abstraites, et par conséquent si vastes, qu'on y ferait tenir tout le possible, et même l'impossible, à côté du réel. (*Œ*: 1253)

For Bergson, then, precision is relative to subject matter.[5] Abstract thinking may be methodologically appropriate for the mathematical sciences, but if philosophy is to have any connection to life, it must root itself in concrete experience in order to avoid missing its subject-matter entirely.[6] Indeed, as Bergson notes in his 'Introduction à la métaphysique' (1903), conceptual analysis is, by definition, incapable of conveying individuality: 'l'analyse est l'opération qui ramène l'objet à des éléments déjà connus, c'est-à-dire communs à cet objet et à d'autres. Analyser consiste donc à exprimer en fonction de ce qui n'est pas elle' (*Œ*: 1395–96). 'Plus de grand système qui embrasse tout le possible, et parfois aussi l'impossible!', Bergson therefore declaims in 1934; 'contentons-nous du réel' (*Œ*: 1307).

Bergson and (subsequently) Marcel thus turned to the human experience of the immediate as the starting point for their philosophical reflection: 'je suis enclin à considérer avec [...] méfiance une pensée philosophique qui oserait se formuler en-deçà d'une expérience authentiquement vécue' (*DH*: 28), declared Marcel in the first of the William James Lectures he delivered at Harvard in 1961 ('Points de départ'). In so doing, Bergson and Marcel contested a more dominant philosophical tradition, which, because of its desire to be universally applicable, assumed that individual personal experience could only bias the thought process, undermining the legitimacy of conclusions intended to be of relevance to anyone other than the author. Bergson voiced such a challenge by positioning himself against Eleaticism,[7] which rejected the epistemological validity of sense experience, instead taking logical coherence and necessity to be the (only) criteria of truth.[8] Marcel, on the other hand, opposed the rationalist certainty boasted by Descartes (1596-1650) — particularly the *cogito*, on which this assurance was initially founded. 'La réalité que le cogito révèle', he contends in the 1925 article 'Existence et objectivité',[9] 'est d'un ordre tout différent de l'existence dont nous tentons ici non point tant d'*établir* que de *reconnaître*, de constater métaphysiquement la priorité absolue' (*JM*: 315).

More specifically, Bergson's and Marcel's opposition to abstraction entailed a renunciation of both idealism and materialism, and Bergson was influential with respect to Marcel's repudiation of both. Although initially drawn to idealism, Marcel came to understand it as banishing the self to regions of ineffable transcendence, so that it rendered individual subjectivity insignificant. As his experiences of the First World War caused him to realize, his concern was with the person and his or her individuality, the 'moi concret' (*RA*: 295). It was Bergson who helped confirm to Marcel the inadequacy of an idealist approach. Marcel thus credits him, in particular, when speaking of 'les abus qui peuvent être faits de l'idée de totalité' in a 1950 Gifford Lecture at the University of Aberdeen: 'Comme l'a vu Bergson avec une admirable clarté, il ne peut y avoir totalisation que de ce qui est homogène; mais n'importe quoi ne peut pas être traité comme une unité susceptible d'être ajoutée à d'autres unités' (*ME II*: 51-52).

A materialist approach, however, was considered equally excessive.[10] It too, Bergson argued, sought to equate perception with absolute knowledge, for while idealism subordinated the domain of science to the (epistemologically superior) realm of consciousness, materialism attempted the inverse, insisting that facts of consciousness could only be derived from science.[11] This reductive approach was something to which Marcel had been averse ever since his school days. He detested the narrow, prescriptive education system that had dominated his formative years, encouraging all truths to be thought of in terms of cold, impersonal sets of facts which were to be committed to memory and regurgitated on demand: 'rien, avec le recul dont je dispose, ne me paraît moins justifiable que l'espèce d'encyclopédisme de pacotille qui commandait alors les programmes [lycéens]' (*EC*: 63), he writes in *En chemin*.[12] Bergson's engagement with meta-psychical aspects of existence — that is, with concrete experiences that surpass conceptualization or rational explanation — thus impressed Marcel considérably. '[Bergson] a été le seul parmi les penseurs français à reconnaître l'importance des faits métapsychiques', he writes in 'Regard en arrière' (1947);

'la pseudo-idée du "tout naturel" a contribué non seulement à décolorer notre univers, mais encore à le décentrer, à le vider des principes qui peuvent seuls lui conférer sa vie et sa signification. [...] l'investigation métapsychique nous aide à remonter cette pente fatale' (*RA*: 313). Hence, Bergson also helped to consolidate Marcel's philosophical reaction against scientific reduction; and Marcel's own investigations into the 'métapsychique' (winter 1916–17), for him, only testified further to a reality beyond the empirically demonstrable and expressible.[13]

Bergson and Marcel, then, were in search of a third, intermediate approach to philosophizing about the human, between these extremes of materialism and idealism: 'Nous soutenons contre le matérialisme que la perception dépasse infiniment l'état cérébral; mais nous avons essayé d'établir contre l'idéalisme que la matière déborde de tous côtés la représentation que nous avons d'elle' (*Œ*: 318), states Bergson in *Matière et mémoire* (1896). The key to this 'third way', for both, lay in taking first-person consciousness seriously, as a deeper, foundational level of reality. Similar to their German contemporary Husserl (1859–1938), whose new discipline of 'phenomenology'[14] was announcing, in Dermot Moran's words, 'a bold, radically new way of doing philosophy, an attempt to bring philosophy back from abstract metaphysical speculation [...] in order to come into contact with [...] concrete living experience' (2000: xiii), Bergson and Marcel (independently) emphasized the importance of subjectivity, arguing that it was not only a legitimate ground for philosophical reflection, but the only basis on which philosophy could investigate the nature of human Being. 'Au départ de cette investigation, il nous faudra placer un indubitable, non pas logique ou rationnel, mais existentiel; si l'existence n'est pas à l'origine, elle ne sera nulle part' (*RI*: 25), affirms Marcel in his 1940 essay 'L'Être incarné'.[15] Not only did they advocate a distinction between (third-person) objectivity and metaphysics — an argument which may be attributed to all the European 'philosophies of existence';[16] Bergson and Marcel also underlined the specific need to recognize the temporality of human existence (or 'la durée', in Bergsonian terms) for a genuine philosophy of Being (ontology).[17] As Marcel maintained in his lecture 'La Responsabilité du philosophe dans le monde actuel' (published 1968), the responsibility of a philosopher is to be engaged, which means situating his or her philosophizing in the world.[18] Philosophy is relevant to life, and must be shown to be such. This cannot be done in an abstract atemporal realm: 'Je doute qu'il y ait un sens à s'interroger sur la responsabilité du philosophe *urbi et orbi*, j'entends par là dans une perspective intemporelle ou détemporalisée. Une analyse [...] ne peut s'exercer que dans la durée, plus exactement dans un contexte temporel' (*PST*: 49).

Although Husserl and Heidegger (1889–1976) also emphasized the importance of time with respect to understanding human experience, Bergson's approach, which Marcel embraced, can be seen as marking a decisive moment within the

French philosophical tradition, for, unlike later French thinkers such as Sartre or Merleau-Ponty (1908–1961), neither was especially influenced by contemporary German philosophy. 'C'est avec Bergson et G. Marcel que s'inaugure en France une nouvelle manière de philosopher', Sumiyo Tsukada affirms (1995: 19).[19] Nevertheless, many parallels can be drawn between Husserlian time and time in Bergson and Marcel, some of which are outlined below. Human time for Heidegger found its true meaning through a confrontation with mortality, so that authentic existence was characterized in terms of Being-toward-death (*Sein zum Tode*).[20] This contrasts sharply with Bergson's and Marcel's position, where human temporality is conceived in relation to the here and now, the immediacy of which, this section suggests, they describe in very similar terms:[21] 'Qu'est-ce que philosopher concrètement? [...] c'est philosopher *hic et nunc*' (*RI*: 85), insists Marcel in 'Appartenance et disponibilité' (1940); and Bergson similarly equates the experience (or 'intuition') of *la durée* with consciousness of the immediate when he writes: 'Intuition signifie donc d'abord conscience, mais conscience immédiate, vision qui se distingue à peine de l'objet vu, connaissance qui est contact et même coïncidence' (Œ: 1273).

If (the here and now of) human reality is temporal and dynamic, as Bergson and Marcel contend, the fact that it is possible to conceive of things in static atemporal terms nevertheless requires explanation. Paralleling Husserl's 'bracketing' of the 'natural standpoint', so as to come into contact with 'conscious experiences *in the concrete fullness and entirety* with which they figure in their concrete context' (1931: 116),[22] Bergson therefore distinguishes between real, lived time (*la durée*; this can be equated with Husserl's *immanente Zeit*)[23] and 'le temps homogène' (the ordinary, objective conception of time which relates primarily to an awareness of succession, analogous to Husserl's *Raum-Zeit*).[24] For Bergson, this second form of time is not really time at all. Rather, it is a confused, 'spatialized' conception of time[25] that freezes the motion of reality so that fixed observations can be made.[26] In his doctoral thesis and first major philosophical work, *Essai sur les données immédiates de la conscience* (1889), he affirms this radical distinction between time and space proper using the example of following the movement of a clock's hands and pendulum:

> Quand je suis des yeux, sur le cadran d'une horloge, le mouvement de l'aiguille qui correspond aux oscillations du pendule, je ne mesure pas de la durée, comme on paraît le croire; je me borne à compter des simultanéités, ce qui est bien différent. En dehors du moi, dans l'espace, il n'y a jamais qu'une position unique de l'aiguille et du pendule, car des positions passées il ne reste rien. Au dedans de moi, un processus d'organisation ou de pénétration mutuelle des faits de conscience se poursuit, qui constitue la durée vraie. C'est parce que je dure de cette manière que je me représente ce que j'appelle les oscillations passées du pendule, en même temps que je perçois l'oscillation actuelle. (Œ: 72)

Whereas time is change and movement, then, space is static simultaneity. There is never more than one spatial arrangement of things at any one time; if anything is re-arranged, the previous arrangement is no more. As a consequence, no link exists between these separate states of affairs without an observer to make such a connection. Conceiving the movement of the hands and the pendulum depends on my temporal duration, for it is only my consciousness of a multiplicity of states, which have been strung together in my memory over time, that creates any relation between these spatial arrangements. Thus, for Bergson — and as will be seen, for Marcel — lived human experience is not a one-dimensional succession of discrete events; its many facets are all lived at once, intermingled to the extent of indissociability in the complex flux of *la durée*.[27] Our everyday conception of time, however, treats time as a homogenous space in which events in our life are located, encouraging us to spectate our life, to view it one frame at a time.[28] Bergson and Marcel wanted to correct this understanding of temporality, and demonstrate that there was another more authentic way of conceiving of ourselves.

Bergson deliberately avoids visual or spatial metaphors in his descriptions of *la durée*, so as not to encourage its structure to be understood in terms of graspable, atomic parts that can be conceived of in isolation from one another and at a distance.[29] Indeed, language, with its fixed objectifying concepts, can be misleading, and Bergson often warns against becoming 'dupe du langage' (Œ: 109). In the 1888 foreword to his *Essai* he writes:

> Nous nous exprimons nécessairement par des mots, et nous pensons le plus souvent dans l'espace. En d'autres termes, le langage exige que nous établissions entre nos idées les mêmes distinctions nettes et précises, la même discontinuité qu'entre les objets matériels. Cette assimilation est utile dans la vie pratique, et nécessaire dans la plupart des sciences. Mais on pourrait se demander si les difficultés insurmontables que certains problèmes philosophiques soulèvent ne viendraient pas de ce qu'on s'obstine à juxtaposer dans l'espace les phénomènes qui n'occupent point d'espace. (Œ: 3)

Of course, as Bergson recognizes in his 1911 lecture 'L'Intuition philosophique', 'il n'y aurait pas place pour deux manières de connaître si l'expérience ne se présentait à nous sous deux aspects différents [...]. Dans les deux cas, expérience signifie conscience; mais, dans le premier, la conscience s'épanouit au dehors, et s'extériorise par rapport à elle-même [...]; dans le second elle rentre en elle, se ressaisit et s'approfondit' (Œ: 1361). So both modes of understanding are just as real experientially — and as mentioned above, engagement with our surroundings on a practical level actually requires us to conceive of things in detached, fixed terms. If we are to come to any deeper, metaphysical understanding of existence, however, the expressive limitations of language need to be recognized. It is not sufficient to base conclusions on this spectatorly, instrumental form of understanding, for this is a spatialized and consequently atemporal construction[30] that ignores the richer underlying reality of *la durée*. In his *Essai*, therefore, Bergson creates a distinction between 'le moi superficiel' and 'le moi intérieur' or 'moi profond':

> notre moi touche au monde extérieur par sa surface; nos sensations successives, bien que se fondant les unes dans les autres, retiennent quelque chose de l'extériorité réciproque qui en caractérise objectivement les causes; et c'est pourquoi notre vie psychologique superficielle se déroule dans un milieu homogène sans que ce mode de représentation nous coûte un grand effort. Mais le caractère symbolique de cette représentation devient de plus en plus frappant à mesure que nous pénétrons davantage dans les profondeurs de la conscience: le moi intérieur, celui qui sent et se passionne, celui qui délibère et se décide, est une force dont les états et modifications se pénètrent intimement, et subissent une altération profonde dès qu'on les sépare les uns des autres pour les dérouler dans l'espace. (Œ: 83)

Both are experiences of one and the same self: 'ce moi plus profond ne fait qu'une seule et même personne avec le moi superficiel' (Œ: 83); but the *moi profond* is more ontologically real,[31] and it is in defence of this reality that Bergson criticizes spectatorship.

Marcel launches a similar critique — in fact, William Cooney believes that 'Marcel's warning and reaction against optical or spectacular thinking forms the vehicle through which his philosophy can best be understood' (1989: iii).[32] Although Marcel does not, for the most part, present his opposition to spectatorship in relation to time, as I will illustrate, he was not unaware of its relevance and does, on a number of occasions, apply his thoughts to time more explicitly. First, however, it is useful to outline Marcel's objection to spectatorship in general, so that his reflections on time can be appreciated in relation to his philosophy as a whole.

For Marcel, human existence is not an object I can behold from the outside; 'je ne puis concevoir l'existence d'un [...] observatoire' (RI: 8–9), he insists in the introduction to *Du refus à l'invocation* (1940). This distinction between existence and objectivity, which Marcel described in the lecture 'L'Homme devant son avenir' (published 1968) as 'le point de départ de tous mes écrits' (PST: 220), is first introduced in the 1925 article 'Existence et objectivité';[33] and here, supporting Kant's (1724–1804) denial that existence is a predicate, Marcel argues that existence is an irreducible 'immédiat pur' (JM: 319), rather than a simple definable property about which logical deductions can be made and which entities can be said to possess. In other words, the 'is' of identity, which we can relate to Marcel's notion of existence, should not be confused with the purely grammatical 'is' of predication, which is relevant to objectivity. If Kant maintained that the (logical/grammatical) predicate of existence gives us no further information about that to which it is attributed,[34] this only confirms that existence, understood as a property, is an empty notion with respect to identity. If identity is one's concern, the significance of existence needs to be reconceived. Thus, Marcel gives primacy to existence over objectivity, consolidating his philosophical turn to the experience of the here and now.

In the diary entries published in *Être et avoir* (1935; the entries themselves date from 1928 to 1933), Marcel develops the distinction between existence and objectivity into a broader division between two kinds of understanding, 'être' and 'avoir', following his reflections on the self and body in 'Existence et objectivité'.[35] The (linguistic) fact that I can talk about 'having' a body, he observes, makes it possible to understand my body as a tool at my disposal. However, if I can talk about the body that 'I have', I can also omit the intermediate terms and simply refer to 'my body'. The status of my body is therefore ambiguous; but, Marcel insists, while the mode of *avoir* makes it possible to adopt an impersonal, wholly instrumental attitude toward my body, this nevertheless rests on the underlying reality of *être*.[36] Just as for Bergson *le moi superficiel* depends on *le moi profound*, then, for Marcel, possession and instrumentality already presuppose the body as mine — or rather, presuppose the lived body that I am. 'Je ne me *sers pas de mon corps*, je *suis* mon corps' (*JM*: 323), he contends in 'Existence et objectivité'.[37] *Être* is thus the deeper mode of existence that needs to be recognized if human identity is the subject of enquiry;[38] and Marcel demands that we stop and question the very language we use when talking about human existence, so as not to fall prey to its equivocity.[39]

Marcel's reaction against the instrumentality of *avoir* articulates a much more general concern with the increasing dominance of technology in society,[40] an anxiety that was widespread amongst French non-conformist thinkers during the 1930s — including Bergson, although it is less visible in his philosophical writings.[41] Jean-Louis Loubet del Bayle explains:

> la guerre avait ébranlé la foi dans le Progrès et la confiance en la Raison qui avaient guidé le XIXe siècle. [...] Des phénomènes convergents vinrent nourrir chez beaucoup la crainte de voir la civilisation écrasée par ses propres productions, l'homme mécanisé par ses machines, l'individu absorbé par la masse. [...] Nombre d'esprits commencèrent alors à s'interroger sur le sort de l'homme menacé dans sa vie personnelle par l'oppression grandissante d'appareils politiques et étatiques envahissants [...,] par l'angoisse de voir se produire, selon le mot de Bergson, 'au lieu d'une spiritualisation de la matière, une mécanisation de l'esprit'. (1969: 21)

For Marcel, the ever-increasing importance society places on technology favours a purely functional world view: 'L'âge contemporain me paraît à se caractériser par ce qu'on pourrait sans doute appeler la *désorbitation* de l'idée de fonction [...]. L'individu tend à s'apparaître et à apparaître aussi aux autres comme un simple faisceau de fonctions' (*HP*: 192–93), he observed in the 1933 lecture, 'Position et approches concrètes du mystère ontologique'. Further contextualizing his argument, he continued:

> Il m'arrive souvent de m'interroger avec une sorte d'anxiété sur ce que peut être la vie ou la réalité intérieure de tel employé du métropolitain par exemple; l'homme qui ouvre les portes, ou celui qui poinçonne les billets. Il faut bien reconnaître qu'à la fois en lui et hors de lui tout concourt à déterminer l'identification de cet homme et de ces fonctions, je ne parle pas seulement de sa fonction d'employé, ou de syndiqué, ou d'électeur, je parle aussi des fonctions vitales. L'expression au fond assez affreuse d'*emploi du temps* trouve ici sa pleine utilisation. Tant d'heures consacrées à telles fonctions. Le sommeil aussi est une fonction dont il faut s'acquitter pour pouvoir s'acquitter des autres fonctions. Et il en est de même du loisir, du délassement. Nous concevons parfaitement qu'un hygiéniste vienne déclarer qu'un homme a besoin de se divertir tant d'heures par semaine.[42]
> (*HP*: 194)

Marcel is therefore prompted to describe the world around him as 'un monde cassé',[43] from which '[une] impression d'étouffante tristesse [...] se dégage' (*HP*: 195). However, precisely because of the inauthenticity of this functional existence, Marcel insists that it is possible to become aware of its insufficiency. In his 1933 lecture he therefore declared:

> il n'y a pas seulement la tristesse de ce spectacle pour celui qui le regarde; il y a le sourd, l'intolérable malaise ressenti par celui qui se voit réduit à vivre comme s'il se confondait effectivement avec ses fonctions; et ce malaise suffit à démontrer qu'il y a là une erreur ou un abus d'interprétation atroce.
> (*HP*: 195–96)

According to Marcel, this tendency to translate human experience into functional or possessive terms is equally reflected in our interpretation of the temporality of existence: we are tempted either to understand time in impersonal terms of succession, or to talk about it as something we simply 'have', forgetting that it fundamentally defines our being. 'Malgré nous, nous imaginons cette durée concrète comme une essence qui serait saisie intemporellement' (*PI*: 88), he observes in his diary on 1 June 1942. Like Bergson (whose influence is clear from Marcel's use of the term 'durée'), Marcel therefore rejected the idea that life is a chronological chain of discrete events, from which we can distance ourselves as we might from a film: 'toute tentative pour se figurer ma vie [...] sur le modèle d'une succession cinématographique doit être résolument écartée', he asserted in a 1949 Gifford Lecture;

> pour autant qu'il y a une substance de ma vie, [...] il est impossible qu'elle se réduise à des images, et par conséquent que sa structure soit purement et simplement celle d'une succession. [...] notre expérience intérieure, telle que nous la vivons, serait impossible pour un être qui ne serait que succession d'images [...].[44] (*ME I*: 204–05)

And in the 1967 conference paper, 'Mon temps et moi (Temps et valeur)', he related a possessive attitude toward time to a state of despair ('la conscience désœuvrée' (*MT*: 13)) analogous to that described in relation to *le monde cassé*. For Marcel, the phrase 'mon temps' describes 'un temps distribué et fragmenté. [...] Je parlerai volontiers d'un éclusage du temps, les journées sont assimilées à des bassins de retenue, à des vases communicants provisoirement à sec. Il s'agit de les remplir' (*MT*: 11). But as he explained in 1967:

> ce n'est pas le temps qui peut être retenu, mais, dans une lutte continuelle, menée contre ce que nous appellerons très approximativement une puissance ou un élément réfractaire [...,] je procède [...] à une sorte de détemporalisation qui me donne l'illusion de contrôler cette puissance ou cet élément: c'est l'éclusage dont j'ai parlé plus haut. (*MT*: 11–12)

To attempt to appropriate time in such a way — that is, to conceive of it in terms of *avoir* — is not to think about the reality of human time at all, for time cannot be controlled. With this realization, Marcel notes, time can appear 'sauvage' or 'untamed [*sic*]', and the person in question 'en proie à l'irréversible ou à l'irréparable' (*MT*: 13). However, just as he argued in relation to *le monde cassé*, Marcel suggests such a feeling to be symptomatic of an inauthentic understanding of one's (temporal) existence. 'Il existe une connexion secrète et rarement discernée entre la façon dont le moi se centre ou non sur lui-même — et sa réaction à la durée, plus précisément à la temporalité' (*HV*: 52), he remarked in the 1942 lecture 'Esquisse d'une phénoménologie et d'une métaphysique de l'espérance'. Consequently, Marcel stresses the need to hold a non-possessive, non-reductive conception of time (referred to as 'la conscience œuvrante' (*MT*: 13; also *PI*: 169–70)), which recognizes the irreducible complexity of human temporality and the undetermined creativity that characterizes Being's freedom:[45]

> Nous avons [...] noté, sous le temps éclusable que nous aménageons de notre mieux, la présence d'un temps sauvage dont nous pouvons dire qu'il est comme l'aspiration de notre être par la mort qui finalement, au moins en apparence, l'engloutira.
> Il nous a paru cependant que ce point de vue n'était pas ultime. Il se peut que chacun de nous soit appelé, au cours même de cette vie, à tisser en quelque sorte les premières mailles d'une durée non homologue par rapport au temps utilisé ou aménagé comme à la surface du gouffre [...]. Dans cette ligne de pensée, le temps prend une tout autre valeur; il apparaît, en effet, comme l'épreuve au travers et à la faveur de laquelle se crée ou s'élabore un mode d'existence auquel ne sauraient convenir les mesures temporelles que nous sommes tentés de dire normales. (*MT*: 18–19)

As Bergson and Marcel both emphasize, then, the way things appear to us can differ, depending on our particular reflective mode. 'We must recognize the importance for Bergson of the word "sens"; like the German "Sinn", sens means not only meaning and sense, but also direction', writes Leonard Lawlor (2003: 3);[46] and the same must be emphasized for Marcel.[47] As philosophers concerned with grasping the unadulterated, temporal character of Being, both therefore argue for *une conversion de l'attention* away from practical utility or function. 'Le rôle de la philosophie ne serait-il pas ici de nous amener à une perception plus complète de la réalité par un certain déplacement de notre attention?', asked Bergson in his 1911 Oxford lecture 'La Perception du changement'. 'Il s'agit de *détourner* cette attention du côté pratiquement intéressant de l'univers et de la *retourner* vers ce qui, pratiquement, ne sert à rien. Cette conversion de l'attention serait la philosophie même' (Œ: 1373–74).[48] This requires a non-objectifying act of reflection, which Bergson refers to as 'intuition'[49] and contrasts with the intellectualizing human faculty of 'intelligence'.[50] Marcel, on the other hand, distinguishes between the spectating mode of 'réflexion primaire' and the more ontologically authentic 'réflexion seconde', defined in 'Position et approches' as 'un mouvement de conversion' (*HP*: 214) — 'ce qu'il y a de moins spectaculaire dans l'âme'[51] (*HP*: 212).

So unlike the neo-Kantian Paul Natorp (1854–1924), who argued that it was impossible to access and investigate true subjectivity directly because of the objectifying and intellectualizing nature of self-reflection,[52] Bergson and Marcel can be aligned with Heidegger in that, although they would accept Natorp's argument to a certain extent, neither believes that subjective life is completely inaccessible or inexpressible. Rather, all three contended that self-understanding was not dependent on conscious reflection, and sought guidance from (temporal) life-experience itself when attempting to disclose this non-objective form of acquaintance.[53] Heidegger valued poetic form for its evocative power (particularly the works of Hölderlin (1770–1843)), the significance of which transcends the literal and the categorical, opening up a space for meaning without delineating (and therefore limiting) it. Bergson and Marcel, on the other hand, privileged music's expressive capacity,[54] seeing, in its non-representational form, an illustration of the non-objective unity they believed intuition and secondary reflection were capable of apprehending, an illustration of *la durée*.[55] As Bergson famously asks in his *Essai*: 'Ne pourrait-on pas dire que, si [les notes d'une mélodie] se succèdent, nous les apercevons néanmoins les unes dans les autres, et que leur ensemble est comparable à un être vivant, dont les parties, quoique distinctes, se pénètrent par l'effet même de leur solidarité?' (Œ: 67–68). 'La preuve en est', Bergson continues, 'que si nous rompons la mesure en insistant plus que de raison sur une note de la mélodie, ce n'est pas sa longueur exagérée, en tant que longueur, qui nous avertira de notre faute, mais le changement qualitatif apporté par là à l'ensemble de la phrase musicale' (Œ: 68).[56] Thus, in 'La Perception du changement', Bergson concludes that as far as human existence is concerned, il n'y a ni un substratum rigide immuable ni des états distincts qui y passent comme des acteurs sur une scène. Il y a simplement la mélodie continue de notre vie intérieure — mélodie qui se poursuit et se poursuivra, indivisible, du commencement à la fin de notre existence consciente. Notre personnalité est cela même. C'est justement cette continuité indivisible de changement qui constitue la durée vraie.[57] (Œ: 1384)

Marcel praises Bergson's analysis of the melody in his 1925 article 'Bergsonisme et musique': 'Une analyse comme celle que M. Bergson esquisse met incontestablement en lumière — ou du moins nous aide à définir — le type tout à fait particulier d'unité qui est celui d'une mélodie: unité à la fois indivise et fluide, unité d'un progrès et non point d'une chose' (*EM*: 36). He himself had a deep appreciation for music and often refers to the impact that certain pieces, or his own piano improvisations, have had on his thought.[58] Bergson's approach struck a chord with him, therefore. 'Idée d'une expression non-représentative, en musique', he writes in his diary on 16 December 1930 — 'un ordre où la chose dite ne peut pas être distinguée de la manière de dire. En ce sens, en ce sens seulement, la musique ne signifie rien, mais peut-être parce qu'elle *est* signification' (*EA*: I 69).[59] Later, in his lecture 'La Musique dans ma vie et mon œuvre' (1958), Marcel even went so far as to say: 'c'est à partir de la musique que j'ai été amené à penser sur l'être ou à affirmer l'être' (*EM*: 93).

For this reason, Marcel also uses musical imagery to describe the form of his philosophy (its 'manière de dire'), and employs musical analogies to express the content of his ideas. In his *lettre préface* to Troisfontaines's 1953 exposition of his philosophy, for example, he describes his thought's development in terms of 'une orchestration vitale de plus en plus riche' (1953: 10). Such a characterization emphasizes the process of Marcel's philosophizing over and above its maxims or conclusions; indeed, this gradual or improvisational development is very evident in Marcel (almost overwhelmingly so, in places). Arguably this conforms to his and Bergson's desire to present time as a fundamental constituent of human reality, and our thoughts and experiences as becoming with and through time. For Bergson, too, to think intuitively — that is, in faithful concordance with Being — is to acknowledge and accept reality's movement, to 'penser en durée' (Œ: 1275) as opposed to confining one's reflections to rigid, ready-made concepts.[60] In his 'Introduction à la métaphysique' he describes an authentic metaphysics as follows:

> les concepts lui sont indispensables, car toutes les autres sciences travaillent le plus ordinairement sur des concepts, et la métaphysique ne saurait se passer des autres sciences. Mais *elle n'est proprement elle-même que lorsqu'elle dépasse le concept*, ou du moins *lorsqu'elle s'affranchit des concepts raides et tout faits pour créer des concepts bien différents* de ceux que nous manions d'habitude, je veux dire *des représentations souples, mobiles, presque fluides, toujours prêtes à se mouler sur les formes fuyantes de l'intuition.*[61] (Œ: 1401–02; my emphasis)

Such a conception is then echoed by Marcel in the 1958 presentation he gave to the *Société française de philosophie*, 'L'Être devant la pensée interrogative': 'j'ai toujours été en garde contre un certain rancissement du langage philosophique; je suis convaincu qu'il est nécessaire de le revitaliser constamment. Mais cette revitalization ne peut s'opérer que par le moyen d'une réflexion vigilante qui se maintient perpétuellement en contact avec l'expérience' (*PST*: 77–78). Accordingly, his own metaphysical terminology is in constant flux, testifying literally to the temporality of existence. Secondary reflection, for example, is also referred to as 'intuition aveuglée' (*EA*: I, 151), 'réflexion récupératrice' (*HP*: 207), 'réflexion à la deuxième puissance' (*HP*: 207), 'le recueillement' (*HP*: 211), 'la pensée pensante' (*RI*: 21–22), and 'la contemplation' (*ME I*: 139)[62] — for, as Marcel explained to his audience in one of the William James Lectures ('Points de départ'), the philosophy he aspired to write was

> une recherche qui, à mesure qu'elle se poursuivait et qu'elle s'éclairait, restait néanmoins et devait rester une recherche — bien loin de se commuer en un ensemble de propositions susceptibles d'être établies une fois pour toutes et reconnues pour vraies, abstraction faite des chemins par lesquels l'esprit avait procédé pour les établir. Aussi n'est-ce aucunement un hasard si ma pensée pendant si longtemps s'est exprimée sous la forme d'un 'journal'. (*DH*: 17)

As the above discussion has demonstrated, then, there are extensive similarities between Bergson's and Marcel's approach to philosophy. 'Le but de G. Marcel est [...] le même que celui de Bergson: retour à l'immédiat', asserts Tsukada (1995: 82). Certainly, both emphasize the temporality of existence, considering an engagement with lived human time to be essential for any investigation into the nature of Being; and for both, this then manifests itself in a critique of a linear, spectatorly understanding of time, which — they argue — fails to recognize the irreducible multiplicity of human experience. Preference is therefore given to a non-objectifying form of reflection, heralded as a more authentic mode of understanding which is capable of appreciating the complexity and free creativity of *la durée*; and as a consequence Bergson and Marcel share a vision of a metaphysics that is faithful to this reality, encapsulating the movement and interconnected nuances of temporal life by means of a more fluid expressive form. What is not so clear, however, is whether this justifies Tsukada's straightforward identification of the two projects: '[leur] démarche est semblable', Tsukada declares; 'ils sont tous les deux partis de l'expérience immédiate de la réalité concrète qu'ils ont analysée et décrite aussi précisément que possible' (1995: 245) — and thus, 'en ce qui concerne [...] l'éternité, Bergson et G. Marcel sont beaucoup plus proches que ne pourrait le faire croire la différence de leurs vocabulaires' (1995: 242). However, as will be seen in the remainder of this chapter, the metaphysics underlying these two philosophies is actually quite different. Although Marcel's engagement with Bergson can be said to have instigated his own philosophical project, this project remained his own, and he moved forward from this *point de départ* in quite a different direction. Surprisingly, as the second section will now reveal, this difference begins with their philosophies of time. The third section will then explore how this affects their understanding of eternity.

Ruptures

If, at first glance, Bergson's *intuition* and Marcel's *réflexion seconde* seem to share the same role, the legitimacy of drawing an analogy between the two became a major subject of contention for Marcel, who rejected Bergsonian intuition as a viable mode of secondary reflection. The very first article published by Marcel, 'Les Conditions dialectiques de la philosophie de l'intuition' (1912), concerns precisely this issue, and illustrates how pivotal Bergson's thought was for the development of Marcel's own distinct position. 'Dans [cet] article [...] je m'étais évertué à montrer que l'intuition, contrairement à ce qu'affirmait Bergson, ne pouvait pas trouver en elle-même sa garantie, et que seule la réflexion pouvait peut-être dans des conditions déterminées lui conférer une valeur' (*DH*: 55–56), explained Marcel in the third of his William James Lectures ('Existence'). For Marcel, the purity of Bergsonian intuition is too passive, and therefore too abstract. In order to remain faithful to the reality of human experience, he argues, intuition's mode of understanding must be conceived of in terms of a 'dialectic' — a constant tension between primary and secondary reflection (or intelligence and intuition). 'La philosophie de l'intuition doit reconnaître à la dialectique un rôle essentiel', Marcel declares in the 1912 article; 'affirmer l'intuition comme indépendante de toute dialectique ce serait nécessairement nier la pensée — nier toute pensée — donc nier l'intuition même' (1912: 639).[63] Indeed, in the 1943 essay 'Grandeur de Bergson' Marcel affirms — again, explicitly in opposition to Bergson — that 'cette "pensée privilégiée" [qui est celle de l'intuition] est un mythe',[64] maintaining instead that 'l'intuition ne prend son sens, sa valeur, qu'à condition de s'insérer dans une dialectique qui la soutient avant de l'exploiter', and asking whether 'toute la philosophie bergsonienne ne devrait pas être repensée de ce point de vue' (1943: 37–38). As Marcel explained to Boutang, 'ce que je veux exclure, c'est l'idée d'une saisie de l'être; je ne crois pas que nous puissions capter l'être' (*GM*: 79). Thus, if the faculty of intuition is to be anything more than an abstract construct, it must be implicated in the struggle of existence and its self-experience, which involves the constant temptation to confine our self-knowledge to primary reflection's intellectualizing mode: 'l'immédiat de la sensation est forcément un paradis perdu' (*JM*: 131), he asserts in his diary on 14 April 1916 — for, he affirms in his next entry, 'la dialectique, le drame de la sensation, c'est qu'elle doit être réfléchie, interprétée' (*JM*: 131; 4 May 1916). He consequently rejects the Bergsonian dichotomy between intelligence and intuition, insisting that their conflictual relationship be taken into account.[65]

The metaphysical differences between the two philosophers become all the more evident in their analyses of 'ruptures' — that is, breaks in continuity — of *la durée*.[66] For Marcel, these are ontologically significant and manifest what he calls 'le tragique' of existence, whereas for Bergson they are not representative of anything more than a lapse in our attention, where we are distracted from the reality of *la durée*: 'L'apparente discontinuité de la vie psychologique tient [...] à ce que notre attention se fixe sur elle par une série d'actes discontinus' (*Œ*: 496), he writes in *L'Évolution créatrice* (1907). In the 1929 article 'Note sur les limites du spiritualisme bergsonien', Marcel states: 'Cette urgence dramatique du problème spirituel, M. Bergson me paraît l'avoir profondément méconnue [...]. La résonance tragique si constante chez un Nietzsche, malgré la faiblesse du système, fait ici défaut' (1929b: 269–70). The tragedy to which Marcel is referring is intimately connected to the fundamental dialectic of self-experience he wishes to emphasize — an existential dialectic which Marcel feels Bergson evades, prompting him to criticize Bergson's impatience in wanting to 'pénétrer *hic et nunc* le sens de l'épreuve incompréhensible dans laquelle elle se trouve jetée' (1929b: 269). In addition to the hermeneutic battle between primary reflection's objectification, and secondary reflection's efforts to recuperate existential experience, Marcel wants metaphysics to take seriously experiences such as suffering, despair, and bewilderment, in spite of how they decentre — that is, detemporalize — our experience of self and plunge us into what he calls 'le temps gouffre' (*EA*: I, 100) or '[le] temps vide' (*JM*: 230). More than mere distractions from *la durée*, for Marcel, these experiences are an integral and inevitable part of (temporal) human existence, and are just as instructive with respect to the human ontological condition. A continuous experience of any *moi profond*, he contends, is impossible.

In *En chemin* Marcel recalls a revelatory moment when this disjointed reality of existence became clear to him:

> Je garde le souvenir précis d'une excursion solitaire au lac Champex, quelques semaines plus tard, et je me revois distinctement au cours de la longue descente vers Martigny, prenant une conscience intense de l'*ici-maintenant*, qui était le mien à cet instant, et du fait qu'il deviendrait plus tard un *alors*, mais sur la base de quel autre *ici-maintenant*? Il se présentait à moi comme caché dans une brume impénétrable. C'était là vraiment reconnaître le mystère du temps, faut-il dire de la distorsion temporelle, dans une dimension qui n'était sûrement pas celle de la durée personnelle. Ce qui s'imposait à moi, c'est bien plutôt la discontinuité des *présents*. Mais combien ce pluriel nous heurte! Le présent n'est-il pas voué par définition

à la singularité? Peut-être est-ce le lieu de dire que ce mystère du temps est certes au cœur de tout ce que j'ai pensé. (*EC*: 71–72)

Thus, although Marcel's opposition to Bergson does not present itself specifically in relation to time, it now becomes clear that the two philosophers' understanding of time is crucial to its appreciation. 'Ce domaine [du temps] comporte des chevauchements et aussi ce qu'on appelle, de façon bien significative, des temps morts' (*MT*: 18), affirmed Marcel in 'Mon temps et moi'. In Bergson's 'durée indivisible, au contraire', writes Tsukada, 'on ne perçoit que la présence et pas l'absence' (1995: 167). Indeed, as Mark Muldoon notes,

> Bergson never reckons with the discordance of life revealed through the language in laments and other elegies of the human condition. There is no need for testaments of struggle and the recognition of suffering. Apart from his last work, the misery of time is only rejected and never confronted. In fact, outside of the need to refer always to duration at the expense of discontinuity, there is little overlapping between the modes of discourse that describe mortal time and duration.[67] (2006: 116)

Rather, in *Les Deux Sources de la morale et de la religion* (1932), Bergson writes of human suffering:

> la souffrance physique n'est-elle pas due bien souvent à l'imprudence et l'imprévoyance, ou à des goûts trop raffinés, ou à des besoins artificiels? Quant à la souffrance morale, elle est au moins aussi souvent amenée par notre faute, et de toute manière elle ne serait pas aiguë si nous n'avions surexcité notre sensibilité au point de la rendre morbide: notre douleur indéfiniment prolongée et multipliée par la réflexion que nous faisons sur elle. (*Œ*: 1197)

Hence, for Marcel, if Bergson argues for a deeper intuition of Being he has failed to engage with its immediate reality — that is, with the complexities and conflicts of its experience and with its tragedy. His notion of intuition can therefore only, at best, be deemed an indirect presupposition of Being; it does not explore its lived nature, its actual temporal experience.

However, as Marcel stated in a William James Lecture ('Existence'), he himself did not fully appreciate *le tragique* at the time he wrote the 1912 article: 'ce qui me frappe surtout aujourd'hui, c'est que toutes ces investigations se développaient à partir de ce que j'appellerais peut-être aujourd'hui une certaine sécurité existentielle. [...] le monde qui était le mien — le nôtre à nous artistes, écrivains et philosophes — ne paraissaient pas encore sérieusement ou vitalement menacé' (*DH*: 56).[68] Marcel first discovered its true extent during World War I: 'avec l'événement [la guerre de 14], ce qui s'est introduit dans ma pensée d'une manière définitive, c'est le tragique' (*GM*: 62), he explained to Boutang. It is in the second part of his *Journal*, therefore, that he decisively equates 'le point de vue dialectique' with 'le point de vue de l'expérience' (*JM*: 144; 23 July

1918), a perspective which, thereafter, he describes in terms of interrogation and response. 'Il me semble que nous sommes forcés de questionner sur l'être' (*PST*: 78),⁶⁹ he declared in his presentation 'L'Être devant la pensée interrogative' — a remark which, significantly, he coupled with a re-affirmation of his rejection of intuition: 'lorsqu'on me parle d'une intuition de l'être, ces mots ne correspondent rigoureusement à rien pour moi' (*PST*: 78).⁷⁰

If Marcel was opposed to the abstract, reflective purity of Bergsonian intuition, he felt that it also failed to grasp human reality on a second, more fundamental level: it did not take into account the intersubjectivity of the human condition. 'Chez moi, l'essence est plutôt musicale et mélodique qu'intuitive' (*GM*: 61), he stated in an interview with Boutang. The musicality to which Marcel is referring is not simply a fluid and indivisible intrasubjective reality; its unity is intersubjective: 'En moi, c'est [...] un certain "nous-tous" que l'œuvre musicale la plus haute semblait atteindre directement' (*DH*: 45), he informed the audience in his second William James Lecture ('Participation'). In addition to neglecting the dialectical character of self-experience, then, Marcel felt the way in which intuition presented ontology was too individualistic. In another William James Lecture ('Mystère ontologique'), Marcel elaborated further, explaining how, for him, it was because of its orchestral, polyphonic nature that music came so close to capturing the essence of Being:

> On pourrait [...], pour faciliter autant que possible l'acheminement de la pensée en direction de l'être se référer à une totalité concrète comme celle d'un orchestre exécutant une œuvre polyphonique, chaque exécutant tient bien sa partie dans l'ensemble, mais il serait absurde d'assimiler celui-ci à une somme arithmétique d'éléments juxtaposés, ou plutôt cette représentation irait de pair avec la méconnaissance absolue de ce qu'est un ensemble musical. (*DH*: 108)

As such, for Marcel, Bergson's analysis of the melody is insufficiently developed; and in 'Bergsonisme et musique' he criticizes it for neglecting the participatory nature of the musical experience, contending that the listener is involved in the creation of a melody (*le devenir musical*) in addition to the musician.⁷¹ As Parain-Vial summarizes: 'Si la comparaison avec la durée musicale [bergsonienne] ne trahit pas la pensée de Gabriel Marcel sur l'être, c'est à condition de la concevoir, non comme la mélodie d'une conscience solitaire, mais comme essentiellement *orchestrale: esse est coesse*' (1976: 192). Indeed, 'to Marcel', Collins writes, 'the reality of [Bergson's] internal duration is too narrow[; he conceives it to be ...] more global and indeterminate [...,] an affirmation of existing being which prescinds from the distinction between the inner and the outer' (1968: 136). Thus, despite all the similarities that appeared to exist between the two philosophies, Marcel's understanding of Being, and the reality of its self-experience, is not the same as Bergson's. As the third and final section of the chapter will now demonstrate, this difference impacts on the way Marcel speaks of eternity, in addition to time.

Memory, Time, and Eternity

In the second part of the *Journal*, in which Marcel reacts against his earlier work and begins to reconceive his philosophical approach, Marcel's move away from a Bergsonian model of time is much more apparent. Furthermore, it becomes increasingly evident that if Bergson and Marcel hold different philosophies of time, these are bound up with antagonistic notions of eternity; and as will now be suggested, the time–eternity relation that Marcel sides with in opposition to Bergson consolidates a potential problematic in his philosophy, whereby time and eternity seem to compete against one another for primacy, rendering the consistency of his position questionable.

Discussions of time in the first part of the *Journal* (January–May 1914) are often situated in relation to an experience of eternity, and, significantly, this seems to be treated as a more authentic experience of Being. In an early entry (27 January 1914), for example, when reflecting on how to conceive of the relation between my conscious self and my empirical self, Marcel describes an authentic experience of Being in terms of an eternal 'actualité', taking the existential experience of the saint ('pour qui tout est actuel' (*JM*: 48), 'pour qui tout est actualité pure' (*JM*: 49)) as exemplary. It is Marcel's belief that there are truths which can be legitimately affirmed, despite being objectively unverifiable in the empirical world of the spatio-temporal. He thus wishes to find grounds on which, for example, a saint's religious faith might be regarded as justified, even if the truth of religion cannot be scientifically proven. At the same time, however, he wants to avoid instituting a dualism between appearance and reality — a difficulty Marcel describes as being 'la plus grave que j'aie encore rencontrée' (*JM*: 48). The solution, he suggests, is to understand consciousness of reality as a progression: one that moves away from absolute dependence on the certainty of temporally-bound objectivity toward a (more authentic) consciousness of Being, which recognizes an eternal form of assurance that transcends any opposition between the objectively verifiable and the objectively unverifiable:

> chez le saint, [... une] base historique [de sa foi] n'est pas nécessaire; l'éternité fait corps avec l'actuel, elle est l'actuel. Chez celui qui ne participe pas au contraire de la sainteté que d'une façon tout empirique, l'histoire, le temps, prennent une valeur, et avec eux s'accuse ce rapport à un élément objectif qui recèle en soi tous les germes de mort. En d'autres termes, la difficulté se résoudrait pour qui saurait voir que l'état de choses qui la provoque est lui-même relatif à un moment qui ne peut à aucun degré être regardé comme ultime du développement de la conscience religieuse (développement que

> la pensée pose du moment où elle conçoit le saint). Dans la mesure où le dualisme de l'apparence et de la réalité n'est pas entièrement surmonté [...], il est naturel, il est nécessaire que subsiste cet élément inéliminable d'objectivité, de position existentielle qui est la tare de la religion chez le fidèle. Le croyant véritable est celui qui n'a pas besoin de toucher (c'est-à-dire de savoir) pour croire.[72] (*JM*: 48–49)

The above passage demonstrates how, for Marcel, time is identified with objectivity[73] — that is, with something that is opposed to the eternity of *actualité* and must therefore be transcended. This is then reiterated in his entry on 17 February, which, following a similar discussion that seeks to establish grounds on which belief in miracles might be intelligible, argues for a transcendent eternal present '[qui] n'est que pour la foi' (*JM*: 83). In this entry, historical and scientific frames of truth (*cadres de vérité*) are described as 'moules intemporels, [...] dont la pensée n'en devient pas moins [...] captive' (*JM*: 82). While this may seem to cohere with Bergson's argument that an objective, scientific account of reality, although seductive and influential for our understanding of the world, fails to capture the essence of human reality, Marcel's response to this observation is decidedly un-Bergsonian. Not only does Marcel argue that 'si l'idée de miracle est susceptible de présenter un contenu philosophique, c'est à condition de ne pas être dissoute dans la notion d'histoire'; he comes to the further conclusion that it is 'par l'intermédiaire de l'idée de présent (de présent absolu) que l'idée de miracle peut recevoir un contenu', where 'présent' is defined as 'le lieu de la pensée religieuse — [...] ce qu'on a appelé l'éternité' (*JM*: 82). Strangely, then, Marcel seems to be arguing both for the necessity of engaging with temporal reality, as illustrated by his dismissal of the *moules intemporels* of science and history, and for the necessity of transcending the temporal in favour of eternity, so as to avoid conceiving of reality in (inauthentic) dualistic terms: 'Le présent est transcendant à cette durée, à la rigueur même elle ne se conçoit qu'en fonction de lui' (*JM*: 83). Is this not a contradiction?

Bergson too was opposed to (Cartesian) dualism;[74] and yet Bergson did not feel the need to argue for an eternal present. Although he does accept a notion of eternity, it is not the immobile conception that Marcel appears to hold. For Bergson, the only notion of eternity that is applicable to reality is that of time's movement and change. 'Le mouvement est la réalité même' (*Œ*: 1378), he affirmed in 'La Perception du changement'; and in his 'Introduction à la métaphysique' Bergson writes:

> nous allons à une durée qui se tend, se resserre, s'intensifie de plus en plus: à la limite serait l'éternité. Non plus l'éternité conceptuelle, qui est une éternité de mort, mais une éternité de vie. Eternité vivante et par conséquent mouvante encore, où notre durée à nous se retrouverait comme les vibrations dans la lumière, et qui serait la concrétion de toute durée

comme matérialité en est l'éparpillement. Entre ces deux limites extrêmes l'intuition se meut, et ce mouvement est la métaphysique même. (Œ: 1419)

Immobility and invariability have traditionally (since Plato) been considered more perfect than change and movement; but Bergson inverts this relation, analysing immobility as composite and relative, merely a constructed relation between movements. The error of other philosophies, he writes,

> consista à s'inspirer de cette croyance si naturelle à l'esprit humain, qu'une variation ne peut qu'exprimer et développer des invariabilités. D'où résultait que l'Action était une Contemplation affaiblie, la durée une image trompeuse et mobile de l'éternité immobile, l'Ame une chute de l'Idée. Toute cette philosophie qui commence à Platon pour aboutir à Plotin est le développement d'un principe que nous formulerions ainsi: 'Il y a plus dans l'immuable que dans le mouvant, et l'on passe du stable à l'instable par une simple diminution'. Or, c'est le contraire qui est la vérité. (Œ: 1424-25)

So although we often speak of change, as Bergson insisted in 'La Perception du changement', 'nous n'y pensons pas. Nous disons que le changement existe, que tout change, que le changement est la loi même des choses [...]; mais ce ne sont là que des mots, et nous raisonnons et philosophons comme si le changement n'existait pas' (Œ: 1367). Thinking of things in terms of immobility is certainly useful as far as our practical relations with the world are concerned. To apply such an understanding to metaphysics, however, is to falsify reality:

> L''immobilité' étant ce dont notre action a besoin, nous l'érigeons en réalité, nous en faisons un absolu, et nous voyons dans le mouvement quelque chose qui s'y surajoute. Rien de plus légitime dans la pratique. Mais lorsque nous transportons cette habitude d'esprit dans le domaine de la spéculation, nous méconnaissons la réalité vraie, nous créons, de gaieté de cœur, des problèmes insolubles, nous fermons les yeux à ce qu'il y a de plus vivant dans le réel. (Œ: 1379)

Bergson's arguments against the legitimacy of any form of immobile, eternal present therefore raise questions about the extent to which Marcel actually engages with time — the extent to which time is of consequence to his philosophy. One might object here that, given Marcel's own criticism of his early philosophy, which he came to regard as abstract and detached from existential experience, we should not be so astonished if the significance of time is marginalized in the first part of his *Journal*. However, as will now be demonstrated, in spite of the evolution Marcel identifies in his thought, the concrete focus of which he credits Bergson for helping him to reinforce, the second part of the *Journal* only continues to position time in relation to an eternal present.

Part Two of the *Journal* commences on 15 September 1915, and the very first question to be considered is that of time:

> Il n'y a pas et ne peut y avoir d'autre origine du temps que le présent, seule limite qu'on lui puisse assigner... Illusion d'un temps qui serait *donné* avant d'être *duré* (comme l'espace qui *est là* avant d'être parcouru). [...] Le temps n'est en rien assimilable à un milieu dans lequel des consciences s'inséreraient et par rapport auquel ces 'insertions' seraient contingentes; il est *la négation même de cela*.
>
> Pourtant cette limite interne du temps qui est sa réalité même se présente à l'imagination comme mouvante — au sein de quoi? ainsi naît l'idée d'un temps-milieu. (*JM*: 129)

In this first entry time seems to be presented in a Bergsonian manner, with Marcel's assertion that time is only authentically experienced in the immediate present of *le temps duré*, and its reference to the fundamental differences between time and space. Indeed, in his next entry (18 September 1915) Marcel affirms: 'Le temps ne peut être pensé comme objet qu'avec de l'espace; mais l'espace ne peut être donné que dans le temps' (*JM*: 129). Again, this appears Bergsonian, for Marcel's acknowledgement of how time and space — although different — are nevertheless used to symbolize one another in our minds, supports Bergson's arguments concerning the entangled conceptions we hold of these notions, whereby we understand time through spatial analogies and vice versa. Nevertheless, Marcel then reintroduces his conception of *actualité*, which, as in the first part of the *Journal*, seems to override the importance of *le temps duré* with respect to the authentic apprehension of one's being:[75] 'Le temps est symbolisé pour nous par le mouvement, qui lui-même est symbolisé par l'espace parcouru. Mais on fait alors abstraction de ce qui est l'essentiel et que faute de mots j'appelle l'*actualité* (correspondant à l'eccéité)' (*JM*: 129). It must be noted that this passage does not, in itself, confirm Marcel's understanding of time to be distinct from Bergson's, for Bergson himself argued that all movement, as conceived in our minds, is only an abstract image of *la durée*.[76] However, the account of memory that the second part of the *Journal* subsequently develops is decisive in establishing Marcel's position as distinct from Bergson, and provides what is perhaps the most complete illustration of how time and eternity mean something different in the two philosophies.

Both Bergson and Marcel attribute ontological significance to memory: 'je vois dans la mémoire un aspect essentiel de l'affirmation ontologique. Combien à cet égard je me sens plus proche de Bergson — et d'ailleurs de saint Augustin' (*EA*: I, 123–24),[77] writes Marcel in his diary on 10 October 1932. In both cases, attention paid to memory is also motivated by a dissatisfaction with empirical reduction and Cartesian dualism. The two philosophers part company, however, as a result of Marcel's insistence that memory's ontological grounding is in the present rather than the past. This lays the foundations for his conception of an eternal present that transcends time, in opposition to Bergson's notion of eternity which is conceived as the movement of time itself.

'Le problème de la mémoire est à nos yeux un problème privilégié' (Œ: 221), writes Bergson in *Matière et mémoire*.[78] Like Marcel, he also wanted to legitimize the empirically unverifiable, for he too believed this to characterize the nature of the human mind (*esprit*). One could demonstrate this, he felt, by engaging with the experience of memory, because memory was something that empirical examination was unable to account for fully: 'Si [...] l'esprit est une réalité, c'est ici, dans le phénomène de la mémoire, que nous devons le toucher expérimentalement. Et dès lors toute tentative pour dériver le souvenir pur d'une opération du cerveau devra révéler à l'analyse une illusion fondamentale' (Œ: 220). The reason for this, Bergson continues, is that memory is not simply a 'weak' form of perception which emanates from the present. It is of a fundamentally different nature — it is past: 'La mémoire ne consiste pas du tout dans une régression du présent au passé, mais au contraire dans un progrès du passé au présent. C'est dans le passé que nous nous plaçons d'emblée' (Œ: 369). Human subjectivity cannot be objectified in the present moment because *la durée* not only relates to the instant in which I find myself, but is also layered with the past memories I have accumulated, any of which might be triggered in response to my present actions. Bergson therefore declares: 'les questions relatives au sujet et à l'objet, à leur distinction et à leur union, doivent se poser en fonction du temps plutôt que de l'espace' (Œ: 218).

In contrast to Descartes's (spatial) dualism then, Bergson proposes a dualism of time, which, as well as accounting for human reality's resistance to materialist reduction,[79] is also able to suggest how the mind (of *la durée*) and the body (of present action) are related:

> Le tort du dualisme vulgaire [de Descartes] est de se placer au point de vue de l'espace, de mettre d'un côté la matière avec ses modifications dans l'espace, de l'autre des sensations inextensives de la conscience. De là l'impossibilité de comprendre comment l'esprit agit sur le corps ou le corps sur l'esprit. [...] A une distinction spatiale nous substituons une distinction temporelle. [...] entre la matière brute et l'esprit [...] il y a toutes les intensités possibles de la mémoire [...]. Dans la première hypothèse, celle qui exprime la distinction de l'esprit et du corps en termes d'espace, corps et esprit sont comme deux voies ferrées qui se couperaient à angle droit; dans la seconde, les rails se raccordent selon une courbe, de sorte qu'on passe insensiblement d'une voie sur l'autre. (Œ: 354–55)

But Marcel takes issue with Bergson's past–present dualism, criticizing the rigid 'past-ness' of Bergson's *souvenir pur*. 'Certainement c'est moi qui rejoins mon passé et non pas du tout mon passé qui m'a suivi et a en quelque sorte fait route avec moi' (*JM*: 177), he asserts on 6 March 1919.[80] According to Marcel, memories are not immanent and carried around in their original form as if they were, in some sense, fixed, eternal 'truths' about the past. To speak in such a way, he contends, is pure abstraction. On 10 December 1918 he writes:

> Il y a un sens où ce que nous appelons un événement est une vérité éternelle (un ensemble de jugements qui se croisent); et en ce sens-là il serait absurde de parler de conservation. Il n'y a conservation que de ce qui dure — de ce sur quoi *le temps mord*. [...] le souvenir immobile de Bergson est pure abstraction, il ne peut durer, il ne peut se conserver.[81] (*JM*: 150)

In Bergson's defence it must be noted that, in practice, pure memory (the past) is never experienced as such, nor for that matter is pure perception (the present): 'la perception n'est jamais un simple contact de l'esprit avec l'objet présent; elle est tout imprégnée des souvenirs-images qui la complètent en l'interprétant' (*Œ*: 276); and conversely: 'le souvenir pur, indépendant sans doute en droit, ne se manifeste normalement que dans l'image colorée et vivante qui le révèle' (*Œ*: 276). As a consequence, we tend to confuse the two notions,[82] conceiving of their differences in terms of intensity rather than nature:

> L'erreur capitale [...] est celle qui consiste à ne voir qu'une différence d'intensité, au lieu d'une différence de nature, entre la perception pure et le souvenir. Nos perceptions sont sans doute imprégnées de souvenirs, et inversement un souvenir [...] ne redevient présent qu'en empruntant le corps de quelque perception où il s'insère. Ces deux actes, perception et souvenir, se pénètrent donc toujours, échangent toujours quelque chose de leur substances par un phénomène d'endosmose. [... Mais] en faisant du souvenir une perception plus faible, on méconnaît la différence essentielle qui sépare le passé du présent. (*Œ*: 214)

This confusion is then reinforced, because 'on tient la perception pour une espèce de contemplation, [...] parce qu'on veut qu'elle vise à je ne sais quelle connaissance désintéressée: comme si, en l'isolant de l'action, en coupant ainsi ses attaches avec le réel, on ne la rendait pas à la fois inexplicable et inutile!' (*Œ*: 215–16). In fact, because we are creatures of action, our perception is necessarily motivated — that is, partial; and the same applies to memory: '[les images du passé] ne se conservent que pour se rendre utiles: à tout instant elles complètent l'expérience présente en l'enrichissant de l'expérience acquise' (*Œ*: 213). For Bergson, therefore, consciousness is in a constant tension between the present and the past, where our experience of *la durée* (the collective totality of our temporal life) can both contract and focus on more immediate concerns, and dilate, so as to reach back to moments in the past. Our everyday experience is thus of neither *le souvenir pur* nor *la perception*, but rather of 'le souvenir-image',[83] which is a mixture of (past) memories and (present) perceptions that relate to our actions.[84] As a result, Bergson argues, the past (our memories of which accumulate layer upon layer) and the present co-exist in us and compete for our attention, creating an eternal dialectic of (temporal) movement between them.[85]

Nevertheless, this account does not satisfy Marcel, who rejects the idea of *le souvenir pur* in principle: 'En parlant de conservations des souvenirs purs,

Bergson postule que [...] le souvenir est là (qu'il s'actualise d'ailleurs ou non)' (*JM*: 130), he writes on 13 April 1916. Marcel interprets Bergson as presenting memory as a kind of directory, where all things past are stored for future reference: 'Je voudrais savoir si chez Bergson les souvenirs purs eux-mêmes ne tendent pas à constituer un répertoire? Tout le problème de la mémoire pure est là' (*JM*: 176; 6 March 1919). He, on the other hand, is strictly opposed to this idea: 'il ne faut pas identifier mémoire et collection: se souvenir, c'est en réalité *revivre* (selon certaines modalités) et non pas extraire une fiche' (*JM*: 163; 22 February 1919; my emphasis).[86] In fact, Marcel seems to have misunderstood Bergson, as Bergson explicitly rejects such a reading;[87] but this is not important for the purposes of this particular section (it will be returned to in the conclusion to Part I). What is important is to understand what Marcel suggests in place of Bergson's analysis.

Over and above the practical realm of action to which Bergson relates memory, in Marcel memory bears witness to values.[88] Conservation of the past, argues Marcel in the second part of the *Journal*, relies on the threat of loss or dispersion: 'N'est-ce susceptible d'être pensé comme conservé que ce qui peut aussi à la rigueur être conçu comme dispersé ou perdu. La conservation implique l'action d'une force qui s'oppose à la dispersion' (*JM*: 147; 10 December 1918). It is the act of valuing, he continues, which opposes such a threat: 'L'idée de conservation implique celles de sauvegarde et de valeur' (*JM*: 147). Rather than proposing a dialectic between (past) memory and (present) perception then, Marcel argues that the ontological structure of memories is itself dialectical, and crucially, he describes this experience in terms of a tension between the temporally finite and the eternal.

Indeed, aspects of my life can appear to me as either finite, contingent, and fleeting, or as necessary and eternal: that which is inconsequential to me I view as finite and contingent; anything that is of genuine consequence, on the other hand, cannot be interpreted as such, since it continues to appear, in some sense, necessary to me. Objects,[89] for Marcel, are not therefore conserved: 'Evidemment la chose est ce qui ne se conserve pas; c'est en tant que choses [...] que nous sommes mortels' (*JM*: 147). And this is because of their passive rigidity — a rigidity that Marcel attributes to memories in Bergson, and considers to be the downfall of his analysis: 'le souvenir immobile de Bergson [...] ne peut durer, il ne peut se conserver' (*JM*: 150; 10 December 1918). That which does survive in memory, however, rises above the temporal order of finitude to a plane of eternity: 'l'expérience mnémonique n'implique-t-elle pas la négation effective et réelle du temps?' (*JM*: 130; 2 April 1916). So, rather than interpreting memories as purely past, Marcel recognizes memories only insofar as they continue to be lived — in spite of the threat of loss — as necessary or eternal to me in the present (this could be either positive, as something valuable, or negative, as trauma or loss).⁹⁰ As such, the past which has survived almost takes on a life of 'my' own, transforming itself in response to the obstacles that challenge it,⁹¹ in order to re-assert itself as 'my' present in spite of the changing flux and finitude of time.⁹²

Memory's eternity of value⁹³ has a further significance for Marcel, which is equally fundamental to understanding his opposition to Bergson: memory is identified with the affirmation of a spiritual realm:

> Les souvenirs ne peuvent être conservés que pour autant qu'ils sont [...] susceptibles d'être dispersés. Mais que veut dire ceci? Y-a-t-il vraiment un sens où on puisse traiter le souvenir comme un élément ou un agrégat susceptible de se défaire? Il semble qu'il faille reconnaître qu'en réalité les idées spatiales de perte, de dissémination impliquent au fond quelque chose de spirituel. (*JM*: 149)

Not only this; in Marcel's view, the spiritual values that are truly capable of surviving loss or dispersion are communal values (and again, these may be positive or negative), which ground an underlying intersubjective relation between all individuals. In order to be recognized as such, though, these values must first be confronted with time's threat. It is then only through a form of collective revolt that the threat of destruction can be overcome: 'La puissance spirituelle nous apparaît — à tort ou à raison — comme ne pouvant triompher que de l'inattention, de la dispersion spirituelle: c'est seulement, croyons-nous, *communément*, de cette façon indirecte qu'elle peut triompher de la dispersion matérielle, des forces matérielles anarchiques' (*JM*: 149; my emphasis).⁹⁴ Hence, Marcel's account of memory confirms the importance of intersubjectivity for his philosophy, previously suggested by certain remarks concerning Bergson's presentation of intuition and music.

For Marcel, then, the 'ruptured' experience of time, with its constant threat of loss, degradation, or dispersion, is an incitement for transcendence — not only of that which is finite and contingent, but also of the division between the self and others; and this then raises the dialectic of existence he affirms to another level, where its tension is between time and (a more essential notion of) eternity. Bergson, on the other hand, categorically refused any notion of going beyond the time of *la durée*, toward an eternity that was conceived to be more authentic. Hence, it seems difficult to support Tsukada's suggestion that the conceptions of eternity in Bergson and Marcel are akin,⁹⁵ for although, as the first section of this chapter observed, Marcel appeared to agree with Bergson about how human existence should be characterized, his metaphysics of time has emerged to be quite different and appears to interlink with a notion of eternity that Bergson would in fact oppose. For Bergson, time is in itself something eternal, and the source of authentic Being. Authenticity in Marcel, on the other hand, is always spoken of in terms of eternity; the (temporal) immediate is merely presented as the starting point for transcendence — that which makes us realize we are something more.[96] But as a consequence, it has been suggested, the consistency of the position he secures against Bergson might be questioned: if he insists that philosophy engage with the temporality of the here and now, and that it recognize a tragic dialectic in the lived reality of human existence, is he not actually contradicting his project by privileging such an eternal present? 'Qu'est-ce qui compte premièrement et principalement pour [ce] philosophe de l'existence', Thành Tri Lê asks, 'l'existence temporelle ou l'existence éternelle de l'être? [...] on attend du Marcel une précision' (1961: 198).

Notes to Chapter 1

1. Cain is of a similar opinion: 'Among thinkers at the turn of the century, Bergson was undoubtedly the closest to Marcel in his attitude and emphases, and played an important role in Marcel's development' (1963: 22). Gilson has also described Marcel as a disciple of Bergson: 'C'est de l'intérieur du bergsonisme même qu'on peut entrer dans cette nouvelle philosophie de l'être [qui est celle de Marcel]' (1947: 5). In general, however, the profound influence Bergson had on Marcel's thought is not reflected in the secondary literature. Alexander (1948) does discuss Bergson's influence in some detail; but only one study to date actually focuses on Bergson and Marcel, namely Tsukada's *L'Immédiat chez H. Bergson et G. Marcel* (1995).
2. Bergson became a professor at the *Collège* in 1900, replacing Charles L'Évêque as the Chair of Ancient Philosophy. In 1904, he transferred his position to the Chair of Modern Philosophy, following the death of Gabriel Tarde. He then continued to hold this chair until he retired from his duties in 1920. For a comprehensive chronology of Bergson's life, see Bergson, Ansell-Pearson, and Mullarkey (2002: viii–xi).
3. Marcel was not the only philosopher to be inspired by Bergson. Jacques Chevalier, for example, also attended Bergson's lectures (from 1901) and describes how 'une foule énorme se pressait autour de la chaire où le maître [Bergson] parlait de l'origine de notre croyance à la causalité, des concepts, de l'idée de temps, de Plotin, de Descartes' (1959: 3); and Sartre attributes his choice of philosophy over literature to Bergson's influence when he writes: 'j'avais lu le livre de Bergson, l'*Essai sur les données immédiates de la conscience*. Et là, j'avais été saisi. Je m'étais dit: "Mais, c'est formidable la philosophie, on vous apprend la vérité"' (1977: 40).
4. 'Ce qui a le plus manqué à la philosophie, c'est la précision' (Œ: 1253).
5. Cf. Marcel: 'une solution en philosophie n'est jamais ce qu'elle est dans le domaine des sciences' (*DH*: 11). Dilthey (1833–1911) argued similarly: 'Starting from the most universal concepts of a general methodology the human studies must work towards more definite procedures and principles within their own sphere by trying them out on their own subject-matter, just as the physical sciences have done. We do not show ourselves genuine disciples of the great scientific thinkers simply by transferring their methods to our sphere; [...] the methods of studying mental life, history and society differ greatly from those used to acquire knowledge of nature' (Rickman 1976: 89).
6. '[L]a métaphysique n'a rien de commun avec une généralisation de l'expérience' (Œ: 1432). Bergson was careful to add that this insufficiency should not be interpreted as a value judgement: 'nous voulons une différence de méthode, nous n'admettons pas une différence de valeur, entre la métaphysique et la science' (Œ: 1285). On the contraire, Bergson writes: 'Nous croyons [que la science et la métaphysique] sont, ou qu'elles

peuvent devenir, *également* précises et certaines. L'une et l'autre portent sur la réalité même' (Œ: 1286; my emphasis).

7. E.g. Œ: 51, 75–76, 156, 326–29, 755–60, 1259.
8. Eleaticism was one of the principal schools of pre-Socratic philosophy, which was founded by Parmenides and flourished in the fifth century BCE. It pursued a strictly logical approach and therefore founded reality in the all-encompassing static unity of Being, dismissing perceptual appearances (for further reading on the pre-Socratic philosophers, see Kirk, Raven, and Schofield (1983)). Bergson's opposition to the Eleatics has wider implications, however: their rationalist methodology and principles greatly influenced the Greek thought that followed; and the Greek philosophical canon has, in general, dominated Western philosophy, setting the standards for (traditional) 'good' thinking.
9. First published in the April–May edition of the *Revue de métaphysique et de morale* in 1925, this article was then reprinted in 1927 as an appendix to Marcel's *Journal métaphysique* (*JM*: 309–29).
10. '[I]déalisme et réalisme sont deux thèses également excessives' (Œ: 161).
11. Œ: 177.
12. 'Je suis tenté de me demander aujourd'hui si mon aversion pour le lycée n'est pas à l'origine de l'horreur croissante que devait m'inspirer l'esprit d'abstraction' (*RA*: 304). See also *EC*: 37–39 and *HV*: 22–23.
13. On Marcel's metapsychical experiences themselves, see *EC*: 100–08, *GM*: 16–17, and Monestier (1999: I, interview 3). For related philosophical reflection see *JM*: 33–36, 151–52, 165–69, 173–75, 233–36, 239, 243–45, 246–48, 262–63, and *RA*: 309–10.
14. Moran defines phenomenology as 'a radical, anti-traditional style of philosophising, which [... attempts] to get to the truth of matters, to describe *phenomena*, in the broadest sense as whatever appears in the manner in which it appears, that is as it manifests itself to consciousness, to the experiencer. As such, phenomenology's first step is to seek to avoid all misconstructions and impositions placed on experience in advance' (2000: 4).
15. Cf. Bergson: 'La vérité est qu'une existence ne peut être donnée que dans une expérience' (Œ: 1292).
16. Associated philosophers include (but are not restricted to): Beauvoir, Berdyaev, Buber, Camus, Heidegger, Jaspers, Kierkegaard, Merleau-Ponty, Nietzsche, Ortega, Sartre, and Unamuno. All of these thinkers reacted against the arrogance of the Enlightenment's exaltation of Man and his all-powerful faculty of Reason, questioning the resultant assumption that logical consistency and deductive reasoning were sufficient for grasping all truths. All turned to personal lived experience in their search for truth, advocating an exploration of — as opposed to objective deductions about — the human.
17. For Barrett, Bergson was 'the first to insist on the insufficiency of the abstract intelligence to grasp the richness of experience, on the urgent and irreducible reality of time, and — perhaps in the long run the most significant insight of all — on the inner depth of the psychic life which cannot be measured by the quantitative methods of the physical sciences' (1990: 15). Indeed, as Capek writes: 'With such few exceptions as Heraclitus, Schelling in his last period, and, to a certain extent, Hegel, there has been a persistent tendency in the philosophy of all ages to interpret reality in static terms and to consider temporal existence as a shadowy replica of a timeless and Platonic universe' (1950: 331).
18. '[L]a philosophie n'a un poids et un intérêt que si elle a un retentissement dans cette vie qui est la nôtre' (*PST*: 37).

19. Bergson analysed human temporality before Husserl, who examined internal time consciousness in a series of lectures delivered in 1928. However, the new generation of French philosophers that emerged after World War II (Sartre and Merleau-Ponty, most notably) was more interested in German thought than recent French philosophy. Bergson consequently lost his intellectual dominance — principally to Husserl and Heidegger. Regarding Marcel's independence, see Cain (1963: 99–100), Ricœur (1976b), Schilpp and Hahn (1984: 495), and Spiegelberg (1994: 467, note 4).
20. Heidegger (1962).
21. In the late lecture 'Ma mort et moi' (published 1968) Marcel in fact criticized Heidegger for failing to appreciate the importance of Bergson: 'Je dirai, quant à moi, que la mort ne m'apparaît comme un terme que si la vie est assimilée à un certain parcours. Mais elle ne se présente à moi de cette manière que si je la considère du dehors — et dans cette mesure, elle cesse d'être éprouvée par moi comme ma vie: ici on rejoindrait Bergson dont Heidegger me semble avoir tellement méconnu l'importance' (*PST*: 181). See also *EC*: 76, and Marcel (1945: 98–99).
22. See, more generally, Husserl (1931:101–16, §27–34).
23. For further detail on Husserl's theory of time, see in particular Husserl (1964).
24. As Marcel does not analyse time so directly, it is not possible to provide equivalent terms in his philosophy. However, as will be seen, many parallels can still be drawn between his position and that of Bergson. As Parain-Vial states: 'Gabriel Marcel admet [...] comme incontestable la description bergsonienne de la première expérience' (1989: 147).
25. The literal meaning of Husserl's *Raum-Zeit* is, similarly, 'space-time'.
26. '[O]n ne saurait établir un *ordre* entre des termes sans les distinguer d'abord, sans comparer ensuite les places qu'ils occupent; [...] si l'on établit un ordre dans le successif, c'est que la succession devient simultanéité et se projette dans l'espace' (*Œ*: 68).
27. '[L]a multiplicité des états de conscience, envisagée dans sa pureté originelle, ne présente aucune ressemblance avec la multiplicité distincte qui forme un nombre' (*Œ*: 80).
28. '[L]e mécanisme de notre connaissance usuelle est de nature cinématographique' (*Œ*: 753).
29. '[A]uditive and kinesthetic images (melody, elan, explosion) definitely prevail in [Bergson's] writings. Once we get rid of a symbolic representation of time as a line, we also escape the tendency to imagine its "parts" geometrically, that is as *points*. In other words, the moments of time are not punctual instants; time is not infinitely divisible' (Capek 1950: 339).
30. 'Un milieu [homogène] de ce genre n'est jamais perçu; il n'est que conçu' (*Œ*: 628).
31. It is thus also referred to as 'le moi fondamental' (e.g. *Œ*: 80).
32. See also Cain (1963: 14).
33. It is however anticipated in his *Journal métaphysique* (*JM*: 236–37, 252–53, 266–67, 273, 292–93, 304–06).
34. Kant (1998: 563–69). Here, Kant is specifically arguing against Saint Anselm's ontological argument for the existence of God, which claims that existence is a perfection, and uses this to define God (who is perfect) into being.
35. *EA*: I, 14, 102–05, 167–68.
36. 'Pour *avoir* effectivement, il faut *être* à quelque degré, c'est-à-dire être immédiatement pour soi, se sentir comme affecté, comme modifié' (*EA*: I, 167–68).
37. See Marcel's earlier reflections in *JM*: 262. See also *DH*: 114, *ME I*: 115–20, and *RI*: 28–32.
38. See also *DH*: 67–70.

39. It is in fact to Bergson that Marcel feels indebted, regarding his discovery of language's complicity with spectatorship. In the 1958 lecture 'La Musique dans ma vie et mon œuvre' Marcel states: 'ce n'est pas à partir d'une donnée visuelle, quelle qu'elle puisse être, que s'est développée en moi la recherche ontologique, mais beaucoup plutôt à partir d'une expérience qu'il est aussi malaisé que possible de traduire en un langage presque toujours élaboré à partir des objets, à partir des choses. [...] un des points sur lesquels la rencontre avec la pensée bergsonienne a été la plus féconde, la plus enrichissante, c'est justement cette sorte de dénonciation méthodique des illusions auxquelles peut donner lieu le langage' (*EM*: 94).
40. See especially *DH*: 199–219, *HH*: 33–76, Marcel (1954a), and *PST*: 151–74.
41. Bergson's *Le Rire* (1900) expresses this concern to an extent: by characterizing humour as an impression of 'du mécanique plaqué sur du vivant' (Œ: 405, 410), he implies that acting like a machine is the furthest one can be from human reality. Other twentieth-century thinkers who are critical of technology include Berdyaev, Heidegger, Jaspers, and the philosophers of the Frankfurt School.
42. And Sartre reflected on the kind of life lived by a waiter: 'Considérons ce garçon de café. Il a le geste vif et appuyé, un peu trop précis, un peu trop rapide, il vient vers les consommateurs d'un pas un peu trop vif, il s'incline avec un peu trop d'empressement, sa voix, ses yeux expriment un intérêt un peu trop plein de sollicitude pour la commande du client [...]. Toute sa conduite nous semble un jeu. Il s'applique à enchaîner ses mouvements comme s'ils étaient des mécanismes' (2004: 94).
43. *Le Monde cassé* is the title of one of Marcel's most well-known theatrical works, but he also uses the term frequently in his philosophy. It is the title of his second 1949 Gifford Lecture, for example, and the (philosophical) lecture 'Position et approches' was first published as an appendix to this play. The relation between Marcel's thought and theatre is examined in Chapter 4.
44. See also *PI*: 56–57 and *RI*: 209–10.
45. '[L]a création n'est possible qu'à condition de nier pratiquement le *réalisme du temps*' (Marcel 1924: 44).
46. Indeed, for Bergson it is crucial to recognize that our engagement with the world is motivated (cf. Husserl's notion of intentionality; this is outlined in Chapter 2, note 12). Our observations are therefore not purely impartial, made only out of a desire to further truth; rather, our perception always occurs in the context of action, necessarily relating it to the specific way in which we happen to be interacting with the world. Bergson thus concludes: 'l'obscurité du réalisme, comme celle de l'idéalisme, vient de ce qu'il oriente notre perception consciente, et les conditions de notre perception consciente, vers la connaissance pure, non vers l'action' (Œ: 362).
47. Indeed, Marcel makes this argument himself in *ME I*: 188–89.
48. Cf. Marcel: 'Plus on s'éloigne de la zone où l'idée est un plan d'action [...], plus on s'enfonce dans le non-pragmatisable, c'est-à-dire au fond de la métaphysique' (*EM*: 44).
49. Husserl (1931) also argues for a form of 'intuition', as the means of accessing pure consciousness.
50. In addition to *intuition* and *intelligence*, Bergson distinguishes a third mode of knowledge: *instinct*. These three notions are discussed, for the most part, in *L'Évolution créatrice* (1907) (see especially Œ: 578–652). Here Bergson suggests that, with the evolution of animals, a dichotomy developed between an innate knowledge of matter (instinct), and a pragmatic, analytical engagement with life (intelligence). When humans evolved, a third form of knowledge developed, namely intuition. Intuition is instinct in its most developed form: it is a reversal of the intellectual, but more than an

attention to matter, it is an attention to the essence of life itself, and as such, in touch with the very becoming (*devenir*) of Being. By the time Bergson wrote *L'Évolution*, this no longer meant *la durée* (of the individual), but the collective generative movement of all human life, which he termed 'l'élan vital'.

51. While the term 'âme' may suggest a religious understanding of self, it is not used systematically by Marcel and is seemingly interchangeable with more neutral terms such as 'le moi', 'le sujet', 'le soi', and 'l'être'. One should not, therefore, be too quick to attribute specifically Christian connotations to the word. See Chapter 5 for further discussion of the 'Christian' status of Marcel's philosophy, which includes discussion of Marcel's non-traditional understanding of life after death.

52. For an exposition of Natorp's argument, see Zahavi (2006: 73–78).

53. 'As [Heidegger] wrote in his 1919–20 lecture course *Grundprobleme der Phänomenologie*: "The aim is to understand this character of self-acquaintance that belongs to experience as such" [...]. Any worldly experiencing [...] is characterized by the fact that "I am always somehow acquainted with myself"' (Zahavi 2006: 80).

54. Bergson and Marcel nevertheless recognize the potential of poetic expression. In 'Musique comprise et musique vécue' (1937), for example, Marcel writes: 'Il est trop clair, au surplus, que la poésie lyrique constitue ici un chaînon essentiel entre la pensée métaphysique et la pensée musicale' (*EM*: 45); and Bergson also relates intuition to poetry: 'Je veux bien aussi que cette part si modique d'intuition se soit élargie, qu'elle ait donné naissance à la poésie, puis à la prose, et converti en instruments d'art les mots qui n'étaient d'abord que des signaux' (*Œ*: 1321).

55. 'This is not incidental to the pervasive post-Hegelian concern with the limits of systematic philosophy. Schopenhauer and Nietzsche are the major figures in the nineteenth century who believed that music was the metaphysical art. There are others in the twentieth century, Adorno most notably, who give some privilege to music. Philosophy, particularly in its logicist forms, can run roughshod over the subtleties, intimacies of being. Music may sing these, as it were, in a manner that forces philosophy to raise the question of the unsayable — the unsayable that yet is sung and so somehow said' (Desmond 1994: 112–13).

56. 'Quand nous écoutons une mélodie, nous avons la plus pure impression de succession que nous puissions avoir, — une impression aussi éloignée que possible de celle de la simultanéité, — et pourtant c'est la continuité même de la mélodie et l'impossibilité de la décomposer qui font sur nous cette impression. Si nous la découpons en notes distinctes, en autant d'"avant" et d'"après" qu'il nous plaît, c'est que nous mêlons des images spatiales et que nous imprégnons la succession de simultanéité' (*Œ*: 1384).

57. Husserl also describes internal time-consciousness using a musical analogy (e.g. 1964: 41, 43).

58. E.g. *EA*: I, 170–71, *EC*: 207–08, and *DH*: 44. On the importance of music for Marcel in general, see for example *EC*: 46–51.

59. See also *EC*: 119.

60. 'Nous voulions surtout protester une fois de plus contre la substitution des concepts aux choses, et contre ce que nous appellerions la socialisation de la vérité. [...] il faut réserver cette socialisation aux vérités d'ordre pratique, pour lesquelles elle est faite. Elle n'a rien à voir dans le domaine de la connaissance pure, science ou philosophie. [...] Nous recommandons une certaine manière difficultueuse de penser' (*Œ*: 1327–28). See also *Œ*: 1292.

61. See also *Œ*: 1288.

62. As a result, Michaud remarks that secondary reflection is 'one of the most elusive [...] notions in Marcel's thought', thereby making it difficult to characterize positively;

'with such a protean development, [...] there has been minimal agreement among Marcel's interpreters as to exactly what secondary reflection is' (1990: 222).

63. Marcel does not provide an explicit definition of 'dialectique', but his understanding of the term seems to share parallels with the Hegel's (as opposed to Marx's, for example), since both Marcel and Hegel examine consciousness' experience of itself, and of its objects, and emphasize the dynamic and conflictual nature of this. For more discussion of Marcel and Hegel see Chapter 2.

64. Marcel was critical of Maritain's Thomistic notion of intuition for similar reasons. See Devaux (1974: 94-98) and Bars (1991: 240-45). Sweetman (2008: 121-34) also discusses relations between Marcel and Maritain. He agrees that the most crucial difference between them concerned 'the respective roles they each assigned to conceptual knowledge in their thought' (2008: 124). Nevertheless, Sweetman feels that the two philosophies still have much in common, and aims to expose these points of agreement as a corrective to the excessively divisive way in which Marcel and Maritain often presented their positions.

65. In Bergson's defence, although he describes intuition as 'un acte simple' (Œ: 1396) and '[une] vision directe de l'esprit par l'esprit' (Œ: 1285), he does not present it as the passive, privileged mode of understanding that Marcel accuses him of advocating. On the contrary, Bergson argues that intuition requires a sustained effort: 'Le travail habituel de la pensée est aisé et se prolonge autant qu'on voudra. L'intuition est pénible et ne saurait durer' (Œ: 1275).

66. Tsukada agrees: 'G. Marcel différait surtout de Bergson par son insistance sur les ruptures de la durée ou sur ce qu'il appelle temps destructeur' (1995: 161).

67. On Bergson's lack of a tragic sense, see also Gutting (2001: 114-15), and Hyppolite (1971: I, 452).

68. See also DH: 36 and EC: 80, 201.

69. See also DH: 57-58.

70. This is also affirmed in the preface to Pour une sagesse tragique et son au-delà (1968): 'tout au long de mon périple philosophique j'ai tenté d'approcher l'Etre. Il convient sous ce rapport d'attacher une importance particulière à la communication faite à la Société de Philosophie sur l'*Etre devant la pensée interrogative*. On est ici aux antipodes de toute philosophie qui entend s'appuyer sur une intuition ou même sur une affirmation préalable de l'Etre. Celui-ci se présente au contraire comme ne pouvant être qu'approché et toujours très imparfaitement dévoilé' (PST: 13). Though, as mentioned above, Marcel does briefly speak of an 'intuition aveuglée', he quickly rejects this expression in favour of others such as 'assurance existentielle', 'exigence ontologique', and 'exigence d'être'. As he states in 'L'Être devant la pensée interrogative', he is 'disposé à admettre [...] une intuition aveuglée, ou obturée. Mais ici le mot intuition peut-il encore convenir? J'estime qu'on peut en discuter' (PST: 78).

71. EM: 36. In a diary entry dated 8 March 1933, Marcel also comments on the universal scope of music's expressive capacity: 'A propos d'une phrase de Brahms qui m'a hanté tout l'après-midi [...], j'ai brusquement compris qu'il y a une universalité qui n'est pas d'ordre conceptuel, et là est la clef de l'idée musicale. Seulement, comment c'est difficile à comprendre! L'idée ne peut être que le fruit d'une certaine gestation spirituelle. Analogie intime avec l'être vivant' (EA: I, 170-71). Significantly, in a 1934 footnote to this entry, Marcel references Bergson, again suggesting his analysis of music to be incomplete: 'Il est certain que Bergson a raison, que nous sommes ici dans un ordre où la durée s'incorpore en quelque façon à cela même qu'elle prépare et fait venir à terme. Mais c'est peut-être sur ce qu'on pourrait appeler l'aspect structural de cet ordre de réalité que malgré tout Bergson n'a pas suffisamment projeté la lumière' (EA: I, 171, note 1).

72. This use of the saint as model is problematized in Chapter 5.
73. Marcel also identifies time with instrumentality, with *avoir* as opposed to *être* (e.g. *JM*: 263).
74. As will be seen later in this section, although Bergson's philosophy is dualistic (and he himself acknowledges this, e.g. *Œ*: 161) this is not the same kind of dualism as Descartes's, which Bergson describes as 'le dualisme vulgaire' (e.g. *Œ*: 318, 354).
75. In his 1941 lecture 'Moi et autrui', Marcel also equated the term 'actualité' with the (ontologically authentic) notion of 'la personne' (*HV*: 15–35). Here too, *actualité* was identified with transcendence of the immediate present.
76. E.g. *Œ*: 79–80.
77. Indeed, for Augustine memory was more than the mere act of recalling something: 'Great is the power of memory, an awe-inspiring mystery, my God, a power of profound and infinite multiplicity. And this is mind, this is I myself' (1998: 194, X. xvii.26), he writes in his *Confessions*. Henry Chadwick's editorial commentary to this work explains further: '*Memoria* for Augustine is a deeper and wider term than our "memory". In the background lies the Platonic doctrine of *anamnesis*, explaining the experience of learning as bringing to consciousness what, from an earlier existence, the soul already knows. But Augustine develops the notion of memory by associating it with the unconscious [...], with self-awareness, and so with the human yearning for true happiness found only in knowing God' (Augustine 1998: 185, note 12). Locke (1632–1704) famously reasserts the importance of memory with respect to personal awareness and identity (1997: Book II, chap. 27).
78. Unless otherwise indicated, citations from Bergson in this section are from *Matière et mémoire*.
79. '[L]'esprit déborde le cerveau de toutes parts' (*Œ*: 858).
80. See also *JM*: 189.
81. See also *EA*: I, 44, 160–61, and *PI*: 46, 54–56 (p. 56 references Bergson).
82. Bergson's distinction between time and space might, then, be mapped directly onto that between memory and perception, since perception, on Bergson's account, takes place in the space of the present instant, whereas memories are the 'substance' of *la durée*.
83. See especially *Œ*: 276.
84. Parallels may be drawn between Bergson's account of memory and psychologists' (increasing) focus on the context of recollection, that is, on the interaction between present circumstances and memory traces. Engel argues, for example, that 'one creates the memory at the moment one needs it, rather than merely pulling out an intact item, image, or story' (1999: 6). See also Schacter (1996).
85. '[A]ll the doublings or dualisms in Bergson derive from the coexistence of the past with the present' (Lawlor 2003: 55). Lawlor's book offers the most thorough and authoritative account of *Matière et mémoire* (on which it is specifically focused) that I have encountered.
86. See also *EA*: I, 160–61, and *JM*: 164, 176, 197, 242–44.
87. See for example *Œ*: 298–90, 856–57.
88. 'La mémoire comme indice ontologique. Liée au témoignage. L'essence de l'homme ne serait-elle pas d'être un être qui peut témoigner?' (*EA*: I, 120; 7 October 1932).
89. Or things which have been objectified. Here it is possible to relate the ontological dialectic of memory to the hermeneutical dialectic between (objectifying) *réflexion primaire* and (recuperative) *réflexion seconde* — for Marcel, the latter dialectic is instructive with respect to the former.
90. '[L]es souvenirs ne sont pas extérieurs à l'activité psychique' (*JM*: 149; 10 December

1918); 'le passé n'est pas séparable de la considération qui porte sur lui' (*EA*: I, 161; 7 February 1933).
91. 'Il semble que le passé ne puisse se conserver que dans la mesure où il se transforme en se survivant à lui-même' (*JM*: 150; 10 December 1918).
92. '[L]e souvenir est vraiment le mode selon lequel une réalité me demeure présente, c'est en lui et par lui qu'elle transcende sa contingence' (Marcel 1973b: 376).
93. '[L]'éternel ne peut être défini en dehors de tout rapport à la valeur' (*JM*: 151; 18 December 1918).
94. Marcel also compares this aspect of memory with telepathy (e.g. *JM*: 168; *EA*: I, 120).
95. Tsukada (1995: 245).
96. '[U]n être ne réalise pas dans l'immédiat, dans le pur maintenant "la plénitude de ce qu'il est"' (*JM*: 194–95; 4 July 1919).

CHAPTER 2

~

Phenomenological Time

Chapter 1 introduced Marcel's and Bergson's shared desire to engage with the time of lived experience as opposed to thinking in abstraction. However, Marcel's philosophy emerged as very different from that of Bergson and, as such, was suggested to be problematic: it appeared to subordinate time to an eternal present, and this seemed at odds with his assertions concerning (temporal) existence's dynamic, dialectical nature. This potentially problematic relation between time and eternity was identified in the first part of Marcel's *Journal* (January–May 1914), before his thought evolved in reaction against his idealist leanings; but strangely, in spite of his subsequent vow to ground his philosophy in concrete human reality, his presentation of time remained unaffected in the *Journal*'s second half (1915–23). Although Marcel's encounter with Bergson had been pivotal in stimulating the reconfiguration of his philosophical approach, his ensuing efforts to differentiate his newly reformed philosophy from Bergson's position as well seemed, in fact, only to reinforce his affirmation of an authentic eternal realm.

Marcel's efforts to dissociate himself from Bergson were most clearly illustrated by Chapter 1's comparison of memory in the two philosophies, which revealed Bergson's concern to be with action within time, and Marcel's to be with eternal values that in some sense surpassed time. Marcel, here, referred to time only in terms of contingency and finitude; he did not understand it as the eternal motion Bergson proposed. And it was in relation to this contingency and finitude that he postulated an eternal present, allowing him to testify to foundational values, via memory's dialectical experience, as well as to an underlying unity between the self and others.

A desire to affirm Being as intersubjective, and its temporal experience as dialectical, thus emerged as the driving force behind Marcel's break from Bergson; these two motivations were pivotal to Marcel's rejection of both Bergson's account of memory, and his (more general) description of *la durée*. Such points of contention remained central to Marcel's philosophy as he continued to shape his reappraised position — with intersubjectivity's importance, in particular, coming to the fore. This chapter therefore engages with Marcel's 'concrete' approach in more depth, and examines the way in which he defends his position in order to establish whether his presentation of time might still be justified. As will be seen, this raises questions about Chapter 1's reading of time in Marcel. The first section suggests that — contrary to Bergson — it is a mistake to interpret Marcel's references to time in directly ontological terms, despite his clear interest in the metaphysics of Being. His 'concrete' interest in time is instead revealed to be phenomenological; he is not philosophizing about time itself. Yet as begins to emerge in the second section of this chapter, such an argument may not be sufficient to absolve Marcel from Chapter 1's charge of inconsistency, for a difficult relation nonetheless exists between his phenomenological interest in time and the broader ontological concern of his philosophy — a difficulty that is made particularly apparent in his arguments for intersubjectivity. The third and final section of the chapter continues to debate the coherence of Marcel's overall project, and in so doing, moves on to discuss the ethical dimension to his thought.

The Irrelevance of Time?

In the first part of the *Journal*, Marcel makes an argument that has the potential to exonerate him from Chapter 1's critique: his argument for the 'irrelevance' of time. According to this, there can be no genuine conflict between time and eternity in his work because, he states, he is not actually concerned with time (or indeed eternity) as such. The argument begins, on 13 January 1914, by questioning the very intelligibility of the idea of time — or more specifically, by observing how beginnings can be attributed to things within time, but not, curiously, to time itself: 'Poser un commencement du temps, c'est certainement ne rien penser réellement; [...] le temps lui-même, comme forme idéale, n'a pas de commencement, mais ce qui est dans le temps peut en avoir un' (*JM*: 8). One could try to make sense of this perplexity by interpreting the idea of time as something which exists outside of temporality; and yet, Marcel notes, such a solution is empty, for it does not help us to understand what time is: 'le temps peut-il être réellement pensé comme purement idéal? comme n'étant durée de rien, duré par rien?' (*JM*: 8). Thus, concludes Marcel on 14 January, we should not attempt to define the nature of time in rational, objective terms: 'la pensée n'a pas le droit de réaliser le temps, de le traiter comme un objet sur lequel on peut raisonner'. Instead, he states in Kantian fashion:[1] 'Le temps est une condition formelle sous laquelle des objets peuvent nous être donnés, il n'est pas lui-même un objet. Le problème est insoluble parce qu'il est illégitime' (*JM*: 8).[2]

Marcel further supports this conclusion by refuting both realist and idealist theories of time, both of which recognize a distinction between the way time appears to us in our experience and time itself, and aim to account for this dualism. In realism's case, it is suggested that what is real is just what appears to us. Hence: 'Il y a solidarité entre le temps et la réalité temporelle' (*JM*: 8). However, Marcel observes, 'on ne peut inférer directement du caractère formel du temps son caractère apparent[; ...] il faut avoir recours comme intermédiaire à l'idée de la solidarité du temps et de ce qui est dans le temps' (*JM*: 9). Consequently, such a position fails to resolve the dualism between time's phenomenality and time as such, since, from the point of view of someone 'in' time, the identity between the experience of time and time itself is not immediately apparent: 'la réponse (négative) [à ce dualisme] claire pour le temps est au contraire très obscure pour ce qui est dans le temps' (*JM*: 9). Marcel thus turns to idealism to consider its alternative proposition — namely, that the world is, in itself, atemporal, and the experience of time simply an imperfect conception of such atemporality that results from consciousness's finitude:

> le seul moyen d'éviter la contradiction paraît bien être de dire que le monde véritable est hors du temps; et que les contradictions auxquelles nous nous heurtons sont liées à l'inadéquation de la forme du temps par rapport à une réalité elle-même (en soi) intemporelle. La subjectivité du temps servira d'explication; on dira que c'est à cause de l'infirmité de notre conscience finie que nous ne pouvons saisir cette réalité que temporellement. (*JM*: 9)

This still does not elucidate the relation between the phenomenality of time and time itself, however: 'Si [...] on pose le dualisme de ce qui est dans le temps et de ce qui est hors du temps, ce ne doit pas être avec l'espoir d'unifier ensuite ces deux ordres' (*JM*: 10; 15 January 1914). And so, Marcel concludes, since neither theory is capable of accounting for this dualism on a metaphysical level, 'le dualisme ne doit et ne peut être conçu en un sens ontologique' (*JM*: 12).

Returning to the question of the ideality of time on 16 January, Marcel therefore affirms that it is not his concern to philosophize about what time is:

> il résulte de tout ce qui précède que, si le temps peut être posé comme idéal, ce n'est pas du tout par opposition à une réalité qui serait soustraite au temps et dont l'ordre temporel serait une apparence subjective; c'est en ce sens seulement que le problème de la réalité ne peut *même pas* être posé à propos du temps. (*JM*: 12)

For this reason, Marcel situates himself in explicit opposition to Hegel (1770–1831), whose philosophical dialectic — he alleges — institutes a temporal hierarchy that judges certain historical moments as ontologically 'inferior' to others, when considered in relation to historical progress as a whole. Marcel, on the other hand, has no interest in making such judgements about the ontological status of temporal moments; he does not believe it is even possible for philosophy to make this kind of judgement:

> la pensée [...] se trouve amenée à se poser certaines questions qui semblent d'abord pouvoir porter sur le monde et sur ses conditions, mais qui, à l'analyse, se révèlent non seulement insolubles, mais mêmes inapplicables

à ce plan. Poser l'idéalité du temps et du monde en tant qu'il est soumis à un développement temporel [...], c'est donc dire seulement: *les problèmes métaphysiques ne peuvent se poser que dans un ordre où il est fait abstraction de tout rapport [...] du temps*. (*JM*: 12)

Rather, his concern is with the way in which the possibilities of (philosophical) affirmation change according to point of view, and thus he is interested in investigating what the process of reflecting on time reveals about reflection itself:

> Le progrès que nous décrivons n'est [...] pas le progrès synthétique hégelien [...], c'est le progrès d'une réflexion qui transcende ses propres positions. Alors que le progrès défini à la façon hégélienne prétend nécessairement avoir une signification ontologique, mais que par là même se trouve soulevée la question insoluble (et qu'on ne peut éviter, mais seulement escamoter) de la réalité indépendante des moments inférieures — la dialectique telle que je la conçois étant purement idéale, portant exclusivement sur des modes d'affirmation, et ne prétendant pas, par suite, valoir ontologiquement, ne se heurte pas à cette difficulté. (*JM*: 12–13)

Here, Marcel also moves away Kant's position; again, this is too dualistic. As he writes a little later, on 13 February:

> l'idée d'une raison organisatrice de l'expérience, quelque vraie qu'elle soit au fond, a ce défaut de paraître tout au moins invinciblement dualistique. Il me paraît que c'est dans un sens différent qu'elle doit être infléchie. Il faut admettre que la pensée (la raison) ne se constitue comme pensée pour elle-même qu'au fur et à mesure qu'elle se réalise dans l'expérience. Et ceci ne doit pas être entendu en un sens purement psychologique [...]; ceci veut dire [... qu'il est] arbitraire de prétendre définir l'expérience en fonction de catégories antérieures à elle ou de faire dépendre la pensée d'une expérience indûment réalisée. Il faut comprendre que la pensée ne se connaît et ne se saisit que dans l'expérience, au fur et à mesure que celle-ci se définit comme intelligible.[3] (*JM*: 75)

Marcel is not simply arguing, like Kant, that reflection relates only to things as they appear to us (*phenomena*), rather than to things in themselves (*noumena*);[4] Marcel also believes that reflection itself has different modes, and that things appear differently to us according to the kind(s) of object that each mode of reflection is disposed to recognize. This 'dialectique d'une réflexion qui s'élève à des plans d'intelligibilité de plus en plus élevés' (*JM*: 13) can still be instructive and allow for progress, Marcel insists, for 'le rapport de la pensée à son objet lui apparaît à elle-même comme différent suivant la façon dont elle conçoit l'objet' (*JM*: 13). And crucially, it has the advantage of avoiding commitment to an appearance–reality dualism: 'Alléguera-t-on enfin la difficulté de comprendre le rapport de la dialectique au temps? mais [*sic*] cette difficulté est précisément nulle, puisque nous ne posons pas la réalité de l'intemporel, mais seulement la transcendance du réel par rapport à l'opposition du temporel et de l'intemporel, c'est-à-dire l'"irrelevancy of time"' (*JM*: 13).⁵

If time is 'irrelevant' to Marcel's philosophical investigations, then, it is as something understood objectively in a direct ontological sense. Passages in the *Journal* which reject time should therefore be understood as making a statement about the way in which it is proper to conceive of ourselves in reflection, not as claiming the ontological transcendence of the self over time.⁶ Indeed, in a diary entry on 9 January Marcel argues against a purely temporal understanding of self, because he considers this to be detrimental to human action. For Marcel, to think and act for oneself requires an unmediated form of self-reflexivity, a relation that transcends any division between 'ce que je suis en tant que pensée (qui veut le bien, qui veut l'être) et cette matière absolument contingente que je suis en tant qu'empirique' (*JM*: 6). Thinking of oneself as temporally-bound makes the self appear contingent and thus invites passivity, because the self's actions seem to be entirely determined by the (empirical) time of causation; and this then encourages a dualistic understanding of self where matter and mind appear utterly distinct, leaving one to question one's relation to the world and thereby compromising one's ability to act. 'Je dois me refuser à me penser comme contingent (c'est-à-dire comme déterminé à l'infini)', Marcel insists.⁷ 'Ceci revient à dire que je dois me penser comme voulu par un acte intemporel qui est lié à moi-même sans intermédiaire'; '[l'unité de moi-même] ne peut être pensé qu'en dehors du temps' (*JM*: 6).⁸

As his philosophy became more established, Marcel began to express his argument for the irrelevance of time in terms of a phenomenological interest. In the discussion following his 1967 paper 'Mon temps et moi', for example, Marcel underlines his intention to restrict his analyses of time to 'la considération phénoménologique' as opposed to 'le temps en soi, physique ou cosmologique' (Hersch and Poirier 1967: 25). Although 'phenomenology' was first (explicitly) defined by Husserl, the application of the term is now much wider. As Moran explains, 'the phenomenological movement understood in its broadest terms includes not just the work of Husserl, but also the work of many original practitioners of phenomenology, who did not feel bound to Husserl's methodology' (2000: 2). One such practitioner was Heidegger, who denied that it was possible to provide, as Husserl attempted to do, a pure (formal) description of what 'is', of the immanent essences (*Wesen*) or fundamental ontological categories underlying the experience of consciousness.⁹ 'The expression "phenomenology" signifies primarily a *methodological conception* [as opposed to a philosophy in itself]', asserts Heidegger in the introduction to *Being and Time* (1962: 50); and 'the meaning of phenomenological description as a method lies in *interpretation*. [...] The phenomenology of Dasein[10] is a *hermeneutic*' (1962: 61–62). Marcel's phenomenology of (temporal) lived experience is similarly hermeneutical. 'Ce qui m'intéresse, c'est de considérer la manière dont nous parlons du temps, et l'expérience que sous-entend notre façon d'en parler', he explained in the discussion following 'Mon temps et moi', adding that if he had left aside the question 'qu'est-ce que le temps?', it was 'parce que je me demande [...] jusqu'à quel point nous pouvons arriver à la poser et même à lui donner un sens' (Hersch and Poirier 1967: 28). In fact, one might compare Heidegger's opposition to Husserl with Marcel's refusal of Bergsonian intuition, for this can also be seen to perform a Husserlian 'eidetic reduction' to the extent that it 'brackets' the more worldly, 'natural standpoint' of the *moi superficiel*, in order to gain pure consciousness, via a form of intuition, of the essential being of the *moi profond*. For Marcel (versus Bergson) and Heidegger (versus Husserl), on the other hand, Being cannot be accessed so directly: 'the thing itself is deeply veiled', argues Heidegger (1962: 49) in direct response to Husserl's claim to return to the 'things' or 'matters themselves' (1931: 256).[11] Whereas Husserlian intentionality[12] analyses the object-directedness of consciousness, Marcel (as this chapter's second section will demonstrate in more detail) and Heidegger understand consciousness' intentionality as relating to something broader and more existential. 'We are ourselves the entities to be analysed', contends Heidegger. 'These entities, in their Being, comport themselves towards their Being[, ... and thus the essence of any such entity] lies in its "to be"' (1962: 67). For both Marcel and Heidegger, each of us is already in the midst of Being.[13] This is the reason they believe an indirect, pre-ontological study of Being's experience in the world is necessary for any ontology.[14] As a result, their phenomenologies of time present lived reality as continual interpretation rather than as a discoverable essence,[15] and, as David Wood writes of Heidegger, are 'essentially participatory[, ... treating] subjects both as embodied — and thus from the beginning "in the world"' (1989: 326), and claiming that 'the most basic temporal patterns affecting us are not those that organize the persisting objects around us, but those that involve our actions and our self-understanding as finite beings' (1989: 326–27).[16]

Chapter 1's challenge to Marcel's philosophical consistency thus seems misplaced. Although Marcel agreed with Bergson that synchrony reduced time to space, and thereby failed to reflect lived reality, he did not believe that Bergson was justified in speaking of ontology so directly. In Marcel, therefore, reflection — indeed, philosophy itself — is what is in motion; it is on this indirect, pre-ontological level that he wished to speak of time. Having now established this, the second section will proceed to examine Marcel's *approches concrètes* in greater detail, so as to analyse his discourse on time further. Particular attention will be paid to his affirmation of intersubjectivity, since this was the other major factor motivating his break from Bergson.

Approches concrètes

Marcel's *approches concrètes* are pivotal to the position he establishes post-World War I, which aims to restore philosophical weight to human existence in all its lived experience.[17] The phrase 'approches concrètes' is first employed in the 1933 lecture 'Position et approches concrètes du mystère ontologique' — Marcel's first public attempt at outlining (and in so doing, consolidating) his position following the evolution he identifies in his thought.[18] Thereafter, Marcel continues to characterize his work in this way, and, in a 1961 William James Lecture, he describes the 1933 presentation as being 'au centre de ma production philosophique' (*DH*: 94).[19] It is only through an engagement with reality as lived, he insists, that one can begin to reveal the actual nature of Being. As a result, phenomenological method becomes pivotal to his reformed philosophy: 'je crois que si l'on ne se place pas sur le plan phénoménologique, on ne peut pas comprendre ce que j'ai écrit' (*GM*: 87), he remarked to Boutang in 1970.[20] In one of the 1949 Gifford Lectures ('L'Exigence de transcendance'), therefore, the purpose of which was to clarify the central themes and trajectory of his work,[21] Marcel underlined the importance of situating his philosophical discussions 'par rapport à la vie telle qu'elle est concrètement vécue' (*ME I*: 49), explaining, in another ('Réflexion primaire et réflexion seconde: le repère existentiel'), how his arguments would be based on 'des exemples les plus simples afin de montrer comment la réflexion s'enracine dans la vie courante' (*ME I*: 91). Furthermore, he argued at the beginning of the 1950 Gifford series (in 'Qu'est-ce que l'être?'), the more one understands one's own life and experience, the more one will be in a position to understand others' being: 'Plus que je m'élèverai à une aperception vraiment concrète de ma propre expérience, plus je serai par là en mesure d'accéder à une compréhension effective d'autrui, de l'expérience d'autrui' (*ME II*: 10). This is not simply a matter of empathy. Rather, the more I engage with the reality of my own experience, the more I will recognize the absurdity of any attempt to dissociate my being from that of others: 'La croyance à la solitude est la première illusion à dissiper, le premier obstacle à vaincre — dans certains cas la première tentation à surmonter' (*EA*: II, 53), he stated in the 1934 lecture 'Réflexions sur la foi'; 'réfuter [...] la position qui consiste à insulariser le moi' (*PI*: 146), he reasserts in his diary on 13 April 1943.[22]

So, according to Marcel, my being is not only incarnate (grounded in the world) but also intersubjective:[23] 'pour moi, le coesse est en effet de la première importance' (*GM*: 87), he affirmed in an interview with Boutang; and this inextricable link between myself and others is confirmed by an 'assurance existentielle', he argues, which is experienced when one is in touch with the concrete reality of one's existence. As he declared in the late lecture 'L'Humanisme authentique et ses pré-supposés existentiels' (published

1968): 'L'assurance existentielle porte [...] sur les conditions structurales qui permettent à l'être individuel de s'ouvrir aux autres' (*PST*: 68). More specifically, this section will show, these structural conditions are temporal. In Marcel's philosophy not only is time crucial for self-understanding; the self's relation to time equally determines the self's relation to others, since the self is not a discrete, isolatable entity. Before demonstrating this specifically, however, it will be necessary to introduce some further details concerning Marcel's philosophy in general.

Having distinguished his position from that of Bergson, Marcel situates his reoriented philosophy in relation to different thinkers, selecting Descartes's thought as his primary counterpoint.[24] More specifically, he situates the (non-objective) indubitable feeling expressed by his *assurance existentielle* in opposition to the (objective) certainty of self announced by Descartes's *cogito*. 'En admettant même que je sois, comment puis-je être assuré que je suis? Contrairement à l'idée qui doit se présenter ici naturellement, je pense que sur ce plan le *cogito* ne peut nous être d'aucun secours' (*HP*: 201–02), asserted Marcel in 'Position et approches'. Marcel does not object to the affirmation 'je suis'; it is the derivation of this assurance from 'je pense' that he challenges — the 'donc' or 'ergo' of the cogito.[25] For Marcel, dogmatic conclusions such as those drawn by Descartes about the 'je' only limit, and therefore betray human Being: 'il n'y *a pas de jugement objectivement valable portant sur l'être [sic]*' (*JM*: 97), he writes in the *Journal* on 20 April 1914.[26] To deduce my existence so analytically — that is to say, impersonally — is to distance myself from the reality of my existence, the experience of which is nothing but personal, and thus its nature cannot be reduced to any (generic) universal truth. The most fundamental indubitable with which we are confronted, Marcel contends, is not the truth of the *cogito* but that of sensation: 'La sensation (conscience immédiate) est infaillible' (*JM*: 131), he asserts on 4 May 1914.[27] He therefore argues that the assurance of one's existence emanates from an underlying feeling of 'presence', as opposed to anything that can be affirmed by a detached spectator's attitude. 'J'ai cherché à montrer', he explained in the 1950 Gifford Lecture 'Existence et être', 'que l'*existant*, [...] c'est mon corps considéré non seulement en tant que corps, en tant que chose corporelle, mais en tant que mien, ou encore en tant que *présence* massive et globalement éprouvée qui ne se laisse par conséquent pas réduire' (*ME II*: 27; my emphasis). If Marcel opposed the way in which Descartes responded to his discovery of self-awareness and attempted to 'establish' the nature of the self's reality, then, his reaction was — more fundamentally — a critique of what he felt was a lack of engagement with Being as it is experienced and lived.

Descartes's ability to doubt the existence of his body, combined with his inability to doubt the existence of a thinking 'je',[28] led him to conclude that mind and body were composed of two different substances — *res cogitans* and *res extensa* — that were absolutely distinct; the self (*le moi*; *le je*) was essentially a 'thinking thing' and, as introspection revealed this with a certainty not attributable to the body, the mind was, by its very nature, separate and more immediately knowable than the body. However, echoing an earlier argument made by Nietzsche,[29] Marcel maintains in 'Existence et objectivité' (1925) that 'seule la structure de notre langage nous contraint à nous demander s'il y a *quelque chose* qui possède l'existence: en réalité, existence et chose existante ne peuvent certainement pas être dissociées' (*JM*: 311). In fact, as Marcel argued as early as 27 January 1914, one only has to examine the *cogito* itself to see that it is its own proof that Cartesian dualism does not accurately understand the nature of existence:[30]

> Nous avons vu que le *je pense* pouvait se convertir en une forme, nous avons même nié la légitimité de cette conversion. Mais le fait même que cette conversion est possible est instructif [...]. Cette conversion n'était possible que par l'intermédiaire d'une pensée qui, confrontant le *je pense* et la matière empirique, interprétait leur relation d'une certaine façon. Déjà par conséquent le cogito était en quelque façon transcendé. (*JM*: 43)

It is impossible for introspection to reveal my 'thinking self' as a whole, for there will always be this transcendent spectator-thought which, dissociated from the rest of my thought so as to make contemplation of 'my thought' possible, will never be contained within 'my thought'; a definable 'je (qui pense)' is not there to be discovered.[31] Since this transcendence is necessary for the affirmation of the *cogito*, postulating a straightforward dualism between the *je pense* (subject) and the world (the object of the *je*'s thoughts) — that is, understanding the self in terms of objective, primary reflection — does not constitute a satisfactory understanding of existence because, contrary to the simplicity of the *cogito*'s form, we cannot actually grasp what such a dualism would consist of. In his 'Ébauche d'une philosophie concrète' (1940), Marcel writes:

> A vrai dire, la séduction que le *cogito* a exercée sur les philosophes réside précisément dans sa transparence au moins apparente. Mais il y aura toujours lieu de se demander si cette transparence n'est pas une *prétention* à la transparence. [...] si le *cogito* est réellement transparent à lui-même, nous n'en tirerons jamais l'existentiel, par quelque processus logique que ce soit. (*RI*: 91)

For Marcel, then, the inadequacy of how experience is presented to us in primary reflection is a negative affirmation of the need for secondary reflection. It is in light of this, as Herbert Spiegelberg observes, that Marcel's interest in phenomenology has to be seen (1994: 460).[32]

If secondary reflection is phenomenological, it is more frequently discussed in relation to the metaphysical — that is, ontological — project which Marcel has in view.³³ Indeed, Marcel's concern when addressing questions of selfhood is to correct the ontological conception of subjectivity suggested by thinkers such as Descartes — and indeed, Bergson. Furthermore, as a result of its intersubjective emphasis, Marcel's metaphysics of *être* is also immediately ethical: for Marcel, the desire to pick out a 'self' to which 'je' refers is ethically detrimental, breeding isolating individualism in taking the 'I' (or the 'eye') as its first point of reference. 'Si j'admets que les autres ne sont que ma pensée des autres, mon idée des autres, il devient absolument impossible de briser un cercle qu'on a commencé par tracer autour de soi. — Si l'on pose le primat du sujet-objet — de la catégorie du sujet-objet — [...] l'existence des autres devient impensable' (*EA*: I, 130–31), he writes on 11 November 1932. True contact with reality, on the other hand, involves 'participation',³⁴ and, as Marcel contended in a 1950 Gifford Lecture ('Qu'est-ce que l'être?'), is distinctly '*non*-optique, *non*-spectaculaire' (*ME II*: 18).³⁵ An inauthentic 'oubli de l'être' is, at the same time, '[un] oubli donc de fraternité', Parain-Vial (1966: 72) observes; and as such, *être* already implies what Marcel calls a 'disponibilité' (often translated as 'availability') toward others,³⁶ which recognizes them as subjects rather than objects.

Crucially, though, this ethico-ontological thesis could not have been reached without phenomenology. It is in fact the phenomenological experience of love that, for Marcel, demonstrates the possibility of escaping the solipsism that a (Cartesian) subject–object conception of self seems to imply, and thereby refutes its ontological foundation. 'C'est dans l'amour que nous voyons le mieux s'effacer la frontière entre l'*en moi* et le *devant moi*' (*HP*: 207), he announced in 'Position et approches'. This importance of love is further reinforced by Marcel's opposition to Sartre — the other major counterpoint to his concrete philosophy. According to Marcel, Sartre's objectifying *regard* excludes the possibility of knowing the Other as a subject (or *toi*,³⁷ in Marcellian terms) because of the pessimistic way in which it defines human relations.³⁸ His outrage at this failure to acknowledge the positive potential of *être* is particularly well illustrated in his fierce criticism of Sartre's play *Huis clos* (1944):

> Ce nihilisme moral est à mon avis tout à fait apparent dans *Huis clos*; il me paraît évident que l'auteur, en mettant exclusivement l'accent sur la dépendance de l'individu par rapport au regard et au jugement d'autrui, escamote systématiquement le 'nous' véritable qui est celui de l'amour ou de l'amitié. (1958a: 47)

For Marcel, love and friendship are what reveal the fundamental inadequacy of subject–object dualism. Philosophy should therefore seek to determine the ontological conditions that make such experiences possible.³⁹

Love and friendship are of course difficult to characterize; but this is pivotal to their transcendent status with respect to primary reflection.⁴⁰ 'On demandera: quel est donc le critère de l'amour réel? Il faut répondre qu'il n'y a de critériologie que dans l'ordre de l'objet ou du problématisable' (*HP*: 208), pronounced Marcel in 'Position et approches'.[41] 'Un problème', as Marcel begins to define in the diary entries of *Être et avoir*, in preparation for the 1933 lecture, 'est quelque chose que je rencontre, que je trouve tout entier devant moi, mais que je puis par là même cerner et réduire' (*EA*: I, 146; 23 December 1932). But, he insists, human affairs are not of the same order as these 'problems', which can be contemplated at a distance and resolved through the implementation of a certain device or procedure. It is thus inappropriate to conceive of them in such terms, Marcel argues; they should instead be understood as 'un mystère',[42] defined as 'quelque chose en quoi je suis *moi-même engagé* [...]. Au lieu qu'un problème authentique est justiciable d'une certaine technique appropriée en fonction de laquelle il se définit, un mystère transcende par définition toute technique concevable' (*EA*: I, 146; my emphasis).[43]

Accepting the reality of 'le mystère' (or 'le métaproblématique', as Marcel also refers to it)[44] does not mean that we must resign ourselves to being left in the dark about how to understand or approach the question of Being.[45] Rather, it means that we must refrain from thinking of human existence in the 'problematizable' terms of objectifying, primary reflection, which treat the question in complete isolation from our own existential situation:[46] 'quel accès puis-je avoir à l'ontologique comme tel? La notion même d'accès est ici évidemment inapplicable. Elle n'a de sens qu'au sein d'une problématique. [...] Impossibilité de traiter l'être de cette façon' (*EA*: I, 125), Marcel writes in his diary on 22 October 1932. Only the indirect (phenomenological) mode of secondary reflection can recover the concreteness of experience that is filtered out by primary reflection (or indeed, by Bergsonian intuition), and thereby approach, through a more global act of 'recollection' (*recueillement*) or 'recuperation' (*réflexion récupératrice*), the reality of the ontological 'mystery' in which I am always already embedded.[47]

Attempting better to illustrate the difference between problem and mystery in his 1933 lecture, Marcel discussed what it was like to experience a significant — perhaps spiritual — encounter with another,[48] and emphasized the impossibility of objectively accounting for such a feeling:[49]

> Vous faites une rencontre qui se trouvera avoir sur votre vie un retentissement profond, indéfini. [...] Qu'on me dise [...], 'vous avez rencontré telle personne à tel endroit, parce qu'elle aime les mêmes paysages que vous ou que sa santé l'oblige à subir le traitement que vous suivez vous-même', on voit aussitôt que la réponse est inexistante. Il y a à Florence ou en Engadine en même temps que moi une foule de personnes qui sont censées partager mes goûts; il y a dans la ville d'eaux où je me soigne un nombre considérable de malades atteints de la même affection que moi. Mais l'identité présumée de ce goût ou de cette affection ne nous rapproche pas, au sens réel du mot; elle

> est sans rapport avec l'affinité intime, unique en son genre, dont il s'agit ici. [...] je me trouve en présence d'un mystère, c'est-à-dire, d'une réalité dont les racines plongent au-delà de ce qui est à proprement parler problématique. [...] moi qui m'interroge sur le sens et la possibilité de cette rencontre, je ne peux me placer réellement en dehors ou en face d'elle, je lui suis en quelque façon intérieur, je dépends d'elle, elle m'enveloppe et me comprend — si moi je ne la comprends pas. (HP: 208–09)

Nevertheless, the specificity of love and friendship is related to a certain experience of 'presence' in Marcel's phenomenological studies,[50] defined in 'La Fidélité créatrice' (1940) as 'le fait [...] de me donner à sentir [que quelqu'un] est *avec* moi' (RI: 201),[51] and which, as Marcel noted in a 1961 William James Lecture ('Fidélité'), 'ne peut être [...] évoquée qu'à la faveur d'expériences directes et irrécusables' (DH: 94).[52] 'Une présence est une réalité, un certain influx; il dépend de nous de rester ou non perméables à cet influx' (HP: 231), stated Marcel in 1933, before explaining that it was 'la fidélité [... qui] consiste à se maintenir activement en état de perméabilité' (HP: 232). As will now be seen, it is the way in which the fidelity of love and friendship perpetuates presence[53] that makes visible the link between Marcel's *approches concrètes* and his engagement with human time.

Crucial to Marcel's conception of love is a notion of unconditional commitment; and in his William James Lecture 'Fidélité', he critiqued the modern-day attitude (as he interprets it) that dismisses such a notion, contending that 'ce qu'il faut [...] mettre en lumière, c'est le fait que l'amour, au sens le plus plein et le plus concret du mot, [...] semble prendre son point d'appui sur l'inconditionnel: je continuerai à t'aimer quoi qu'il arrive' (DH: 103). Marcel even goes so far as to present this form of commitment in terms of immortality: 'Aimer un être, c'est dire: "toi, tu ne mourras pas"' (HV: 194),[54] he affirmed in his 1943 lecture 'Valeur et immortalité'. As he explains in the essay 'La Fidélité créatrice', 'le problème de la mort coïncide avec le problème du temps saisi au plus aigu, au plus paradoxal de lui-même. J'espère réussir à faire voir comment la fidélité, appréhendée dans son essence métaphysique, peut nous apparaître comme le seul moyen dont nous disposions pour triompher efficacement du temps' (RI: 199).[55] For Marcel then, time and mortality are identified with conditionality; true value on the other hand, he insists in 'Valeur et immortalité', 'ne peut être pensée comme réalité [...] que si elle est référée à la conscience d'une destinée immortelle' (HV: 211).[56] Because of this, he believes it is possible to feel a person's presence even after death, because those who have had true influence on me, in some sense, continue to be with me and impact on my worldly experience whether or not they are (or even can be) present in the flesh.[57] In a 1950 Gifford Lecture ('Existence et être'), Marcel therefore rejected any analogy between a person who has died (he uses the examples of Victor Hugo and Napoléon) and '[un] appareil [qui] ne fonctionne plus' (ME II: 28). 'Si nous considérons *concrètement* Victor Hugo ou Napoléon, c'est-à-dire si nous ne pratiquons pas [cette] grossière réduction', he continued, '[il est infiniment douteux qu']il y ait un sens quelconque à dire que Napoléon ou Victor Hugo n'existe plus' (*ME II*: 28–29; my emphasis). And this is because such a question relates to the mystery of Being, not to a determinable problem.[58]

So, just as Marcel rejects the idea that life is simply a linear timeline of events, he refuses an (objective) temporal understanding of love and introduces, instead, a (transcendent) notion of eternity: 'cette réalité de l'aimé ne peut être maintenue que parce qu'elle est posée par l'amour comme transcendante à toute explication, à toute réduction. En ce sens il est vrai de dire que l'amour ne s'adresse qu'à ce qui est éternel' (*JM*: 63), he writes on 7 February 1914. Notice, however, that Marcel qualifies his statement with the words 'En ce sens'. His concern here is phenomenological (in a Heideggerian sense) — that is, hermeneutical; he is not referring to eternity as something ontological. This is reaffirmed on 7 March 1929, where, after writing in his diary that 'on commet une grave erreur en traitant le temps comme mode d'appréhension' (*EA*: I, 19), he then explains that this is not a question of grounding the self in an eternal, unchanging whole (*totum simul*), but merely of transcending a representational understanding of one's (temporal) life as linear succession:[59]

> transcender le temps, ce n'est pas du tout s'élever [...] à l'idée vide en somme d'un *totum simul* — vide parce qu'elle me demeure extérieure, et par ce fait même se trouve en quelque façon dévitalisée — mais participer d'une façon de plus en plus effective [...]; en d'autres termes, s'élever à des plans où la succession apparaît comme de moins en moins donnée, où une représentation [...] est de plus en plus inadéquate, et cesse même à la longue d'être possible. (*EA*: I, 19–20)

Yet, as I will now demonstrate, Marcel's characterization of love's fidelity, although phenomenological, is nevertheless linked to his more general position concerning human ontology. This is what complicates any attempt to interpret time in his philosophy.

In the diary entries published in *Être et avoir* Marcel has the idea of defining 'l'être comme lieu de la fidélité' (*EA*: I, 49; undated, but *c.* 1930),[60] and sees this as a way of accessing the ontological ('Accès à l'ontologie' (*EA*: I, 49)). On 6 November 1930 he also identifies the structure of the promise as metaphysically significant,[61] and equates this question of personal engagement with the question of fidelity.[62] Continuing to emphasize fidelity's ontological import, he links this act of faith with time,[63] describing it as 'une façon de transcender le temps' because 'la fidélité est liée à une ignorance fondamentale de l'avenir' (*EA*: I, 57; 8 November 1930). Although reflection on a promise he himself has made causes him to question the unconditionality he wishes to accord the promise, and to debate the legitimacy of an act of faith in general,[64] Marcel decides that these difficulties can be avoided if fidelity is conceived not as fidelity to a certain future state that I desire, but as fidelity to an underlying unity that grounds my being:[65] 'La fidélité comme reconnaissance d'un permanent' (*EA*: I, 118), he writes on 5 October 1932.

In keeping with his argument concerning the intersubjectivity of Being, Marcel insists that the promise of love or fidelity can never simply be confined to an isolatable dyad, but extends far beyond this, expressing fidelity to a broader, interconnected community. Thus after asserting, in his lecture 'Valeur et immortalité', that 'il n'y a pas d'amour humain [...] qui ne constitue aux yeux de celui qui le pense [...] une semence d'immortalité' (*HV*: 212), for example, Marcel qualified this by adding:

> il n'est sans doute pas possible de penser cet amour sans découvrir qu'il ne peut pas constituer un système clos, qu'il se dépasse en tout sens, qu'il exige au fond, pour *être* pleinement lui-même, une communion universelle hors de laquelle il ne peut se satisfaire, et est voué en fin de compte à se corrompre et à se perdre.[66] (*HV*: 212)

This fidelity and *disponibilité*[67] to a wider, intersubjective community is also (more frequently) discussed in relation to an attitude of hope, where, as Marcel stated in his 1942 lecture 'Esquisse d'une phénoménologie et d'une métaphysique de l'espérance', 'j'espère en toi pour nous' (*HV*: 81).[68] Since Marcel's discussions of hope in fact provide the most developed illustrations of how time and eternity relate to his ontological arguments,[69] it is to this theme that I will now turn.

As with love, time, in Marcel's phenomenological explorations of hope, is associated with reductive objectification that does not even allow for hope's (genuine) possibility, whereas eternity signifies a *récollection*,[70] a more global appreciation of human reality capable of recognizing the unconditional values that ground such a disposition.[71] When referring to time and eternity explicitly, it does again seem that Marcel's reflections are restricted to the phenomenological. Similar to his discussion in 'Mon temps et moi', Marcel describes how an understanding of time in objective terms of succession can lead to despair, while a mode of understanding that transcends this limited view encourages a more hopeful outlook. In a diary entry on 15 March 1931, for example, he identifies a certain experience of temporality — 'l'angoisse de *se sentir* livré au temps' — with 'inespoir' or 'unhope' (*EA*: I, 91; my emphasis),[72] whereas he states that 'la liaison est intime entre l'espérance et *une certaine affirmation de* l'éternité' (*EA*: I, 94; my emphasis).[73] And in his 1942 lecture he explained how 'le désespoir, c'est en un certain sens *la conscience du* temps clos, ou plus exactement encore, du temps *comme* prison', whereas, 'l'espérance *se présente comme* percée à travers le temps; tout se passe alors *comme si* le temps, au lieu de se refermer sur la conscience, laissait passer quelque chose à travers lui' (*HV*: 71; my emphasis).⁷⁴ However, these analyses of hope are also used to ground Marcel's metaphysics of Being. Indeed, the fact that hope is possible, for Marcel, tells us something about human ontology.⁷⁵ Furthermore, on 17 March 1931 Marcel describes hope as 'l'étoffe même dont notre âme est faite', reasoning, on this basis, that 'désespérer d'un être, [c'est ...] le nier en tant qu'âme' (*EA*: I, 100).⁷⁶ As such, (the eternity of) hope also seems to be presented as ontologically authentic, whereas (the temporality of) despair is considered a treacherous temptation that can only blind us to this reality.⁷⁷

Hope and despair are not unrelated, though; on the contrary, remarked Marcel in 'Position et approches', 'ils ne me paraissent pas séparables'. 'La structure du monde où nous vivons permet et en quelque façon peut sembler conseiller un désespoir absolu' (*HP*: 219), he observed in the 1933 lecture. This is the 'tragedy' of existence that Marcel felt Bergson had failed to recognize. As Marcel argued in his 1941 lecture 'Moi et autrui', our initial situation in the world is 'inéveillé[e]', and, as such, 'en marge du réel' (*HV*: 28). The tragedy of temporal existence then awakens us, however, through the protestation it incites: 'Un esprit est métaphysique pour autant que sa position par rapport au réel lui apparaît [...] comme foncièrement inacceptable', he writes on 17 October 1922.⁷⁸ 'Il s'agit de se redresser [...]. La métaphysique *est* ce redressement' (*JM*: 279).⁷⁹

Despair is thus intimately connected to the ontological possibility of hope: 'ce n'est que dans un monde semblable qu'une espérance invincible peut surgir' (*HP*: 219), argued Marcel in 'Position et approches'; 'il ne peut y avoir à proprement parler espérance que là où intervient la tentation de désespérer, l'espérance est l'acte par lequel cette tentation est activement ou victorieusement surmontée' (*HV*: 49), he re-affirmed in 'Esquisse d'une phénoménologie et d'une métaphysique de l'espérance'.⁸⁰ Time, as he stated on 7 March 1929, is considered to be 'la forme même de l'épreuve' (*EA*: I, 19);⁸¹ and hope itself, as Marcel explained in 'Esquisse d'une phénoménologie', 'se situe bien dans le cadre de l'épreuve, à laquelle elle ne correspond pas seulement, mais constitue une véritable *réponse* de l'être' (*HV*: 40).⁸² That is, an ontological response that is equated with eternity — for as Marcel writes on 8 March 1919: 'découvrir l'être, ce serait s'élever à un mode d'expérience ou de vie sur lequel l'expérience critique cesserait d'avoir prise. Cela revient à reconnaître, je crois, qu'il n'y a d'être qu'en l'éternité' (*JM*: 181).

Not only is Marcel's philosophy concerned with 'toute une gamme [de modalités de conscience]' (*PI*: 171; 6 May 1943), then; it is also concerned with different, but nevertheless related, 'modes d'être' (*DH*: 93). Time stands for what Marcel terms, in his 'Esquisse d'une phénoménologie de l'avoir' (1933), 'la déficience ontologique' (*EA*: I, 219–20) and is associated with brute objective existence — that which must be transcended.⁸³ Eternity, on the other hand, is identified with the plenitude of authentic Being. As such, time and eternity map onto a distinction Marcel draws between *existence* and *être*,[84] of which he gave his most developed explanation in the second of his 1950 Gifford Lectures ('Existence et être'):

> tout se passe ici [quand on considère ce que c'est 'l'existant'] comme si nous étions en présence de quelque chose qui est sur *une pente* et qui tend à glisser en bas de cette pente — mais qui en même temps est comme faiblement retenu, peut-être comme par une ficelle, [...] en sorte que cette chose est malgré tout susceptible de *remonter* la pente. [...] nous sommes spontanément portés à traiter l'existence comme le fait pour une chose d'être là [...]. Mais si mon attention se concentre sur ce simple fait: j'existe — ou encore: tel être que j'aime existe, la perspective change [...]. (*ME II*: 29)

Insofar as we understand our existence as a (mortal, contingent) 'thing', Marcel argued, we succumb to '[une] sorte de pesanteur [analogue à la tragique] qui tend à entraîner l'existence en direction de la chose, de la mortalité inhérente à la chose' (*ME II*: 31), and slide down the *pente de l'existence* toward non-being. However if, through our freedom, we undertake a secondary reflection and are able to appreciate something of Being's (eternal) value and intersubjective presence, then the hope and faith this incites will allow us to climb the *pente* toward ontological authenticity or *être* (see Figure 1, below).[85] It is not only my perspective that changes when I cease to focus my attention on the simple temporal 'fact' of my existence; according to Marcel, my very mode of being also changes such that I progress, up this *pente*, from mere *existence* toward *être* — Being proper. Does this not, therefore, mean that Marcel has failed to restrict his examinations of time and eternity to phenomenology, because of how his phenomenological observations slide onto the ontological plane?

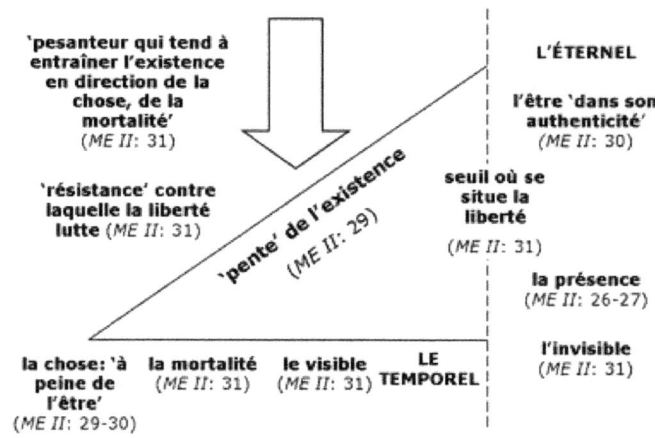

Phenomenology, Ontology, and Ethics: The Indeterminacy of Marcel's Philosophy

This chapter's examination of Marcel's *approches concrètes* has revealed the importance of phenomenology for Marcel, and showed how his explicit discussions of time and eternity conform consistently to this hermeneutical interest, in keeping with his argument for the 'irrelevance' of time. The temporality of existence is characterized by Marcel in terms of several possible perspectives: it can be understood in objective, deterministic terms (which Marcel refers to as 'le temps'), where acts appear contingent and — owing to time's finitude — inconsequential; or it can be understood as something more fluid, which is linked to the lives of others and carries a notion of value ('éternité'). But as the end of the second section reminded us, Marcel does not only speak of human *existence*, but also of *être*. Moreover, *être* is contrasted with *existence*; it is allegedly more authentic. This discourse then seems to cause Marcel to revert to his early use of the term 'existence' (prior to 'Existence et objectivité'), where it is identified solely with objectivity and contingency, not with 'ce que l'esprit doit reconnaître comme proprement insurmontable' (*JM*: xi); and as a consequence, a rigid binary is instituted between the ontologically authentic and inauthentic, to which Marcel relates (with equal consistency) eternity and time respectively.

In practice, then, phenomenology and ontology are not so distinct.[86] Time and eternity both do, and do not, relate to ontology in his writings — and thus the question of the significance of time is raised again. Either the 'temporal interference' in his ontological discussions finds Marcel guilty of a severe lack of philosophical rigour, owing to his use of (non-temporal) temporal terms. Or, through his distinction between *existence* and *être*, Marcel institutes an ontological division between an eternal and a temporal realm, and in so doing contradicts his position on two counts: first, with respect to his argument for the irrelevance of time; and second, by privileging eternity, for this then undermines the alleged importance of time in his philosophy.

Interestingly, Marcel himself struggles to understand what he means by *existence* and *être*, describing their relation as 'enveloppée d'ambiguïté' (*ME II*: 35).[87] His discussions of time and eternity were not, in theory, supposed to relate to an ontological hierarchy, but rather to different understandings of Being. For this reason, *existence* and *être* should not in fact be so separate — and Marcel, on occasion, agrees: 'it does not make sense to claim that existence is different from being' (Schilpp and Hahn 1984: 122), he acknowledges in reply to a 1968 article by Sam Keen. Nevertheless, he continues to employ the binary *existence/être* when speaking of human ontological potential, especially in relation to intersubjectivity. In so doing, he forcibly separates out the two terms — and time and eternity along with them. Thus, the difficult relation between *existence* and *être* can be seen as manifesting a deeper tension between the temporal and the eternal; and the reason this remains unresolved, it might be suggested, is because Marcel has introduced, and yet neglected to consider, time and eternity on an ontological level.

Such a problematic is particularly prominent in a dialogue between Marcel and Parain-Vial at a 1973 symposium in Dijon, at which Parain-Vial asked Marcel whether dualism (which Marcel generally insists he does not want to support) is not reinstated in his work in spite of itself, because of how human beings are presented as both mortal (temporal) and, in some sense, immortal (eternal):

> dire que nous sommes incarnés, c'est dire aussi que nous sommes mortels. Toujours donc, pèse sur nous la conscience de nos limites [...]. Mais alors, dans cette philosophie qui affirme la parenté de l'homme et de l'univers, [...] ne se dessine-t-il pas une sorte de dualité entre l'âme immortelle et le corps périssable? (Marcel 1974: 385)

Marcel is clearly uncomfortable with the question, but Parain-Vial does not allow it to be dismissed. After several re-iterations of her enquiry, Marcel admits that there is perhaps a need to talk of 'une sorte de dégradation qui va d'une inter-subjectivité en un sens extrêmement large, [...] au visible et ensuite au manipulable', so that although he wishes to affirm a certain unity, this is 'une unité en quelque sorte dégradable' (1974: 391). In response, Parain-Vial asks whether it is perhaps 'cette opposition du dégradable et de l'unité qui nous donnerait ce sentiment peut-être fallacieux d'une dualité' (1974: 391). Marcel seems more satisfied with this analysis — but he nonetheless fails to address the central thrust of her enquiry, since a question still remains as to whether unity exists, and if so, in what sense.

The eternal value of Being that Marcel wishes to affirm through his *approches concrètes* would suggest his metaphysics to be essentialist and teleological, as indeed is supported by his numerous references to 'l'universel'.[88] In 'Autour de Heidegger' (1945), for example, Marcel states that 'une description de l'existence qui ne fera aucune place à la protestation qu'éveille en toute conscience vivante la négation de l'universel [...] trahit en fin de compte cette réalité humaine à laquelle pourtant elle se veut adéquate' (1945: 99); and in *Les Hommes contre l'Humain* (1951) he declares: 'L'universel, c'est l'esprit — et l'esprit est amour' (*HH*: 12). Marcel's *approches concrètes* therefore appear to set the (eternal) unity of love up as the model for intersubjectivity,[89] over and above other more asymmetrical, (temporally) conditional interpersonal relations that nevertheless — one might argue — legitimize an intersubjective ontology through the relational space they create (not to mention the question as to whether love itself is ever entirely symmetrical or unconditional).[90] Marcel's description of existence in terms of *une pente* and the conversation with Parain-Vial mentioned above, however, seem to admit that a certain gradation must be recognized,[91] in which case a strict opposition between time (fragmentation, conditionality) and eternity (unity, unconditional value) need no longer be maintained. Occasionally Marcel does consider other, more intermediate intersubjective relations: in a William James Lecture ('Existence'), for example, he discussed the phenomenology of experiences such as asking someone for directions in the street, making an enquiry in a shop, having a banal conversation, and being with others in the sense of spatial proximity.[92] And yet Marcel's central reference point remains that of love, for all these relations are interpreted in terms of a progression toward it. One is therefore left wondering whether any form of ontological 'arrival' is possible, whether *être* can actually be achieved. Marcel's opposition to a clear-cut dualism between *existence* and *être*, and acknowledgement of the need to postulate dynamic degrees of Being,[93] suggest such an idea to be crude and misleadingly abstract. 'Ce qui est essentiel, c'est cette fluctuation possible de l'expérience; elle n'est en rien assimilable à une donnée opaque et constante [...]. La métaphysique n'est imaginée comme étant en dehors de l'expérience que là où celle-ci est arbitrairement stabilisée' (*PI*: 146), he writes on 17 April 1943. But the all-or-nothing terminology he employs then negates this assertion: unity and eternity can never be partial.

The intersubjectivity Marcel embraces in opposition to Bergson thus threatens to come into tension with Marcel's second reason for refusing his philosophy — namely, his insistence that the reality of Being cannot directly be known or affirmed. Marcel's phenomenological ontology does not, as Wood writes of Heidegger, present us with 'a subject whose worldliness is in doubt, but [...] a person whose Being is in question' (1989: 327);[94] and this question, Marcel seemed to suggest contra Bergson, is necessarily hermeneutical (a constant dialectic between primary and secondary reflection), because Being's mystery, into which we are always already thrown,[95] prevents us from approaching it as one might a determinate problem. Yet, if Marcel is then willing to draw definitive conclusions about *être*, does this not simply problematize his ontological method (in the Marcellian sense), such that it becomes a question of what the subject 'is' and 'is not', and thereby negates the indirect approach he claimed to adopt as a corrective to Bergson?

Marcel felt that his early idealist tendencies had prevented him from engaging the question of ontology: 'sous l'influence persistante de l'idéalisme, je n'ai pas cessé d'éluder le problème ontologique proprement dit. J'ai toujours eu, je m'en rends compte, une répugnance intime à penser selon la catégorie de l'être' (*EA*: I, 34), he writes in his diary on 17 July 1929.[96] His expressed opposition to Hegel in the argument for the irrelevance of time is representative of such dissatisfaction with post-Kantian idealism. Philosophers such as Schelling and Hegel, he maintained, went too far in arguing for the dependence of the individual on the collective (in Hegel's case, the historical collective), neglecting concrete, individual experience in favour of the wider whole. Yet although Marcel explicitly opposes Hegel, the philosophical position he proceeds to establish is, in certain important respects, very similar. Like Hegel Marcel also believed that the reality of Being manifests itself concretely in the phenomenal world. He did not regard metaphysics to be impossible as Kant did, and thus reacted against the break between epistemology and ontology (*phenomena* and *noumena*), seeking to revive metaphysics, to affirm that some form of knowledge of Being itself was possible.[97] Like Hegel, Marcel attempted to reconcile his faith in metaphysics with what Kant saw as reason's unavoidable tendency to pose insoluble antinomies for itself; and again, like Hegel, Marcel's solution to this problem was to regard dialectical process as phenomenally real (as opposed to confused, subjective illusion), and grounded in the constant flux of time. Marcel may have been more concerned with individual spiritual progress rather than with the progress of the collective, but crucially, as far as time and eternity are concerned, Hegel's 'Spirit' (*Geist*) is in many ways analogous to Marcel's *être*: both transcend time, and it is in this absolute (omni-temporal) presence that the unity of Being is grounded.[98] As such, both their philosophies appear, in effect, to negate the movement and development (through conflict) of becoming, condemning this to relative unreality because it falls short of their absolute (eternal) standard for existential authenticity — in Marcel's case, that of intersubjective presence.

Thus Marcel's discourse on time and eternity, in particular, suggests that his *approches concrètes* might still be guided by certain idealist assumptions — especially that of the unity of Being — in spite of the experiential dialectic he discovers as a result of his phenomenological investigations. It is therefore not clear that Marcel does take the phenomenality of individual experience seriously, leading critics such as Spiegelberg to conclude that Marcel 'had little interest in phenomenology as such' (1994: 466), for 'his stake in phenomenology was merely subordinate to his major [ontological] concerns' (1994: 460). There is considerable evidence in Marcel's writings to support Spiegelberg's interpretation: on 13 October 1933, for example, Marcel describes phenomenology as simply 'une introduction utile à une analyse renouvelée de l'être' (*EA*: I, 189); and in his 'Ébauche d'une philosophie concrète' (1940) Marcel makes the even more radical assertion that 'la philosophie ne pourra se contenter [...] d'employer la forme [phénoménologique] "comme si".[99] [...] toute philosophie authentique est l'active *négation* du "comme si"' (*RI*: 107; my emphasis). The 1944 work *Homo viator* then seems to confirm this marginalization of phenomenology in practice. Although its method is predominantly phenomenological, the subtitle of the work declares its primary concern to be metaphysical, announcing the collection to be 'prolégomènes à une métaphysique de l'espérance'. 'While phenomenology thus serves as an approach', writes Spiegelberg in response to this, 'it clearly is not considered adequate without additional steps of a more metaphysical nature' (1994: 458).[100] Indeed, surely if phenomenology were so important for Marcel, he would have devoted more time to explaining why, and — more crucially — how, it allows for (ontological) metaphysical discoveries, as opposed to bypassing this and focusing primarily on metaphysics?

Marcel is clearly aware that (at least in some respects) he is straddling the two fields: the central essay in *Homo viator* (a published lecture, in fact), for example, is boldly entitled 'Esquisse d'une phénoménologie et d'une métaphysique de l'espérance'. Yet how phenomenology and metaphysics actually relate is never explained;[101] and in most instances Marcel focuses on either one or the other, so that the need to explain their relation can be avoided. Exceptionally, in a diary entry dated 11 November 1932, Marcel does acknowledge that the shift from phenomenology to metaphysics might be problematic. The issue is raised in specific relation to the difference between treating another as *toi* as opposed to *lui*, and Marcel attempts to legitimize the metaphysical distinction he draws between these two modes by suggesting that it is not simply a question of private phenomenological experience. Rather, he argues, how I conceive of things directly affects my interactions with others and the world, and for this reason the purely phenomenological is surpassed:

> On me dira encore: 'Mais cette distinction du *toi* et du *lui* ne porte que sur des attitudes mentales; elle est phénoménologique au sens le plus restrictif. Prétendez-vous fonder cette distinction métaphysique, conférer au toi une validité métaphysique?'
>
> Mais en réalité c'est le sens de la question qui est extrêmement obscur et difficile à élucider. Tentons de la formuler plus clairement; par exemple de la façon suivante:
>
> Lorsque je traite un autre comme un toi et non plus comme un lui, cette différence de traitement ne qualifie-t-elle que moi-même, mon attitude envers cet autre, ou bien puis-je dire qu'en le traitant comme un toi je pénètre plus avant en lui, que j'appréhende plus directement son être ou son essence? (*EA*: I, 131–32)

Intersubjectivity thus appears to be the key to defending Marcel's position; but this matter is not pursued in any further depth, and Marcel does not develop or clarify this argument in subsequent works. At the very least, therefore, Marcel's justification for his philosophy of intersubjectivity is incomplete. At worst, however — and as the above discussion has suggested — it seems that (as Wood, again, observes with respect to Heidegger) the existential temporality in Marcel

> takes for granted as an ontological premise the value of unity. [...] the pervasive distinction between the authentic and the inauthentic [...]

> ultimately serve[s] to redefine and re-establish notions like personal identity and responsibility [...]. If it does not suppose that personal [or in Marcel's case, interpersonal] wholeness and integrity is something given or guaranteed, it does suppose it to be achievable.[102] (1989: 328)

In so doing, Marcel appears to subordinate temporal change and flux to an eternal constant, and the *tâtonnement* of his indirect approaches to (the alleged mystery of) Being to the comfortable certainty of direct affirmation.

The fact that Marcel himself questions the status of his conclusions makes it difficult to formulate a straightforward critique of his work, though. Marcel's confessed uncertainty concerning the relation between *existence* and *être* is the most frequent example of this; but on 7 March 1929 Marcel even questions the meaning of eternity in his philosophy. After stating that 'je ne peux pas ne pas m'apparaître comme contemporain de l'univers (*coaevus universo*), c'est-à-dire comme éternel' (*EA*: I, 21), he adds: 'Seulement, de quel ordre est cette appréhension de soi comme éternel? Là est, sans doute, le point le plus difficile' (*EA*: I, 21–22). Instances such as these lead one to question whether one might be wrong to assume that Marcel's main conclusions are ontological; the status of his philosophy is indeterminate.

Such indeterminacy is also visible in Marcel's fluctuating use of the terms 'assurance existentielle' and 'exigence ontologique' or 'exigence d'être'. 'Assurance existentielle' appears fairly non-committal, and could conceivably be interpreted as referring only to the phenomenological.[103] 'Exigence ontologique', however, is a much stronger, more univocal term, which can only be related to ontology. Confusingly, though, the two terms seem to be used interchangeably (although an article will often employ one or the other), in spite of the different senses one might be tempted to assign them. And yet Marcel does provide us with some justification for wanting to distinguish between them: one cannot doubt the ontological interest conveyed by 'exigence ontologique'; but in contrast to this, during a discussion at a 1973 symposium Marcel describes 'assurance existentielle' in a way that seems to correspond only to the purely phenomenological (*EAGM*: 86). Is there a reason why Marcel employs two distinct terms? If so, what determines his usage of one over the other, and what justifies their synonymous treatment in certain circumstances?

The other major complication regarding the status of Marcel's philosophy is the ethical concern that his arguments also appear to convey. As Ricœur remarked in a 1975 conference paper: 'un accent *éthique* est inséparable de la description phénoménologique et de son intention purement épistémologique. Gabriel Marcel ne peut réfléchir sans s'indigner' (1976a: 66). Indeed, in conversation with Monestier in 1970, Marcel declared that 'une philosophie sans éthique n'est pas une philosophie' (Monestier 1999: I, interview 2);[104] and as stated at the end of the *Journal*, Marcel believed that his focus on experiences such as presence had uncovered a way in which he could legitimately move from metaphysics to ethics. After noting that 'il y a une valeur infinie dans le fait de sentir un autre comme présent, une valeur infinie du contact comme tel' (*JM*: 292; 1 March 1923), he begins to wonder

> si ce n'est pas par la présence qu'on peut passer de l'existence à la valeur. Ce qui a de la valeur, n'est-ce pas ce qui accroît en nous le sentiment de la présence (qu'on dise qu'il s'agit de la nôtre ou de celle de l'univers — cela n'importe pas). Il y a dans ces réflexions ceci d'essentiel qu'elles semblent rendre possible un passage de la métaphysique à l'éthique: nous vaudrons d'autant moins que notre affirmation de l'existence sera plus restreinte, plus pâle, plus hésitante. (*JM*: 306; 24 May 1923)

Not only does intersubjective presence ground Marcel's transition to metaphysics, then; it serves as the foundation for his ethical conclusions as well.[105] 'Ce mot de vérité doit être pris ici dans une acception à quelque degré normative. Nous pourrions d'ailleurs remplacer le mot vrai par le mot authentique' (*PST*: 60), he declared in his lecture 'L'Humanisme authentique et ses pré-supposés existentiels'. But similar to the difficulty concerning phenomenology and ontology, Marcel does not explain what the relation between the ontological and the ethical is. If 'nous vaudrons d'autant moins que notre affirmation de l'existence sera plus restreinte', what ontological condition makes this the case?

In fact, the relation between ontology and ethics is all the more obscure than that between ontology and phenomenology, because the two seem to be situated on the same plane (phenomenology, on the other hand, appears subordinate to ontology). As Gillman therefore remarks, there appear to be two 'disparate' agendas in Marcel's work. First is the search for an indubitable point of departure for philosophical exploration, as illustrated, in particular, by the privileging of sensation in 'Existence et objectivité'. Second is Marcel's preoccupation with the increasing functionalization of humanity, the emergence of which Gillman relates to 'Position et approches', and which seems to gain ever increasing importance thereafter (1980: 173). Again, Marcel's thesis of ontological intersubjectivity is responsible. As Treanor comments: 'Love implies and depends upon a foundation of justice' (2006a: 87). Indeed, in *Les Hommes contre l'Humain* Marcel makes it clear that he sees 'un lien infrangible' between his philosophical thought and socio-political life (*HH*: 7),[106] and proceeds to identify love with peace, over and against social conflict or war.[107] However, these arguments are more polemical than they are demonstrative. Treanor's analysis is thus accurate to the letter, for if love's ontological foundation of intersubjectivity coincides with the foundation of justice in Marcel's philosophy, such a relation is only ever implied. In practice, intersubjectivity's dual status simply seems to grant Marcel a licence to discuss the ontology and ethics of intersubjectivity simultaneously, making it almost impossible to separate the two and leaving one to wonder whether Marcel does not, at times, simply confuse or conflate the two issues. How, for example, should one understand references to Marcel's 'volonté d'intersubjectivité' (1969: 257; *EC*: 133) and the need for the philosopher to act as a 'veilleur' (1965: 49–50; 1969: 259; *EC*: 150), when in other contexts Marcel asserts that the role of philosophy is one of 'explication' (1937: 180)?

Time and eternity are not exempt from such difficulties. In addition to its function as the symbol of ontological tragedy, time is referred to as a form of 'tentation' (e.g. *EA*: I, 27) — a term that is also identified with the desire to 'convertir le mystère en problème' (e.g. *EA*: I, 125). Eternity, on the other hand, is equated with ethical harmony and virtue, in addition to the ontological mystery. On 7 November 1932, for example, Marcel defines 'le métaproblématique' as 'la paix qui passe tout entendement, l'éternité' (*EA*: I, 128); and André Devaux describes how Du Bos, Maritain, and Marcel were all in absolute agreement

> sur l'*éternité* des valeurs inspiratrices de l'action véritablement humaine. Une identique volonté d'actualisation de l'éternel dans le temporel détermina, par exemple, leur commune attitude face à la crise internationale de Munich, en 1938, au cours de laquelle 'la volonté de paix a prévalue sur la volonté de justice'. Faire du temps 'le vestibule de l'éternité', telle fut sans doute leur plus commune aspiration.[108] (1974: 99)

When confronted with these alternative characterizations of time and eternity, then, how is one to interpret Marcel's argument for the transcendence of time in favour of eternity? Does Marcel reject time for eternity on the basis of this ethical judgement (in which case, are Marcel's arguments more rhetorical than philosophical), or on the basis of something relating to ontology, which is never quite made clear?

It therefore seems that, as Ricœur declared in 1973, 'la difficulté fondamentale, avec laquelle l'ontologie existentielle de Gabriel Marcel n'a jamais fini de s'expliquer, concerne le statut de ses propres énoncés' (1976b: 70). Ricœur admits that '[cette critique] a un allié dans cette œuvre [de Marcel] même', namely in secondary reflection, which, by definition, cannot be definitively characterized without collapsing into primary reflection (1976b: 72–73). Nevertheless, Marcel's terminology lacks rigour,[109] obscuring matters further than is warranted; and his philosophical language is particularly nebulous where time and eternity are concerned. Although it is evident that Marcel distances himself from Bergson's conception of time and eternity in order to affirm human intersubjectivity, neither his own position, nor his method, is justified in any satisfactory manner — indeed, the domain in which he seeks to legitimize his position is not even clear, for his arguments for intersubjectivity span the phenomenological, the ontological, and the ethical, without offering any explanation as to how transitions between these fields are possible. It therefore seems ironic that

Marcel declared his interest in time to concern 'la manière dont nous parlons du temps' (Hersch and Poirier 1967: 28), for it is precisely his failure to consider the manner in which he himself talks about time that renders his philosophy so problematic.

Notes to Chapter 2

1. In the 'Transcendental Aesthetic' of his *Critique of Pure Reason*, Kant argues that time is 'the *a priori* formal condition of all appearances in general' (1998: 163) — that is, a necessary condition of sensibility which, along with space, makes experience of the world possible in the first place, by providing a structural framework through which all other sensory experiences can be individuated and thus interpreted. For this reason: 'Time is not something that would subsist for itself or attach to things as an objective determination' (ibid.). 'It is only of objective validity in regard to appearances, because these are already things that we take as objects of our senses; but it is no longer objective if one abstracts from the sensibility of our intuition, thus from that kind of representation that is peculiar to us, and speaks of things in general. Time is therefore merely a subjective condition of our (human) intuition [...], and in itself, outside the subject, is nothing' (1998: 181).
2. See also Parain-Vial's account of a 1973 interview with Marcel and Simonne Plourde (1976: 200–01).
3. See also *JM*: 105.
4. Though the term 'idéale', which Marcel uses to describe his dialectic, is presumably meant in a Kantian transcendental sense; it is clearly not meant in the strong philosophical sense, as Marcel has just declared such an idealist position to be inadequate.
5. It is not clear who Marcel is quoting here. Possibly, it is a reference to Russell (1872–1970), who, around the same period of time, was writing and lecturing on the irrelevance of time for philosophy and theoretical science (Marcel at least had some familiarity with him, as he refers to him in *EA*: II, 9–10 and *RI*: 88). In his 1914 essay 'Mysticism and Logic', for example, Russell writes: 'The importance of time is rather practical than theoretical, rather in relation to our desires than in relation to truth. [...] to realise the unimportance of time is the gate of wisdom' (1925: 21–22). In 'The Ultimate Constituents of Matter' (1915 lecture) Russell also declared: 'The one all-embracing time, like the one all-embracing space, is a construction' (1925: 141; see also 167); and in 'On the Notion of Cause' (1912 lecture), Russell argued that 'this principle of the irrelevance of the time may be extended to all scientific laws. In fact we might interpret the "uniformity of nature" as meaning just this, that no scientific law involves time as an argument, unless, of course, it is given in an integrated form, in which case *lapse* of time, though not absolute time, may appear in our formulæ' (1925: 205).
6. Marcel affirms this hermeneutical interest on 18 January 1914, when he writes: 'Il me semble que je peux regarder comme établi que l'on ne peut parler de l'existence qu'à propos d'objets donnés dans un rapport immédiat à une conscience (au moins posée comme possible). Comme on peut concevoir une multiplicité de façons dont un même objet (un même contenu) pourrait être donné à la conscience dans un rapport immédiat, il faut concevoir une série infinie de plans d'existence relatifs aux modes possibles d'appréhension' (*JM*: 18; see also 179–80). His critical reading of Royce's theory of time and eternity (published 1918) is likely to have consolidated his thoughts on temporal experience and intentionality, while reinforcing his opposition to a higher eternal reality (Marcel 2005: 118–30, 229–32).

7. See also *JM*: 256. Sartre (2004) also establishes the (free) self against contingency, in terms of its lived (phenomenological) awareness of its being in time.
8. See also *JM*: 105, 108–09. Here Marcel analyses time as a mode of exteriority that compromises authentic self-understanding, concluding that although it is a necessary starting point for reflection, the temporal must be transcended for an authentic conception of self (*l'esprit*; *la pensée pure*). Troisfontaines writes: 'Il ne peut s'agir évidemment de nous affranchir à proprement parler du temps, car dès que nous écartons toute idée de succession, nous risquons de retomber sur la représentation statique d'un invariant qui ne serait qu'un pur abstrait. Il n'est pas plus raisonnable de concevoir au principe de la vie quelque chose qui s'apparenterait à une notion hypostasiée, qu'il ne l'est de figurer la vie comme un pur déroulement d'images. Nous devons seulement nous dégager d'un certain schématisme temporel qui ne s'applique en réalité qu'aux choses. Dans la mesure où nous ne nous assimilons pas à des "choses", nous dépassons l'opposition du successif et de l'abstrait, nous découvrons que nous appartenons à une dimension entièrement différente du monde' (1953: I, 256).
9. In *Ideas I* (1913), Husserl describes his phenomenology as a 'science of Essential Being' (1931: 46), and in Draft A of his *Encylopædia Britannica* article, 'Phenomenology' (1927), he calls it a 'a philosophy of self-reflection at its most original and its most universal', which, while joining with Kant in the 'battle against the shallow ontologism of concept-analysis, [...] is itself an ontology' (1997: 102).
10. Defined as 'this entity which each of us is himself and which includes inquiring as one of the possibilities of its Being' (Heidegger 1962: 27).
11. See also Husserl 1931: 82–83, §19. Here, the German term 'Sachen' is rendered as 'facts', but more recent versions have translated it as 'things'.
12. Husserl argues that all mental acts are characterized by 'intentionality', such that (following Brentano's 'act psychology') consciousness is essentially an act. As such, consciousness is never directed toward itself, but is rather always consciousness 'of' something, that is, it is always directed toward an object in the world (1931: 120, §36).
13. '[R]ien de plus difficile que de penser l'ordre naturel parce que c'est, d'une façon générale, *à partir de lui* que nous pensons' (*PI*: 171; 7 May 1943).
14. 'If to Interpret the meaning of Being becomes our task, Dasein is not only the primary entity to be interrogated; it is also that entity which already comports itself, in its Being, towards what we are asking about when we ask this question. But in that case the question of Being is nothing other than the radicalization of an essential tendency-of-Being which belongs to Dasein itself — the pre-ontological understanding of Being' (Heidegger 1962: 35).
15. '[W]henever Dasein tacitly understands and interprets something like Being, it does so with *time* as its standpoint. Time must be brought to light — and genuinely conceived — as the horizon for understanding Being and for any way of interpreting it' (Heidegger 1962: 39).
16. Cf. Wolff (2010), who argues that Marcel facilitated the positive reception of Heidegger in France.
17. 'Sans doute ai-je été retenu par une crainte qui s'est toujours fait sentir en moi depuis que je me suis dégagé des abstractions auxquelles je m'abandonnais inconsidérément pendant mes années de formation. La crainte de devenir le prisonnier d'un système qu'il s'agirait de gérer ou d'exploiter le plus adroitement possible: d'où mon souci permanent de coller, si j'ose dire, aux situations concrètes qu'il m'était donné d'observer, ou dans certains cas, d'imaginer' (Marcel 1969: 258). Marcel's gravitation toward these concrete experiences can be traced in his diaries. See especially *EA*: I, 149 and *JM*: 202.

18. '[L]es thèses fondamentales que je vais proposer sont l'aboutissement de toute l'évolution philosophique et spirituelle qui se poursuit à travers le *Journal*' (*HP*: 192).
19. See, in addition, *EC*: 151–52, where he states that it was also from this point onward that he came to regard his thought as 'existential'.
20. See also *ME I*: 70.
21. See Marcel's first lecture, where he describes his task as being to 'reprendre mon œuvre entière dans une lumière renouvelée, [...] en marquant surtout l'orientation générale' (*ME I*: 10).
22. See also Marcel (1976: 17).
23. In a diary entry dated 22 November 1928, Marcel states that incarnation is the 'donnée centrale de la métaphysique' (*EA*: I, 11). Intersubjectivity is emphasized increasingly after this point as the full-blown implication of incarnation, though Marcel did not actually use this term until his Gifford Lectures (*DH*: 61). On intersubjectivity as foundation in his thought, see for example *EC*: 112–14.
24. Marcel opposed Descartes's position from the beginning. However this opposition becomes all the more pronounced once Marcel has distanced himself from Bergson.
25. '*Je suis*; voilà une affirmation qui ne se fonde sur rien, mais qui par définition exclut tout fondement, qui, séparée de la conscience que nous en avons, perd toute signification; elle n'est sans fondement que parce qu'elle fonde elle-même toute certitude' (*EC*: 134).
26. This is a constant in Marcel's work. As Du Bos has commented, from the very beginning, Marcel exhibited a marked 'méfiance à l'égard de l'objectivité' (1931: 153; see for example *JM*: 196, 215). The primacy of existence (in its modified sense) over objectivity can be considered the main conclusion of the *Journal*. 'Existence et objectivité' then confirms the importance of this for Marcel, asserting 'la priorité absolue de l'existence' (*JM*: 319).
27. In a 1949 Gifford Lecture Marcel states that he is almost tempted to propose an alternative to the *cogito*: '*sentio, ergo sum*'; but he then retracts the suggestion, rejecting the maxim's deductive structure and declaring that '"*j'existe*" est d'un ordre différent; elle est d'ailleurs en marge de toute inférence quelle qu'elle soit' (*ME I*: 105).
28. In the *Discours de la méthode* (1637), in which his *cogito* first appears, Descartes employs his famous 'method of doubt', doubting everything he possibly can 'afin de voir s'il ne resterait point, après cela, quelque chose [...] qui fut entièrement indubitable' (1951: 61). The one thing that remains impossible to doubt is the existence of something — a 'je' — that is doing this doubting: 'je pris garde que, pendant que je voulais ainsi penser que tout était faux, il fallait nécessairement que moi, qui le pensais, fusse quelque chose' (1951: 62). Hence Descartes declares that 'cette vérité: *je pense, donc je suis*, était si ferme et si assurée que toutes les plus extravagantes suppositions des sceptiques n'étaient pas capables de l'ébranler', and that 'je pouvais la recevoir, sans scrupule, pour le premier principe de la philosophie que je cherchais' (1951: 62).
29. Nietzsche objected to the way in which Descartes's conclusion was conditioned by grammar: 'that there must be something "that thinks" when we think, is merely a formulation of a grammatical custom which sets an agent to every action' (1913: 484).
30. 'Le *je suis* se présente, me semble-t-il, comme un tout indécomposable. [...] la position cartésienne [...] n'est pas séparable d'un dualisme que pour ma part je rejetterais sans hésitation' (*HP*: 202).
31. 'La pensée se crée en se pensant; elle ne se trouve pas, elle se constitue' (*JM*: 32; 26 January 1914).
32. 'La démarche métaphysique essentielle consisterait dès lors en une réflexion sur cette réflexion [primaire], en une réflexion à la seconde puissance' (*EA*: I, 147; 23 December 1932).

33. '[J]e suis conduit à me demander quel est le statut ontologique de ce moi par rapport à l'être qui l'investit' (*EA*: I, 176; 12 March 1933).
34. 'La reconnaissance d'une participation [...] fonde ma réalité de sujet' (*HP*: 204).
35. See also *RA*: 294-95.
36. For discussion of *disponibilité*, see: *DH*: 22; *EA*: I, 85-86, 90, 155; *HP*: 234-39; *HV*: 28-29, 31-32; Marcel (1969: 258); *RI*: 55-80.
37. In Marcel's philosophy, the *toi* is encountered by the *moi* as an other with whom a real, concrete relation can be established, rather than as a third-person *lui* or *elle* who lacks presence for the *moi*. This is achieved through the *moi*'s and *toi*'s mutual *disponibilité* for one another. Thus, for Marcel, the other does not inevitably reinforce individual alienation, but may also provide a glimpse of a richer mode of Being. This may then stimulate further reflective transcendence, which extends beyond the individual *nous* experience and opens the subject up to the intersubjectivity of human Being as a whole.
38. '[C]'est une métaphysique du *nous sommes* par opposition à une métaphysique du *je pense*. [...] Sartre, en reprenant à son compte un cartésianisme d'ailleurs décapité (puisqu'il a retiré son couronnement théologique), s'est condamné lui-même à ne saisir l'autre que comme menace à ma liberté ou, à la rigueur, comme une possibilité de séduction qu'il est bien difficile de ne pas interpréter dans un sens sadiste ou masochiste. [... Sartre] rend impossible l'inter-subjectivité' (*ME II*: 12–13). See also Marcel (1976: 10–11; 1981). It is mainly on the basis of *La Nausée* (1938), and the sections concerning *le regard* and *les relations concrètes avec autrui* in *L'Être et le Néant* (1943), that Marcel forms his judgement about Sartre's philosophy. However, as soon as one has any acquaintance with Sartre's other writings, or even the other sections in *L'Être et le Néant*, the superficiality of Marcel's judgement becomes rather evident: firstly, Marcel has failed to understand the exposure of bad faith; and secondly, if Sartre is dualistic, his dualism is of a different, more nuanced nature from that of Descartes.
39. 'Plus je vais et plus je suis frappé de ce fait que les préoccupations épistémologiques tendent à mettre l'esprit hors d'état de résoudre et même de poser le problème de l'amour' (*JM*: 294; 5 March 1923).
40. '[L]'intersubjectivité est trahie dans son essence même lorsqu'elle est traduite en un langage d'objet' (*DH*: 9).
41. See also *JM*: 157-58, 215, 226-28.
42. 'Le "problème de l'être" ne sera donc qu'une traduction en un langage inadéquat d'un mystère' (*EA*: I, 147; 23 December 1932).
43. In 1940 Marcel describes his distinction between problem and mystery as 'la distinction [...] centrale[, ...] qui m'apparaît aujourd'hui comme en réalité présupposée par tout l'ensemble de mes écrits philosophiques, mais qui ne s'est formulée d'une façon expresse qu'en octobre 1932' (*RI*: 94).
44. Especially in the diary entries published in *Être et avoir*, e.g. *EA*: I, 128, 140, 142, 151.
45. *DH*: 113.
46. 'Le point de départ d'une philosophie authentique — et j'entends par là une philosophie qui est l'expérience transmuée en pensée, c'est [...] la reconnaissance aussi lucide que possible de cette situation paradoxale qui non seulement est la mienne, mais *me fait moi*' (*RI*: 39).
47. 'Je suis convaincu pour ma part qu'il n'y a d'ontologie possible, c'est-à-dire d'appréhension du mystère ontologique, [...] que pour un être capable de se recueillir' (*HP*: 211).
48. 'Chacun de nous, s'il veut bien se donner la peine de procéder à cette sorte de discrimination, sera [...] amené à reconnaître que dans sa vie il y a des présences et des fidélités qui diffèrent radicalement des rapports mondains ou professionnels et des obligations qui en résultent' (*DH*: 95).

49. See also *EC*: 111, 198–99.
50. Note, here, how the affirmation of the other's existence is analogous to that of the self.
51. In 'Position et approches', Marcel emphasized the metaphysical import of the preposition 'avec' (*HP*: 233). This is prefigured in *JM*: 169, but see also *DH*: 62–63 and *ME* I: 193–94.
52. 'Le propre d'un sujet [...] consiste précisément à ne pouvoir être donné [comme les données objectives]. Il n'est un être pour moi que s'il est une présence, et ceci veut dire que d'une certaine manière, il m'atteint par *le dedans*, et me devient, jusqu'à un certain point, intérieur. C'est en cela que la présence en tant que telle relève du mystère' (*DH*: 8). See also *DH*: 15.
53. '[L]a fidélité c'est la présence activement perpétuée' (*HP*: 229). See also *DH*: 94.
54. Here, Marcel is citing a line from one of his plays (*La Mort de demain*; Marcel 1931: 161).
55. '[L]a seule victoire possible sur le temps participe, selon moi, de la fidélité' (*EA*: I, 14; 28 February 1929).
56. 'Rapprochement entre l'existence et la valeur: il y a une valeur infinie dans le fait de sentir un autre comme présent, une valeur infinie du contact comme tel' (*JM*: 292; 1 March 1923).
57. As Cain writes: 'Value is thus bound up with immortality and eternal destiny. If death is ultimate, if it is the annihilation of being, then value becomes meaningless, reality becomes empty, and human communion is broken at its very core. Love, the "pledge and seed of immortality", involves "the recourse to absolute transcendence" and requires a going beyond all finite closedness and an opening up to that "universal communion"' (1963: 86).
58. '[D]ès le moment où il y a présence, nous sommes au-delà du problématique' (*EA*: I, 143; 18 December 1932).
59. Alexander (1948: chap. 7) provides an excellent discussion of this relationship between eternity and time, whereby eternity is not timelessness but instead 'timefulness' (1948: 398) — a deepening of temporal experience that allows the self to '[transcend] the successivity characteristic of objective thinking', to '[emerge] from Objectivity into Existence' (1948: 417).
60. '[À] partir d'une certaine date que je ne peux pas préciser absolument, mais qui se situe sans doute en 1930 ou peut-être avant, ce thème de fidélité prend pour moi une valeur centrale, comme en témoigne cette note non datée d'*Etre et Avoir*: "De l'Etre comme lieu de la Fidélité"' (*DH*: 91).
61. *EA*: I, 49. Ricœur describes this as 'l'expérience concrète qui mêt la quête [de Marcel] en mouvement' (1989: 157).
62. *EA*: I, 51.
63. *EA*: I, 52, 117–18, 149. This is prefigured in the *Journal*, where, on 22 February 1919, Marcel writes: 'je suis d'autant moins mon passé que je le traite davantage comme une collection d'événements enregistrés ou sériés dans le temps [...]. Si donc par *mon passé* on entend le passé que je *suis* en tant que je le suis par opposition à celui que j'*ai*, il faut dire que ce passé-là ne saurait en aucune manière être pensé sous forme de collection, mais au contraire comme objet de foi ou d'amour' (*JM*: 163–64).
64. *EA*: I, 57–69; undated. See also *DH*: 97–100.
65. '[L]'acte par lequel ce privilège de mon être futur se trouve ainsi consacré fait partie de mon présent: voici donc une valeur de l'avenir en tant qu'avenir qui est attachée à mon état actuel, et qui cependant s'en distingue, puisqu'elle en commande en quelque façon le rejet. [...] Admettrai-je que c'est mon état présent lui-même qui se nie et prétend se dépasser? Comment ne pas voir là un arrangement suspect, et qui du reste, à supposer

que je le sanctionne, implique je ne sais quelle vérité transcendante au devenir et susceptible de le fonder? Mais s'il en est ainsi il ne peut plus être question pour moi de me prêter sans résistance au flux de mes dispositions momentanées. Quelque chose qui n'est aucune d'elles, une loi peut-être, en règle les caprices. Et c'est à cette loi, à cette unité qu'il s'agit pour moi de demeurer fidèle. Le langage cependant une fois de plus menace de m'induire en erreur. Cette unité c'est moi, précisément: c'est un même et unique principe — forme ou réalité — qui exige sa propre permanence. Fidélité non plus à un devenir, ce qui n'a point de sens, mais à un être que je ne vois pas la possibilité de distinguer de moi. J'échappe ainsi aux mirages d'un lendemain qui se décolore à mesure qu'il se précise' (*EA*: I, 63-64; undated).

66. See also *ME II*: 156.
67. 'La notion de disponibilité n'est pas moins importante pour notre sujet que celle de présence' (*HP*: 234).
68. '[T]oute espérance est, en son fond, *chorale*. Mais il s'agit ici d'une évidence mystérieuse, qui, dans le monde qui est le nôtre, risque de s'altérer en se rationalisant' (*PST*: 209).
69. Parain-Vial agrees: 'l'espérance manifeste encore plus peut-être que l'amour et la foi, notre double appartenance à l'éternel et au temps' (1989: 81).
70. '[S]i le temps est par essence séparation et comme perpétuelle disjonction de soi par rapport à soi-même, l'espérance vise au contraire à la réunion, à la récollection, à la réconciliation' (*HV*: 72).
71. '[L'espérance] enveloppe l'affirmation de l'éternité, de biens éternels' (*EA*: I, 93; 13 March 1931).
72. Marcel explains that 'l'expression [unhope] se rencontre dans un poème de Thomas Hardy' (*EA*: I, 91).
73. '[D]ans la mesure où je conditionnalise mon espérance, [...] je livre une part de moi-même à l'angoisse' (*HV*: 62).
74. See also *HV*: 80.
75. 'Il faudra reprendre ce que j'ai dit du fait que la structure de notre univers permet l'espérance et voir sa signification ontologique' (*EA*: I, 91; 15 March 1931).
76. Marcel anticipates this in the second part of his *Journal* when he writes: 'Triompher de l'épreuve ce sera se maintenir en tant qu'âme, sauver son âme. A approfondir' (*JM*: 198; 16 October 1919).
77. This is echoed, in a broad sense, by existentialist thinkers such as Nietzsche, Sartre, and Camus.
78. For Marcel this explains 'l'attrait qu'exerça longtemps sur moi l'hégélianisme — car, malgré les apparences, Hegel a fait un admirable effort pour sauvegarder le primat du concret, en marquant avec la plus grande force que celui-ci ne peut en aucun cas se confondre avec l'immédiat' (*HH*: 8).
79. See also *HH*: 119, *PST*: 32-33, and *RI*: 88-89. A comparison might be made here with Camus's (1951) notion of revolt.
80. See also *EA*: I, 92 and *RI*: 100.
81. 'Il n'y a épreuve que si la situation dans laquelle je suis engagé fait surgir en moi une véritable dualité entre ce que je suis comme donné et ce que je suis comme réagissant' (*JM*: 228-29; 25 February 1920); 'il faut que la foi triomphe de l'état de division d'avec soi qui est lié aux conditions d'existence d'un être fini' (*JM*: 231; 29 February 1920).
82. This is suggested as early as Marcel's *Journal*. On 6 March 1919 Marcel writes: 'L'être se définit en somme pour moi comme ce qui ne se laisse pas dissoudre par la dialectique de l'expérience [...]. Être, ai-je dit c'est résister à l'épreuve de la vie' (*JM*: 179-80; see also 181, 198-99).
83. See also *JM*: 264-65 and (especially) 284-85.

84. 'Les recherches de Gabriel Marcel sur le temps sont inséparables de ses recherches sur l'existence, celle-ci étant temporelle, ce qui la différencie de l'être' (Parain-Vial 1985b: 480, note 1).
85. '[M]a faculté d'attention donne la mesure de ma liberté' (*PST*: 12).
86. Vigorito's investigation of time in Marcel's philosophy fails to confront this difficulty. Vigorito considers the conflict between Marcel's disparagement and accentuation of time to be a terminological difficulty, which, he claims, can be easily resolved 'once we distinguish between Marcel's concept of temporality and what he takes to be inadequate theories of time' (1984: 414, note 1). In effect, then, he defends Marcel in accordance with the argument for the irrelevance of time, attributing this apparent problem to Marcel's synonymous use of the terms 'time' and 'temporality': 'Marcel is well aware of [contemporary philosophy's distinction between these two terms]', he writes, 'since he never confuses his own questions concerning temporality with those of objective time', although he uses the terms interchangeably (ibid.). What Vigorito neglects to ask, however, is whether Marcel's references to the 'ontological' do not in fact introduce a dimension of objectivity which warrants discussion of time in addition to temporality, and thereby transforms Marcel's synonymous use of 'time' and 'temporality' into a case of muddy thinking, as opposed to the simple desire for linguistic variety.
87. See also *EA*: I, 44–45.
88. Or to 'un permanent [ontologique]' (e.g. *EA*: I, 118, 149).
89. '[L]e *nexus* où cette exigence prend racine, ne peut être que l'amour' (*EC*: 286). See also *DH*: 110, 195, *HV*: 212, and *JM*: 199, 206.
90. 'Marcel never troubles to sort out the connections between the possible varieties of communal existence. What really interests him is authentic ontological communion', observes Gallagher (1962: 22); but this, Bernard argues, is interpreted by Marcel 'd'une manière trop unilatérale' (1952: 109) — and, he adds, 'il n'y a même pas de communion intersubjective qu'il n'y a jamais intersubjectivité pure dans l'amour si authentique soit-il' (1952: 113).
91. '[I]l faudra distinguer des degrés non seulement dans l'élucidation, mais dans l'intimité avec soi et avec l'ambiance — avec l'univers lui-même' (*RI*: 25).
92. *DH*: 60–62. See also *DH*: 94–95, 110, and *ME I*: 153–60.
93. 'Il y aurait [...] lieu d'admettre une distinction de degré dans l'être' (*PI*: 140; 13 April 1943).
94. In 'Autour de Heidegger' Marcel writes: 'le *Dasein* est un existant dont l'être est [...] toujours mis en jeu ou en question. Il lui est essentiel de se dépasser lui-même, et ce dépassement est l'être de son existence. [...] J'ai tenté [...] d'exprimer une idée analogue lorsque j'ai écrit à propos de la personne que le verbe qui lui convient est *sursum*, et non *sum*' (1945: 90–91; Marcel refers to the person as 'sursum' in *HV*: 32).
95. Just as Marcel's notion of *le mystère* describes human persons as (always already) inextricably involved in Being, Heidegger (1962) speaks of the 'thrownness' of Dasein (literally 'being-there': *da-sein*), where the human subject is already 'there' in the world.
96. See also *EA*: I, 145.
97. '[T]hough the categories, such as unity, or cause and effect, are strictly the property of thought, it by no means follows that they must be ours merely and not also characteristics of the objects. Kant however confines them to the subject-mind, and his philosophy may be styled subjective idealism' (Hegel 1975: 70, §42z). The 'z' indicates that the passage is an addendum (*Zusatz*). These were added by Leopold Henning, and first appeared in the 1840 edition of Hegel's works, published after Hegel's death.
98. Hegel writes: 'Spirit necessarily appears in Time, and it appears in Time just so long as

it has not *grasped* its pure Notion, i.e. it has not annulled Time. It is the *outer*, intuited pure Self which is *not grasped* by the Self, the merely intuited Notion; when this latter grasps itself it sets aside its Time-form [...]. Time, therefore, appears as the destiny and necessity of Spirit that is not yet complete within itself' (1998: 487, §801).

99. 'Comme si' representing the experiential appearances that phenomenology seeks to take seriously.
100. '[L]e phénomène, c'est l'être incomplètement pensé' (*JM*: 202; 17 October 1919).
101. Gallagher comments: 'Of all the "concrete approaches" to the ontological mystery, hope is the one which most unambiguously announces its references to transcendence. So much is this true that a phenomenological analysis of hope passes almost immediately into an elucidation of its hyper-phenomenological roots; the transcendent vector seems to be not eventual but dominant from the beginning' (1962: 73).
102. As Plourde comments: 'Malgré son caractère interrogatif, la philosophie de G. Marcel prend pour point de départ certaines assurances fondamentales. Sa méthode philosophique (réflexion seconde) lui permet de "retrouver ces assurances existentielles fondamentales, constitutives de l'être humain véritable [...]"' (1985: 77).
103. This is how Henri Gouhier might understand it (*EAGM*: 41).
104. See also *EA*: I, 29.
105. Ricœur (1989) also observes the concurrent ethical status of Marcel's ontological arguments, which he feels is manifest, in particular, in the notion of *disponibilité*.
106. See also *EC*: 243.
107. *HH*: 9, 13, 118.
108. The first citation references Marcel in his preface to Du Bos's *Commentaires* (1946), and the second citation references a phrase used by Maritain. On Marcel's position regarding the Munich agreements, see *EC*: 171 and Rémond (1989: 36).
109. This is the main reason why Ricœur (who was heavily influenced by Marcel) moved away from his work. 'La philosophie existentielle ne peut [...] se borner à une critique de l'objectivité, de la caractérisation, du problématique; il lui faut prendre appui sur des déterminations de la pensée, sur un travail du concept' (1976b: 73), he contends in 1973. In a 1995 interview he therefore explains: 'Si je me suis éloigné de sa philosophie, ce n'est pas en raison de ses convictions profondes, mais plutôt en raison d'un certain manque, chez lui, de structure conceptuelle' (1995: 43). Gallagher (1962) too criticizes Marcel for the mutually exclusive relationship he establishes between objective knowledge and metaphysical knowledge. See also Blundell (2003).

CONCLUSION TO PART I

Metaphysics and Presence

If, as a result of his attempt to draw ontological conclusions from phenomenology, one is tempted to accuse Marcel of philosophical inconsistency, it should be noted that this is not in fact so unusual; phenomenology in general has been criticized for this. Wood explains: 'Phenomenology could never have had any interest unless its descriptions of the structures of consciousness had a value that went beyond their being an accurate account of subjective phenomena. That value lay in what was always assumed to be the epistemological and ultimately ontological significance of consciousness' (1989: 324–25).[1] Indeed, in theory, phenomenology was to offer a radically new kind of metaphysics, free from prejudice or dependence on previous philosophies; but as Derrida has argued (with respect to Husserl in particular), the metaphysics proposed by phenomenology still remains within the sphere of traditional metaphysics, because it continues to privilege the mode of presence in its affirmations. In *La Voix et le Phénomène* (1967) he writes of Husserl:

> Les résultats qu'il présente [...] sont, dit-il, 'métaphysiques, s'il est vrai que la connaissance ultime de l'être doit être appelée métaphysique. Mais ils ne sont rien moins que de la métaphysique au sens habituel du terme; cette métaphysique dégénérée au cours de son histoire, n'est pas du tout conforme à l'esprit dans lequel elle a été originellement fondée en tant que philosophie première [...]'. On pourrait faire apparaître le motif unique et permanent de toutes les fautes et de toutes les perversions que Husserl dénonce dans la métaphysique 'dégénérée' [...]: c'est toujours une cécité devant le mode authentique de l'*idéalité*, celle qui *est*, qui peut être *répétée* indéfiniment dans l'*identité* de sa *présence* [...].[2] (1967a: 4)

According to Derrida, then, there has only ever been one kind of metaphysics in the history of Western philosophy: a 'metaphysics of presence': 'De Parménide à Husserl, le privilège du présent n'a jamais été mis en question. [...] La non-présence est toujours pensée dans la forme de la présence [...] ou comme modalisation de la présence. Le passé et le futur sont toujours déterminés comme présents passés ou présents futurs' (1975: 36–37), he asserts in 'Ousia et grammè' (1968). At the heart of this metaphysics, as Wolfgang Fuchs explains, is 'the teaching that the presence to being *is* the metaphysical moment' (1976: 5). More precisely, the metaphysics of presence assumes that the notion of Being as it is in itself coincides with the notion of absolute presence. The living present, where Being can be present as either subject, object, or as their unity, is therefore treated as the ultimate foundation of (and also epistemological point of access to) ontology. As such, the conception of (authentic) Being in Western metaphysics excludes the (nominally inauthentic) notions of temporality, incompleteness, and absence; and thus a binary opposition between the authentic and the inauthentic comes to structure the whole of Western philosophy, with presence as the figurehead of authenticity.[3] Each side of this binary, Derrida argues, tends to be considered as pure and independent of the other. Furthermore, the binary is only a binary insofar as the second term is understood as a negation of the first.[4] Thus, the positivity of presence is considered primary, whereas absence (non-being; nothingness) is treated as a derivative (and therefore inferior) category, because it is defined only as a lack of presence's positivity. However, for Derrida, the idea of such ontological purity and of such a reassuringly stable foundation is a myth. His method of deconstruction therefore aims to think against Western metaphysics' clear-cut opposition between (the foundation of) presence and (the non-foundation of) absence, in order to expose the mutual 'contamination' of the two notions. Presence is always already inhabited by absence, and vice versa; the two are inextricable, equiprimordial.[5]

Although Derrida's critique of phenomenology targeted Husserl in particular, it is no less relevant to Marcel. 'La phénoménologie nous paraît tourmentée sinon contestée de l'intérieur par ses propres descriptions du mouvement de la temporalisation et de la constitution de l'intersubjectivité', writes Derrida (1967a: 5) in *La Voix et le Phénomène* — a statement which resounds strongly with Chapter 2's critique of Marcel's *approches concrètes*. As did Husserl, Marcel wished to break from the traditional, logocentric conception of ontology proposed by Western philosophy. But as Chapter 2 has demonstrated, Marcel also privileged presence in his characterization of authentic (intersubjective) temporality, and, by means of his separation of (temporal) *existence* and (the eternal presence of) *être*, seemed to found his metaphysics on a binary opposition, which in turn appeared to confound his plans to reconceive ontology. The error, Chapter 2 suggested, lay in the decision to treat phenomenology as ontologically instructive: 'la ressource de la critique phénoménologique est le projet métaphysique lui-même', asserts Derrida (1967a: 3) in *La Voix et le Phénomène*; and this temptation, Derrida states in 'Ousia et grammè', can be symbolized by the desire to link phenomenological time with time as such:

> le concept de temps appartient de part en part à la métaphysique et il nomme la domination de la présence. Il faut donc en conclure que tout le système des concepts métaphysiques, à travers toute leur histoire, développe

la dite 'vulgarité' de ce concept [...], mais aussi qu'on ne peut lui opposer un *autre* concept du temps, puisque *le temps en général appartient à la conceptualité métaphysique.* (1975: 73; my emphasis)

In other words, Derrida is claiming that to philosophize about time is, in itself, to be complicit with the metaphysics of presence, because to conceptualize time is already to postulate a metaphysics, and, he insists, there can be no conception of metaphysics other than one of presence — precisely because of its conceptual nature, which aims to expose the true character of Being, to present Being as it is. Time itself is therefore in need of deconstruction.

If Derrida is right, then there is something to be said for Marcel's attempted avoidance of the ontology of time. Indeed, it is significant — and arguably commendable — that he explicitly refused to relate phenomenological time to ontological time. Nevertheless, in Marcel's case such praise is double-edged, for if this is true, the consistency of his concrete investigations of human Being is only called into question all the more. Just as for Heidegger, it is Marcel's belief that — as Wood writes — the dialectical experience of lived time 'will provide him with a horizon (*the* horizon, he claims) [... which will] open up the possibility of thinking Being itself' (1989: 138). 'In *principle*', Wood observes, 'the question of Time seems subservient to the question of Being' (1989: 138); and accordingly, Marcel does not elevate its status to the same level. 'But', he continues, 'it soon becomes clear that the two questions are inseparable' (1989: 139). This last remark is especially pertinent with respect to Marcel, for although his philosophical approach can be compared with Heidegger, unlike Heidegger he does not link (ontological) Being and time, but instead attempts to confine his discussion of time to the phenomenological. Yet, it does not, in fact, seem possible to isolate the two domains, for Marcel's phenomenological analyses of time prove inextricably bound to his ontological conclusions. Marcel's error is therefore his failure to recognize the intimate link between the question of time and the question of Being. Time may not exist as an entirely separate ontological and epistemological object from human reality (Marcel's 'irrelevance of time' argument), but this does not mean that it can be considered without reference to ontology. In fact, Marcel's very justification (in the discussion following 'Mon temps et moi') for avoiding the question 'qu'est-ce que le temps', in favour of an investigation into its phenomenological experience, was 'parce que je me demande [...] jusqu'à quel point nous pouvons arriver à la poser et même à lui donner un sens [ontologiquement, en lui-même]' (Hersch and Poirier 1967: 28). Is this not precisely to say that what time 'is' cannot be dissociated from human Being?

The question remains, however, as to whether Derrida is right to declare that any notion of time is inherently metaphysical, or to assume (as he seems to) that there can be no form of metaphysics other than a metaphysics of presence. Lawlor's radical (Deleuzian, post-structuralist) reading of Bergson would suggest the contrary, arguing that Bergson succeeds in challenging phenomenology and the metaphysics of presence for two main reasons: first, because his metaphysics equates Being with memory — that is, it defines Being in terms of what is past, as opposed to what is present; and second, because, owing to Bergson's notion of matter as 'image' (rather than 'thing'), Bergson proposes a new, non-totalizing conception of presence, where presence is not yet presence to consciousness, but remains a non-representational experience of the vital movement between the objective (matter, extension) and the subjective (mind, spirit).[6]

For Bergson, time is not internal to consciousness; rather, consciousness is internal to time. This is crucial to the non-totalizing character of his philosophy: because time cannot be 'contained' within consciousness, time cannot be presentified and philosophized about unproblematically. In fact, because our lives are predominantly past,[7] it is — if anything — what we conserve of the past (memory) that 'contains' the content of the present moment (Lawlor 2003: 46). Furthermore, Bergson's theory of memory (and of presence to consciousness in general) is non-representational, and thus is not concerned with objects. Rather, in *Matière et mémoire* (1896) Bergson defines matter in terms of 'images': 'La matière, pour nous, est un ensemble d'"images". Et par "image" nous entendons une certaine existence qui est plus que ce que l'idéaliste appelle une représentation, mais moins que ce que le réaliste appelle une chose — une existence à mi-chemin entre la "chose" et la "représentation"' (Œ: 161). 'La vérité est qu'il y aurait un moyen, et un seul, de réfuter le matérialisme: ce serait d'établir que la matière est absolument comme elle paraît être' (Œ: 219), Bergson proceeds to argue. The significance of *le souvenir-image* (see Chapter 1) now becomes clear. Just as my perceptions are images, reflective of the possibles of action I might intend, memories are also images, which, through a dilation of my awareness away from the simultaneity of the present instant and into the layers of my past experience, come to inter-penetrate and inform my present experience, directed by my projects of action. Thus, Lawlor explains, 'consciousness does not engender representations but is the selection of images from the whole of images called matter [...]. Insofar as consciousness is the selection from the whole of images called matter, Bergson does not define consciousness as consciousness of something; rather consciousness *is* something' (2003: 27; my emphasis).

So Bergson is not content to talk about what presently exists; or rather, if existence is anything, it is not fixed in any present — be it objective or subjective. Instead, the present of Being is the realm of action, flux, and change. 'Vous définissez arbitrairement le présent *ce qui est*, alors que le présent est simplement *ce qui se fait*' (Œ: 291), he writes. In the present moment, the (present, objective) materiality of the world collides with (past, subjective) memory of my life; but only for an instant, because, for Bergson, 'cette continuité de devenir qui est la réalité même' (Œ: 281) means that matter — and therefore my relation to matter — is constantly becoming. 'Immediately we can see that Bergson is conceiving being as time', writes Lawlor, 'but he is not conceiving time in terms of the present, since the present is nothing more than a "quasi-instantaneous section of continuous becoming". Becoming or time is not, for Bergson, a function of the present; the present is a function of time or becoming' (2003: 49). So, unlike traditional metaphysics, in Bergson's ontology there is no present to privilege, whether it be the presence of consciousness' being, or the presence of Being to consciousness. Being may be defined in terms of time, but Being-in-time is never stable. Instead, it oscillates between the past and the dynamic present, between contemplation and future-directed action; and thus, for Lawlor, Bergson 'twists free' of Platonism and the metaphysics of presence it instituted (2003: ix–x).

If Lawlor is right, it is all the more ironic that, in seeking to distance himself from Bergson, Marcel consolidated his subservience to a philosophical tradition which he thought — thanks to Bergson — he had succeeded in escaping. Bergson appears to hold the key to a radical new ontology that breaks free from the metaphysics of presence, a metaphysics that Marcel too wished to reject. One therefore recalls with pathos how adamantly Marcel refused Bergson's notion of memory as past in favour of a present conception, for this is emblematic of Marcel's blind conviction to uphold ontological idealism, not to dismiss it as he intended. Indeed, as Deleuze (1925–1995) has commented: 'Si nous avons tant de difficulté à penser une survivance en soi du passé, c'est que nous croyons que le passé n'est plus, qu'il a cessé d'être. Nous confondons alors l'Être avec l'être-présent. Pourtant [pour Bergson] le présent *n'est pas*, il serait plutôt pur devenir' (2004: 49). It seems that Marcel experiences precisely this difficulty.

Bergson is certainly more rigorous than Marcel, then; but the interpretative difficulties that the third section of Chapter 2 drew to our attention continue to make any definitive evaluation of Marcel's thought problematic. If Marcel really is concerned with ontology, in the traditional sense, he can indeed be accused of philosophical inconsistency. However, although Marcel clearly states that his preoccupation to be ontological,[8] it is not obvious that his understanding of ontology is the same as what is ordinarily understood by the term. In fact, in a 1970 interview with Boutang he stated: 'Je ne sais pas, au fond, dans quelle mesure j'ai le droit d'user du terme d'ontologie' (GM: 63). I want to suggest, therefore, that Marcel may still have challenged traditional metaphysics on some level — albeit using a highly clumsy and confusing vocabulary. Even if he were aware of the ontological possibilities Bergson's philosophy was offering, he might still choose to take issue with the direct nature of Bergson's method.

Indeed, in the same interview with Boutang cited above, Marcel defines his conception of 'ontology' as something indirect and inescapably hermeneutical: 'pour moi, au fond, il s'agit, surtout et toujours, d'une réflexion sur *ce que nous voulons dire* quand nous parlons de l'être' (GM: 63; my emphasis). So, given the difficulties Part I has exposed, both in declaring Marcel's interest to be unambiguously ontological, and in restricting a reading of his philosophy to the phenomenological, might a thoroughly hermeneutical interpretation in fact prove more fruitful?

Notes to the Conclusion

1. Accordingly, Widmer writes of Marcel: 'Pour prendre tout son sens, la description phénoménologique ne doit pas se considérer comme dernière, mais se prolonger en une ontologie, car la réalité a un statut d'être, une structure qui permet seule de donner aux descriptions phénoménologiques leur pleine signification et leur fondement solide' (1971: 119).
2. Scholars such as Wood (1989) have since defended Husserl (or at least questioned the purported distance between Derrida's position and Husserl's), arguing that Husserl does not always give primacy to presence, but in some instances (intentionally or not), demonstrates the co-primordiality of absence. See also Fuchs (1976), Dostal (2006), and Held (2007). Even more radical is the suggestion that Husserl's project is non-metaphysical, concerned more with its metaphilosophical or methodological conditions of possibility, than with substantial claims about the ontological nature of the world (Carr 1999; Crowell 2001). However, as Zahavi writes: 'it is one thing to make that point and something quite different to claim that phenomenology has no metaphysical implications. [...] Is it really true that Husserl's transcendental phenomenology excludes the actual existing world from consideration, and that it is concerned with meaning rather than with being?' (2003: 11–12).
3. To this we can relate more specific binaries such as the opposition between good and evil, truth and falsity, or sameness and difference: one can be in the presence of Good, Truth, or the Same, or encounter a lack of goodness, truthfulness, or sameness (Evil, Falsity, Difference).
4. '[D]ans une opposition philosophique classique, nous n'avons pas affaire à la coexistence pacifique d'un vis-à-vis, mais à une hiérarchie violente. Un des termes commande l'autre (axiologiquement, logiquement), occupe la hauteur' (Derrida 1972: 56).
5. '[L]a présence du présent perçu ne peut apparaître comme telle que dans la mesure où elle *compose continûment* avec une non-présence et une non-perception [...]. Ces non-perceptions ne s'ajoutent pas, n'accompagnent pas *éventuellement* le maintenant actuellement perçu, elles participent indispensablement et essentiellement à sa possibilité' (Derrida 1967a: 72).
6. A direct link to Deleuze can be made here. See Rodowick (1997: chap. 2).
7. 'Lorsque nous pensons ce présent comme devant être, il n'est pas encore; et quand nous le pensons comme existant, il est déjà passé. Que si, au contraire, vous considérez le présent concret et réellement vécu par la conscience, on peut dire que ce présente consiste en grande partie dans le passé immédiat. [...] *Nous ne percevons, pratiquement, que le passé*, le présent pur étant insaisissable progrès du passé rongeant l'avenir' (Œ: 291).
8. '[M]a préoccupation fut [...] de mettre sur pied une conception excluant toute réification de l'être, mais en maintenant l'ontologique en tant que tel' (RA: 318).

PART II

~

Time and the Problem of Hermeneutics

CHAPTER 3

~

Narrative Time

Part I suggested the theory of time which underwrites Marcel's philosophy to be problematic, for although he plainly declares his lack of interest in its ontology, he nevertheless is concerned with human Being. Owing to the temporal character of human existence, time is thus necessarily bound up with Marcel's ontological investigations. The conclusions he draws from his phenomenological *approches concrètes*, which make reference to both an experience of finitude and of eternity, therefore (indirectly) reify what Marcel only intended to be a phenomenological distinction between time and eternity. As such, Marcel's philosophy of time emerges as paradoxically concerned, and unconcerned, with both ontology and phenomenology; and in seeming contradiction with Marcel's allegedly realist (as opposed to idealist) project, time risks becoming subordinate to eternity.

Marcel's attempt to recognize a dialectical experience of both time and eternity, and his efforts to relate and reconcile the two existentially — albeit on an indeterminate level of analysis — are reminiscent of the account of time given by Augustine of Hippo (354–430 CE) in Book XI of his *Confessions* (397–400 CE). Here, Augustine also comes to understand the temporal notions of past, present, and future as a distension of a (more authentic) eternal present. The contrast between temporal fragmentation and this foundational unity then causes the absence of eternity to be experienced as a lack, analogous to Marcel's *exigence ontologique*. Furthermore, the status of time with respect to ontology is just as puzzling in Augustine as it is in Marcel. As Garrett DeWeese notes: 'Augustine cannot be interpreted as being a thoroughgoing idealist about time; time does bear some relation to objective reality' (2004: 120). But it is nevertheless the phenomenological character of time that is emphasized in Augustine, leaving us to wonder how exactly our human consciousness and experience of time relates to ontology.

However, such a discussion of time and eternity is considered important by Ricœur, who, in *Temps et récit* (1983–85), proposes a reading of Augustine that makes this tension between the phenomenological and the ontological intelligible. Augustine's reflections draw Ricœur's attention to what he calls '[le] caractère *aporétique* de la *spéculation* sur le temps [*sic*]' (*TR III*: 19), which renders reflection about its nature forcibly inconclusive.[1] As Muldoon explains, for Ricœur, 'there is no one time that can adequately account for both ordinary [ontological/cosmological] time and a purely phenomenological time that issues from the one "who" experiences time' (2006: 29); and this is because 'on ne peut penser le temps cosmologique (l'instant) sans subrepticement ramener le temps phénoménologique (le présent) et réciproquement' (*TR III*: 177). Yet, continues Ricœur in *Temps et récit III* (1985):

> insister sur les apories de la conception augustinienne du temps, [...] ce n'est pas renier la grandeur de sa découverte. C'est, bien au contraire, marquer, sur un premier exemple, ce trait fort singulier de la théorie du temps, que tout progrès obtenu par la phénoménologie de la temporalité doit payer son avancée du prix chaque fois plus élevé d'une aporicité croissante. (*TR III*: 19–20)

Thus, Ricœur does not accuse Augustine of failure, but instead points to how Augustine's reflections deepen our understanding of the experience of time, and, drawing further evidence from his detailed analyses of Aristotle, Husserl, Kant, and Heidegger, proceeds to generalize his suspicion that no pure theory of time is ascertainable, whether it be ontological or phenomenological. Rather, the aporias that arise from such attempts to think time (directly) can only be overcome through the mediation of narrative,[2] the mode in which — Ricœur claims — human time is (always) conceived and articulated, for it enables one to make impersonal ontological time one's own, and thereby bridge the gulf separating phenomenological and cosmological time: 'Le temps raconté est comme un pont jeté par-dessus la brèche que la spéculation ne cesse de creuser entre le temps phénoménologique et le temps cosmologique' (*TR III*: 439). Through its constant, creative, and indeed personal mediation of the tension between the phenomenological and the ontological, therefore, narrative time both recognizes, and (furthermore) renders productive, time's apparent inscrutability.

The similarities between the presentation of time in Marcel and Augustine open up the possibility of a parallel resolution of the tensions in Marcel's thought. Part II of this book will thus explore the scope of a Ricœurian reading of time in Marcel, seeking to establish whether, by means of such a hermeneutical interpretation, a more coherent analysis of his philosophy is possible. The first section of this chapter begins by giving a more detailed account of Augustine's study of time in the *Confessions*,[3] and describing the way Ricœur responds to this. It then explains how Augustine's theory of time may be usefully compared to Marcel's, and how Ricœur's analyses may therefore be seen as relevant and illuminating with respect to the difficulties Part I identified in Marcel's philosophy. The second section then reveals that narrative time is in fact important in Marcel's texts themselves, suggesting a Ricœurian reading of time in Marcel to be particularly pertinent; and the third section examines the relationship between narrative time, ontology, and ethics, in an attempt to shed further light on what Pedro Adams refers to as the 'curious blend between metaphysics and morality' (1966: 183) in Marcel's philosophy.

Time, Eternity, and Narrative: Augustine and Ricœur

Augustine's meditations on time in Book XI of the *Confessions* famously commence with an affirmation of time's familiarity, and a simultaneous confession of ignorance regarding its nature:

> What is time? Who can explain this easily and briefly? Who can comprehend this even in thought so as to articulate the answer in words? Yet what do we speak of, in our familiar everyday conversation, more than of time? We surely know what we mean when we speak of it. We also know what is meant when we hear someone else talking about it. What then is time? Provided that no one asks me, I know. If I want to explain it to an inquirer, I do not know. (1998: 230, XI. xiv.17)

Although Augustine does appear to make some progress in the detailed reflections that follow these remarks, it is significant that he never answers the question. Moreover, the question that is being asked is actually rather difficult to determine, so it is unclear whether Augustine is conducting an enquiry into the nature of time itself, or simply its psychological apprehension.[4] It is this interpretative difficulty that interests Ricœur.

Augustine begins his examination of time by observing the paradox of its existence. While intuitively one feels that time must exist, once a philosophical explication of its nature is attempted it appears that it cannot, since none of its parts — the past, the present, and the future — can truly be said to exist:[5]

> Take the two tenses, past and future. How can they 'be' when the past is not now present and the future is not yet present? Yet if the present were always present, it would not pass into the past: it would not be time but eternity. If then, in order to be time at all, the present is so made that it passes into the past, how can we say that this present also 'is'? The cause of its being is that it will cease to be. So indeed we cannot truly say that time exists except in the sense that it tends towards non-existence. (1998: 231, XI. xiv.17)

On this analysis (and just as in Marcel), 'to be' is 'to be present'; and consequently, Augustine insists, wherever the past and future exist, they too must exist as present, otherwise the paradox of their non-existence will reoccur:

> If future and past events exist, I want to know where they are. If I have not the strength to discover the answer, at least I know that wherever they are,

they are not there as future or past, but as present. For if there also they are future, they will not yet be there. If there also they are past, they are no longer there. Therefore, wherever they are, whatever they are, they do not exist except in the present. (1998: 233–34, XI. xviii.23)

His solution, therefore, is to postulate a three-fold present: if the past, present, and future cannot be located outside the mind and said to exist as such, they must instead be considered qualities of the mind, which all exist within the mind's present:

> What is by now evident and clear is that neither future nor past exists, and it is inexact language to speak of three times — past, present, and future. Perhaps it would be exact to say: there are three times, a present of things past, a present of things present, a present of things to come. In the soul there are these three aspects of time, and I do not see them anywhere else. The present considering the past is memory, the present considering the present is immediate awareness, the present considering the future is expectation. (1998: 235, XI. xx.26)

This also allows Augustine to resolve the problem of time's measurement, which vexes him as a result of his reflections on time's paradoxical non-existence.[6] Indeed, we seem to be able to measure time, and frequently speak of periods past, present, and future as differing in length; but if the past, present, and future do not exist to be measured, what is it that we are measuring when we speak of and compare varying periods of time? Augustine's three-fold present leads him to conclude that

> it is in you, my mind, that I measure periods of time. [...] That present consciousness is what I am measuring, not the stream of past events which have caused it. When I measure periods of time, that is what I am actually measuring. Therefore, either this is what time is, or time is not what I am measuring. (1998: 242, XI. xxvii.36)

Augustine thus 'purifies the present from being anything remotely instantaneous', Muldoon explains; 'it is extended rather, in relation to certain intentions (*intentio*). The mind distends itself (*distentio*) actively between present attention, past memory, and future anticipation in constant engagement given what it intends to accomplish' (2006: 33). More specifically, *distentio* denotes the dispersal and fragmentation of the self over the past and future: 'see how my life is a distension in several directions', and how we 'live in a multiplicity of distractions by many things', Augustine (1998: 243–44, XI. xxix.39) proclaims. *Intentio*, on the other hand, denotes the concentration and unification of the self when it is focused toward the true present of Being: God's eternal simultaneity.

Augustine's presentation of time in the *Confessions* is often interpreted as an argument for time's subjectivity,[7] where time is understood as a psychological construct that exists only in human minds, and acts as a fragmentation of and distraction from the eternal reality of God's presence. Closer examination of his text, however, shows that Augustine recognizes time as something objective,[8] this being most evident when Augustine rejects the definition of time as the movement of the heavenly bodies: 'I heard a learned person say that the movements of sun, moon, and stars in themselves constitute time. But I could not agree' (1998: 237, XI. xxiii.29), he writes. If these heavenly bodies ceased to move, but a potter's wheel continued to turn, we would not, Augustine suggests, wish to say that there was no time by which one could measure the revolutions of the wheel (1998: 237, XI. xxiii.29); and if the sun completed its circuit from east to west in the space of one hour instead of twenty four, we would not wish to call this a day. 'Let no one tell me then that time is the movements of heavenly bodies' (1998: 238, XI. xxiii.30), he concludes; 'a body's movement is one thing, the period by which we measure is another' (1998: 239, XI. xxiv.31). For Augustine, then, objects in the external world exist independently of what is thought about them, so the movement of objects themselves cannot be time; time is something independent.[9]

And yet, Augustine then seems to argue that time is mind-dependent, associating its extension with the extension of the mind as opposed to anything external, and thereby presenting time as purely phenomenological. It is at this point that it becomes difficult to determine the question Augustine is seeking to answer. The above citations concerning time's relation to objects in the world all seem to relate to his initial question, 'What is time?'; but the conclusion Book XI draws about time (the three-fold present) is primarily based on Augustine's reflections concerning what we mean when we speak of time (and measurements thereof),[10] the answer to which cannot be assumed to be the same as the answer to the former. This is what Ricœur notices, and in *Temps et récit III* he responds:

> Le moment précis de l'échec est celui où Augustin entreprend de dériver de la seule *distension* de l'esprit le principe même de l'extension et de la mesure du temps. A cet égard, il faut rendre hommage à Augustin de n'avoir jamais vacillé dans la conviction que la *mesure* est une propriété authentique du temps [...]. Pour Augustin, la division du temps en jours et années, ainsi que la capacité, familière à tout rhétoricien antique, de comparer entre elles syllabes longues et brèves, désignent des propriétés du temps lui-même. (*TR III*: 21–22)

Ricœur, then, identifies two kinds of time in Augustine's analyses — the phenomenological time of the three-fold present, and a cosmological time that is distinct from and envelopes all wordly objects[11] — and declares that these are incommensurable. Furthermore, he contends, it is never possible to delineate a purely phenomenological time of the self because, as Augustine's meditations themselves demonstrate, phenomenological time and cosmological time mutually contaminate each other:

> L'échec majeur de la théorie augustinienne est de n'avoir pas réussi à *substituer* une conception psychologique du temps à une conception cosmologique [...]. L'aporie consiste précisément en ce que la psychologie s'ajoute légitimement à la cosmologie, mais sans pouvoir la déplacer et sans que ni l'une ni l'autre, prise séparément, ne propose une solution satisfaisante à leur insupportable dissentiment. (*TR III*: 21)

To support his thesis further, Ricœur identifies a number of problematics that arise as a direct result of Augustine's triple present. First, he asks: 'Comment mesurer l'attente ou le souvenir [...] sans prendre en considération le changement physique qui engendre le parcours du mobile dans l'espace?' (*TR I*: 48). Indeed, surely expectation and memory cannot exist without reference to physical change outside the mind, that is, without some form of cosmological reference point? 'Quel accès indépendant avons-nous à l'extension de l'empreinte en tant qu'elle serait *purement* "dans" l'esprit?' (*TR I*: 48), he enquires. But 'l'énigme la plus impénétrable', according to Ricœur, '[est] celle au prix de laquelle on peut dire que l'aporie de la mesure est "résolue" par Augustin: que l'*âme* se "distende" à mesure qu'elle se "tend"' (*TR I*: 48; my emphasis). For Ricœur, Augustine's three-fold present does not solve the problem of time, but merely displaces it, creating an internal problem instead of an external one. 'Instead of time being split between being and non-being in the sense of external metaphysical substances', writes Richard Kearney, 'it is now our inner soul which finds itself cleft' (2005: 146). Ricœur therefore concludes:

> La trouvaille inestimable de saint Augustin, en réduisant l'extension du temps à la distension de l'âme, est d'avoir lié cette distension à la faille qui ne cesse de s'insinuer au cœur du triple présent: entre le présent du futur, le présent du passé, et le présent du présent. Ainsi voit-il la *discordance* naître et renaître de la *concordance* même des visées de l'attente, de l'attention et de la mémoire. (*TR I*: 49)

An analysis of Augustine's argument is incomplete, however, without consideration of time's broader situation in relation to eternity, for this is what provides the possibility of unity and stability: 'il manque quelque chose au sens *plénier* de la *distentio animi*, que seul le contraste de l'éternité apporte' (*TR I*: 50), Ricœur remarks. But as Kearney notes, 'it is, paradoxically, when we attend to the still and steadying character of the eternal Word that we fully realize just how distended and scattered our temporal lives are' (2005: 149). Thus, Ricœur argues, Augustine's reflections on eternity only intensify the human experience of time, and call for an internal hierarchization of the soul, which layers its temporal experience in an effort to bridge the gap between the distended present and the eternal present.[12] The problem of time's multiplicity has therefore been replaced with a new problem for personal identity: the different levels of this reconceived notion of temporalization fracture the self in order to accommodate a unified conception of time. Hence, Augustine cannot even be said to provide a satisfactory phenomenology of time,[13] because of the further contradictions his theory creates for personal identity. The only way of resolving these tensions, Ricœur argues, is to recognize both the reciprocity of the phenomenological and the cosmological and their radical difference, and to seek a poetic (rather than theoretical) solution to the aporias these create[14] — namely through the mediation of narrative: 'le temps devient temps humain dans la mesure où il est articulé sur le mode narratif' (*TR I*: 105), he asserts.

Marcel of course differs from Augustine in that he does not intend to reconcile the paradoxes of time itself; he considers such an endeavour to be misguided — illegitimate even — as testified to by the antinomies of time themselves. The difficulty in Marcel is in fact the reverse of that in Augustine, for the former's attempt to reconcile (ontologically) the dichotomy between the subject and the object (primarily in opposition to Descartes) succeeds only in displacing the problem, creating, as Chapter 2 proposed, an external tension between time and eternity where an inner conflict existed before. This external division between the temporal and the eternal is the dilemma that ends up troubling Marcel, manifesting itself in the difficult relation between *existence* and *être*. Yet this divide should never have been instituted in the first place, because Marcel only intended to engage with time on a phenomenological level. It therefore seems that, in spite of himself, Marcel has not been able to help but refer to time on a cosmological or ontological level, thereby corroborating Ricœur's theory that cosmological and phenomenological time mutually affirm and occlude one another: neither can be substituted for, nor derived or analysed in isolation from, the other without generating irresolvable aporias; they are incommensurable. 'L'aporie de la temporalité, à laquelle répond de diverses manières l'opération narrative, consiste précisément dans la difficulté qu'il y a à tenir les deux bouts de la chaîne: le temps de l'âme et le temps du monde', writes Ricœur in *Temps et récit III*. 'C'est pourquoi il faut aller jusqu'au fond de l'impasse, et avouer qu'une théorie psychologique et une théorie cosmologique du temps *s'occultent* réciproquement, dans la mesure même où elles *s'impliquent* l'une l'autre' (*TR III*: 24–25). If Augustine's discussions of time and eternity can be related to Ricœur's argument concerning the incommensurability of phenomenological and ontological time and the consequent importance of narrative, then, so in fact can Marcel's.

Narrative Time and Identity: Marcel and Ricœur

Not only do similarities between Augustine and Marcel enable links to be made with narrative time as discussed by Ricœur; as the remainder of this chapter demonstrates, narrative time is pivotal to Marcel's philosophy itself,[15] suggesting a Ricœurian reading of Marcel to be all the more legitimate as an approach to interpreting time in his philosophy. As will be seen, narrative time is in fact the underlying structure of the (temporal) self in both Marcel and Ricœur; and in Marcel's case, this is visible in the form of his philosophy as well as its content, owing to his distinctive first-person style.

Following his detailed study of the temporal dimension of human existence in *Temps et récit*, Ricœur came to a heightened awareness of the importance of narrative, not only as regards the human understanding of time, but with respect to the very constitution of the self. Narrative identity in *Temps et récit*, observes Ricœur in *Soi-même comme un autre* (1990),

> répondait à une autre problématique: au terme d'un long voyage à travers le récit historique et le récit de fiction, je me suis demandé s'il existait une structure de l'expérience capable d'intégrer les deux grandes classes de récits. J'ai formé alors l'hypothèse selon laquelle l'identité narrative [...] serait le lieu recherché de ce chiasme entre histoire et fiction. (*SA*: 138, note 1)

Indeed, here, Ricœur observed that history was never concerned with 'pure' cosmological time directly, nor fiction ever completely dechronologized, so that it bore no relation to the real world. Rather, the time of fiction and history stemmed from the same configuring operation — the narrative function — so that the understanding both modes reached was structured in an analogous way. Furthermore, their temporal reconfigurations always involved a reciprocal interweaving of the historical and the fictional: historical accounts always depended on creative imagination (that of the historian, when connecting the facts, but also that of the reader), inscribing phenomenological time into the cosmological time it sought to narrate; and fictional narratives were historical to the extent that the (unreal) events they presented were nevertheless 'past facts' (*faits passés*) for the narrative voice, thereby inscribing cosmological time into the phenomenological.[16] But what these analyses lacked, Ricœur realized, was 'une claire compréhension de ce qui est en jeu dans la question même de l'identité [...]. La question de l'entrecroisement entre histoire et fiction détournait en quelque sorte l'attention des difficultés considérables arrachées à la question en tant que telle' (*SA*: 138, note 1). Thus, Ricœur refocuses and modifies his argument so as to emphasize that: (i) personal identity itself 'ne peut précisément s'articuler que dans la dimension temporelle de l'existence humaine' (*SA*: 138); and (ii) the configuration of (temporal) personal identity also involves a reciprocal interweaving of the phenomenological and the cosmological. Narrative time, he therefore concludes, is central to the development of a conception of self.[17]

The specific difficulty Ricœur feels a narrative theory of (temporal) identity can help to untangle is that of identity's equivocity.[18] Positioning himself against the reductive simplicity of Descartes's notion of self, Ricœur aims to confront (and in so doing, affirm the reality of) this ambiguity, and, using the Latin distinction between *idem* and *ipse*, he argues for a distinction to be made between two different kinds of identity — *idem*-identity and *ipse*-identity — both of which need to be recognized when considering the lived reality of human time.[19] *Idem*-identity refers to that which remains the same over the course of time: 'la *permanence dans le temps* constitue le degré le plus élevé [de l'hiérarchie de significations déployée par l'identité au sens d'*idem*], à quoi s'oppose le différent, au sens de changeant, variable' (SA: 12–13). On this level of identification, there is essentially no difference between persons and things, because *idem*-identity is only concerned with static, objective similarity, such as numerical or qualitative identity. By definition then, *idem*-identity is not affected by time's passage; it is an eternal constant. *Ipse*-identity, on the other hand, 'n'implique aucune assertion concernant un prétendu noyau non changeant de la personnalité' (SA: 13).[20] Instead, it is grounded in the reflexive structure of human identity, and expresses a self-relatedness that does not rely on observable identity detectable through conscious (objectifying) reflection. It is only on the level of *ipse*-identity that the question *qui suis-je?* can arise. Whereas time was irrelevant for *idem*-identity, time is fundamental to *ipse*-identity, for the question *qui suis-je?* can relate only to individuals, the distinctive identity of which emanates from the unique track they carve through time, reflecting their different — and indeed, ever-changing — temporal situations and experiences.

Although radically different kinds of identity, the eternity of *idem*-identity and the temporality of *ipse*-identity are nevertheless related, in that both orders of identity are lived by the self. In *Soi-même* Ricœur writes:

> Parlant de nous-mêmes, nous disposons en fait de deux modèles de permanence dans le temps que je résume par deux termes à la fois descriptifs et emblématiques: le *caractère* [l'*idem*; le *que*] et la *parole tenue* [l'*ipse*; le *qui*]. En l'un et en l'autre, nous reconnaissons volontiers une permanence que nous disons être de nous-mêmes. (SA: 143)

Moreover, he contends, 'on ne peut penser jusqu'au bout l'*idem* de la personne sans l'*ipse*', for 'le caractère, c'est véritablement le "quoi" du "qui"' (SA: 147). An understanding of personal identity therefore demands an appreciation of both time and eternity. While *idem*-identity founds the self's sense of spatio-temporal permanence and eternity of its character (e.g. firmly established dispositions or habits), it is from *ipse*-identity that the self draws an assurance that it can initiate something new and other, against the objective sameness of the *idem* (le *que*), and that it can nevertheless remain constant as a subject (*qui*) while undergoing such change — hence, why Ricœur identifies its form of temporal permanence with the act of keeping one's word (see the citation above). The term 'assurance' here is key, for Ricœur insists that the lived certainty of personal identity is not demonstrable through empirical verification, but is instead affirmed through an existential feeling of belief or confidence which he calls 'attestation'.[21] Thus, Ricœur does not reject Descartes's testimony to a certainty of self entirely,[22] but he believes self-assurance to be characterized by a certain fragility or vulnerability: 'l'attestation manque [... la] garantie de l'hypercertitude[, lui conférant ...] une fragilité spécifique à quoi s'ajoute la vulnérabilité d'un discours conscient de son défaut de fondation' (*SA*: 34). *Ipse*-identity, therefore, 'met en jeu une dialectique complémentaire de celle de l'ipséité et de la mêmeté, à savoir la dialectique du *soi* et de *l'autre que soi*' (*SA*: 13); and for Ricœur, only a narrative model of personal identity can mediate such a dialectic.[23]

While Cartesian identity affirms the self in the present eternal ('je *suis*'), narrative identity treats 'l'agir humain comme *un mode d'être* fondamental' (*SA*: 32); it is not static, but always in motion. Only by constantly relating difference (over time) to the same (the eternal), and marrying them in the story itself, can narrative introduce a dimension of unity.[24] Through this continuous, yet open-ended, action of unification narrative identity then attests to the self's 'créance' in its permanent individuality, in its ability to act while remaining a constant subject through time and change (*ipse*-identity), as well as to endure in a more objective sense (*idem*-identity). 'La créance [de l'attestation] est [...] une espèce de confiance [...]: l'attestation est fondamentalement attestation *de soi*' (*SA*: 34), writes Ricœur in *Soi-même*. But at the same time, he argues, 'cette confiance sera tour à tour confiance dans le pouvoir de dire, dans le pouvoir de faire, dans le pouvoir de se reconnaître personnage de récit' (*SA*: 34). Thus, much more complex than Cartesian identity, Ricœur's *confiance en soi* — 'le pouvoir de se reconnaître personnage de récit', which gives one the confidence to speak and to act *as* someone — is not something that can be possessed or definitively stated, but is instead a dynamic constancy that is dependent on this saying and doing and on their continuation. David Carr explains: 'La narration, dans ce sens, [...] n'existe pas indépendamment d'une action qui la précède mais constitue précisément l'action. L'action n'existe pas indépendamment de son sens, et c'est la narration qui lui donne son sens en lui donnant sa forme et sa cohérence intérieures et extérieures' (1991: 210–11); '*est* personnage celui *qui fait l'action* dans le récit' (*SA*: 170; my emphasis), declares Ricœur in *Soi-même*.

Although not explicitly presented as such, close inspection of Marcel's analyses of what it means to think about the temporality of one's existence —

what it means to remember one's past, situate oneself in the present moment, and project oneself into the future — reveal narrative to be the underlying model upon which Marcel's account of personal identity rests as well.[25] Marcel also presents narrative time as fundamental to the self's experience and understanding of Being, I shall now argue.

As Marcel observed in his 1949 Gifford Lecture, 'Ma vie': 'Ma vie se présente à ma réflexion comme quelque chose dont l'essence est de pouvoir être raconté' (*ME I*: 169).[26] In fact, he continued, 'ceci est [...] tellement vrai qu'il est permis de se demander si les mots "ma vie" gardent un sens précis, abstraction faite à l'acte narratif' (*ME I*: 169). However, he then proceeds to illustrate the extent to which any 'story of my life' is always a reductive interpretation: 'Si je dis par exemple: mes premières années ont été tristes, j'étais souvent malade — j'*interprète* ce qui ne m'a peut-être pas du tout été donné sous cette forme. Ai-je eu seulement conscience à cette époque d'être triste? Cela est douteux' (*ME I*: 170; my emphasis). He therefore declares:

> Tout se passe comme si rétrospectivement je répandais une certaine couleur sur une trame ou un tissu. Mais cette couleur peut m'apparaître seulement maintenant, à la lumière de ce que j'ai appris ou vécu par la suite. Il est donc probablement impossible — et cela par définition — de prétendre que je raconte ma vie telle que je l'ai vécue. Contrairement à l'idée naïve que nous nous en formons, un récit ne peut en aucune façon être une reproduction de la vie.[27] (*ME I*: 170)

But if, here, Marcel rejects narrative as a way of (authentically) understanding 'my life', this is because Marcel has misunderstood the term 'narrative', interpreting it as meaning simply a story, as opposed to the 'twofold structure' that Ricœur (1980: 178) understands it to be — which includes not only the story, but also the act of its configuration. 'So many authors have hastily identified narrative time and chronological time [...] because they have neglected a fundamental feature of a narrative's temporal dialectic', Ricœur (1980: 177) writes in his article 'Narrative Time' (1980).[28] In fact, what one must recognize is that

> every narrative combines *two* dimensions in various proportions, one chronological [or temporal] and the other nonchronological [or eternal]. The first may be called the episodic dimension, which characterizes the story as made out of events. The second is the configurational dimension, according to which the plot construes significant wholes out of scattered events. [...] To tell and to follow a story is already to reflect upon events in order to encompass them in successive wholes. [...] the humblest narrative is always more than a chronological series of events.[29] (1980: 178; my emphasis)

Closer scrutiny of Marcel's texts, however, actually reveals his description of the structure of self experience to coincide with Ricœur's. This is what allows his notion of personal identity to be interpreted as (Ricœurian) narrative identity, in spite of his explicit rejection of narrative as a mode of self-understanding.[30]

As early as 23 July 1918, Marcel recognized the necessity of asking the question 'qui?', in addition to 'que?', in order to escape an abstract conception of myself:

> le *je* correspond précisément à ce refus de poser la question *qui*, parce qu''il n'y a pas lieu'. Seule une réflexion supérieure mettra en doute l'objectivité même de cette ordre [...]. Mais il est [...] impossible de ne pas poser la question *qui* [dans] l'ordre du témoignage. [...] Le *qui* est [...] essentiellement une signature. (*JM*: 141)

The answer to the question *qui suis-je?*, as Marcel later argues, lies in the past that I continue to live as 'my present' — 'dans l'actualité fulgurante du souvenir' (*ME I*: 170), as he stated in his lecture 'Ma vie'. Part I has already shown memory to be important for Marcel's understanding of subjectivity, although the dialectic between time and eternity that his account involved appeared problematic. If time in Marcel's philosophy is understood as narrative time however, a much more fruitful — and crucially, more coherent — reading becomes possible. As Chapter 1 explained, Marcel refuses to conceive of memories as conserved 'truths' about the past. Rather, memories testify to the indeterminacy of meaning in general. For Marcel, the significance of my past is not directly knowable (as a *que*). A passive, impersonal record of my encounters, thoughts, and actions over time — that is, my past grasped through primary reflection — does not therefore express 'my' life, for my life is always experienced *as*: as individual to me, as personal to me, as mine and not another's.[31] 'Ma vie n'est pas séparable d'un certain intérêt pris à ma vie' (*ME I*: 177), he argued in 'Ma vie'.[32] Thus, for a past truly to be 'my' past, its *que* must continue to hold meaning for me in the present. It is for this reason that Marcel describes a meaningful grasp of my life as occurring 'dans *l'actualité* fulgurante du souvenir'.[33]

This links in with memory's further significance for Marcel — the fact that conservation of the past is also a personal and creative act of valuation:[34] 'Il n'y a conservation que dans l'ordre du créé qui est aussi l'ordre de ce qui *vaut* (la conservation impliquant la lutte active contre la dissolution)' (*JM*: 151), writes Marcel in his diary on 10 December 1918. Memory thus becomes the focal point of the continuity I experience in myself, the locus of personal identity, for to value (positively or negatively) is, at the same time, to assume myself, my individuality, because anything that is of true consequence to me has an impact in spite of the threat of temporal dispersion. For better or for worse, through the act of remembering (valuing), I actively choose and create myself (as a subject, *qui*): 'je ne *suis* que dans la mesure où il y a des choses, disons des êtres qui comptent pour moi' (*JM*: 224), Marcel declares on 17 December 1919 — a statement which might also be said of *ipse*-identity, since Ricœur models this on fidelity to others.

If these analyses are combined with Marcel's other discussions of 'ma vie' it becomes clear that, for Marcel, if the experience of human existence is tensed, the structure of our tensed experience is not objectively given. Instead, I structure my experience, interpreting it in accordance with what is significant, valuable, or felt (eternally) to be necessary to me,[35] in dialectical relation with the (temporal) threat of change and dispersion. It is in so doing that I create my individual identity. Personal identity in Marcel then emerges as homologous to narrative identity in Ricœur,[36] for narrative identity, as a poetic rather than a theoretical solution to the aporetic of time,[37] relies on the productive imagination and therefore, as with Marcel, creatively re-figures time for me. 'Déterminer le statut philosophique de la reconfiguration [du temps par le récit], c'est examiner *les ressources de création* par lesquelles l'activité narrative répond et correspond à l'aporétique de la temporalité' (*TR III*: 12; my emphasis), writes Ricœur in *Temps et récit III*. And as he notes in *Soi-même*: 'Cette fonction *médiatrice* que l'identité narrative du personnage exerce entre les pôles de la mêmeté et de l'ipséité est essentiellement attestée par les *variations imaginatives* auxquelles le récit soumet cette identité. [...] ces variations, le récit [...] les recherche. [...] l'intrigue est mise au service du personnage' (*SA*: 176). So, whereas Cartesian identity is straightforwardly present, Marcel's and Ricœur's narrative notions of identity are self-constructed — or rather self-constructing: always in process, always becoming. Creativity, for both, manifests the irreducible quality of human Being. It is this that they feel the *cogito*'s self-evidence threatens.[38]

Such an interpretation is further confirmed by arguments made in another of Marcel's 1949 Gifford Lectures, 'Le Sens de "ma vie": identité et profondeur'. In this lecture, Marcel addressed the problem of discontinuity in one's personal experience of self, and introduced the notion of 'profondeur' in order to resolve the conflict between my identity as ephemeral succession and discontinuity, and my identity experienced as a unity.[39] This *profondeur* already resembles Ricœur's narrative identity in that it acts as a transcendent mediator between the tensions of unity and difference (eternity and time) that I experience in myself. However, the parallel extends further to include the active, affirmatory creativity that Ricœur also associates with narrative identity, and which we saw Marcel stress in relation to memory. If Marcel's analysis of memory showed that my past is always experienced *as*, Marcel's description of my *profondeur* shows that the same applies for my identity in general.[40] My identity, as meaningful for me, relates to my past, present, and future together, all of which present themselves to me as mine in light of what I value. As with the past, a future can only become 'my' future if my present is oriented toward it, devoted to it in accordance with my values (here the link with *ipse*-identity as keeping one's word becomes clearer). But configuring my present in such a way requires genuine will and creative action on my part — and thus Marcel speaks of the importance of 'une fidélité créatrice', 'une fidélité qui ne sauvegarde qu'en créant' (*EA*: I, 119; 7 October 1932). Experiencing both my past and future (in my present) as affirming the same set of values is what then allows my life to appear unified,[41] and enables me to experience the *profondeur* of my identity. And when this happens, explained Marcel in 'Le Sens de "ma vie"',

> l'avenir se raccorde mystérieusement au plus lointain passé.[42] On dirait, si obscure que soit une pareille notion, que le passé et l'avenir au sein du profond se rejoignent, s'étreignent dans une zone qui est à ce que j'appelle le présent ce que serait un Ici absolu à l'ici contingent; et cette zone où le maintenant et le alors tendent à se confondre, [...] ne peut être que ce que nous appelons l'éternité. (*ME I*: 209)

Marcel's 'Notes sur *Le Discours cohérent* de Julien Benda' (16 December 1930) shed light on what might be understood by 'éternité' here. In these notes, Marcel reacts against Benda's declaration that 'penser le monde comme infini sous le rapport du temps, c'est le penser d'une manière telle que ses distinctions dans le temps n'existent plus, c'est le penser d'une manière telle que les différents y sont indifférents'. Rather than asserting such an extreme opposition between infinity (or eternity) and time, Marcel recommends that we understand temporal distinctions as distinctions '[qui] n'ont pas une signification ultime, mais seulement superficielle; encore que dans un certain registre elles conservent toute leur valeur' (*EA*: I, 73). In order to illustrate his argument, he draws an analogy with the structure of a book:

> Un livre est quelque chose qui comporte une pagination parfaitement déterminée, et celui qui est chargé d'en mettre en ordre les feuillets est tenu de respecter cet ordre unique et irréversible. Mais il est tout à fait clair d'autre part que ce livre comporte d'autres types d'unité infiniment plus profonds que celui qui s'exprime dans la mise en pages. Et ceci ne veut pourtant nullement dire que la mise en pages soit 'illusoire'. (*EA*: I, 73)

The pagination of the book, as a means of successive ordering, can be compared to chronology; and if the analogy is followed through, we can conclude that, for Marcel, temporal order, legitimate as it is, is nevertheless only one form of order, which should not be understood as contradicting other more fluid, global (and according to Marcel, 'profound') orders that can be equated with infinity or eternity.[43] 'La définition en vertu de laquelle l'être infini serait l'être en tant que contradictoire à lui-même repose sur un paralogisme et peut être immédiatement éliminée' (*EA*: I, 74), he insists.

Narrative, for Ricœur, introduces a similar dimension of necessity or eternity, which unifies contingency, thereby mirroring Marcel's 'zone qui est [... au] présent ce que serait un Ici absolu à l'ici contingent' (*ME I*: 209). Indeed, the individual elements in any narrative are contingent; all of these (incidents, speech events, interactions) could have been different, or indeed might not have happened at all. However, once emplotted — that is, structured as a narrative — these elements become necessary to the plot's unfolding, and are essential in configuring the layers of internal significance that hold the narrative together.[44] 'Le paradoxe de la mise en intrigue', observes Ricœur in *Soi-même*, 'est qu'elle inverse l'effet de contingence, au sens de ce qui aurait pu arriver autrement ou ne pas arriver du tout, en l'incorporant en quelque façon à l'effet de nécessité ou de probabilité, exercé par l'acte configurant' (*SA*: 169–70).[45] But if this appears paradoxical, as with Marcel's above analysis of time and eternity, it can nevertheless be reconciled, namely through the intermediary mode of narrative. 'L'histoire racontée [...] constitue une alternative à la représentation du temps comme s'écoulant du passé vers le futur, selon la métaphore bien connue de la "flèche du temps"' (*TR I*: 131), writes Ricœur in *Temps et récit I* (1983); 'en lisant la fin dans le commencement et le commencement dans la fin, nous apprenons aussi à lire le temps lui-même à rebours [...]. Bref l'acte de raconter, réfléchi dans l'acte de suivre une histoire, rend productifs les paradoxes qui ont inquiété Augustin' (*TR I*: 131). Hence, for Ricœur, as with Marcel's notion of *profondeur*, narratives come to signify a non-linear mingling of the past, present, and future.[46] It is this — not chronology — that is the true articulation of human time. In *Temps et récit I* Ricœur writes:

> S'il est vrai que la pente majeure de la théorie moderne du récit [...] est de 'déchronologiser' le récit, la lutte contre la représentation linéaire du temps n'a pas nécessairement pour seule issue de 'logiciser' le récit, mais bien d'en approfondir la temporalité. La chronologie — ou la chronographie — n'a pas un unique contraire, l'achronie des lois ou des modèles. Son vrai contraire, c'est la temporalité elle-même. (*TR I*: 65)

Thus, if we interpret the presentation of eternity in Marcel in a Ricœurian sense, it need not be conceived of as the static, isolated notion we saw Bergson criticize in Chapter 1; it need not be understood in (ontological) opposition to time. On this phenomenologico-hermeneutic reading, time and eternity are both part of the structure of human temporality. This then allows for an experience of *durée* which is 'ruptured' and yet still continuous, and also manifests the creative possibilities of human understanding (mimesis I) and action. Indeed, for both Marcel and Ricœur, it is a mistake to conceive of (any single mode of) reflection as positing the subject directly, on an ontological level (as do Descartes's *cogito* and Bergson's intuition). Rather, reflection continues to create the subject through its engagement with, and transcendence of, chronology's temporal determinations, thereby affirming a certain ontological foundation (through its reference to cosmological time), while, at the same time, placing the possibility of individual action at the centre of personal identity (through its concurrent relation to personal, phenomenological time).[47] 'Il n'y a pas de philosophie ou de vie philosophique sans une sorte de constante réinterprétation qui n'exclut d'ailleurs pas une certaine inaltérabilité du contenu fondamental: mais cette inaltérabilité reste plutôt *visée* et comme *pressentie* que conceptuellement exprimable' (*DH*: 11), declared Marcel in the first of his 1961 William James Lectures ('Points de départ').

However, time is not yet mine until I read and identify with the narrative I create: 'narrative does more than just establish humanity, along with human actions and passions, "in" time; it also brings us back from [...] "reckoning with" time to "recollecting" it', writes Ricœur (1980: 178) in 'Narrative Time'. This dynamic, creative act of recollection is in fact precisely the role that Marcel's notion of secondary reflection must perform (it is even described as 'recollection'),[48] in order for it to be possible to grasp the significance of my life over and above chronology, over and above my life as understood through primary reflection. Indeed, as Marcel argues in 'Ma vie', were I to have documented my every thought and move in a diary, it does not follow that, upon re-reading the diary, every entry will continue to hold significance for me. Indeed, I may have forgotten much of its content, and think myself to be reading someone else's diary by mistake.[49] As both Ricœur and Marcel argue, therefore, the narrative act does not end at the stage of emplotment (mimesis II); I must be able to re-read, synthesize, and identify with the narrative on a global level if the story it tells is to be recognized as 'my' story (mimesis III), if the temporality it configures is to be seen as 'my' life.[50]

Thus, although Marcel does not present personal identity as narrative identity himself, narrative — in the Ricœurian sense of the term — nevertheless seems to be the underlying structure of personal identity in his philosophy. It is therefore possible to understand the 'ruptured' (but also somehow continuous) experience of *la durée* that Marcel argued for over and against Bergson in terms of the dialectic of narrative identity — which must constantly mediate difference and sameness, time and eternity, contingency and necessity — rather than in the (direct) ontological terms that generated such a problematic in Part I.

Interestingly, if the content of Marcel's philosophy (implicitly) presents narrative as the key to constituting and conceiving a unified notion of identity over time, the first-person form of Marcel's writings also mirrors the structure of narrative identity as defined by Ricœur. On two levels therefore, Marcel's philosophy presents narrative as fundamental to the self's experience and understanding of time and eternity. The remainder of this section provides an introduction to the significance of the narrative form of Marcel's writings. The third and final section then further engages with this, in order to demonstrate how it can shed light on the relation between ontology and ethics in his philosophy.

If Marcel's philosophical investigations of personal identity seemed to affirm Ricœur's narrative conception of personal identity, the form of his texts also corroborates such a theory, for his philosophical trajectory over time is constituted by a range of different 'selves', the relation between which he narrates in order to maintain the cohesion and coherence of his own philosophical identity. Indeed, Marcel's theoretical project, though objectifiable as an investigation into human subjectivity,[51] is also continually re-examined and re-considered, often being cast in a different light as Marcel approaches this question from new angles, and in relation to different situations.[52] Frequently, he also voices his difficulty in identifying with his earlier work, which appears to him as if it were written by another,[53] causing him to ask the question *qui?* Thus, in the case of Marcel's shift away from idealism emerge a number of distinct selves: his idealist, pre-World War I self; his transitional World War I self, who, confronted with the tragedy of existence, develops a concern for human individuality; and his 'non-idealist', post-World War I self, who decides to embrace a more concrete approach to philosophy.[54] Each self is discontinuous with respect to characteristics of the other selves, but at the same time continues to manifest Marcel's unchanging philosophical concern with the nature of human subjectivity. One can therefore identify Marcel's concern with subjectivity with Ricœur's *idem*-identity, in that it represents a constant in his philosophical identity. The form of Marcel's philosophy, on the other hand, can be equated with Ricœur's *ipse*-identity, for this dimension of Marcel's thought, which is not continuous over time, expresses Marcel's philosophical identity in its various human contexts, in the form of these different philosophical selves. And crucially, it is only through the over-arching narrative of his break with idealism that Marcel is able to mediate this 'ruptured' experience of his own philosophical self: in spite of Marcel's claim that the second (reformed) part of the *Journal* (1915–23) 'rend à mes oreilles un son infiniment plus familier, plus intelligible que la première', it is this which allows him to claim that the two parts cannot be dissociated (*JM*: x). In this way, a meta-narrativity emerges in Marcel's writings: his approach to his investigation of the self — which is itself explored through the lens of his self, his own individual (phenomenological) experience — is also narrated. This then rescues what may be seen as incoherence and fragmentation in his work (the two parts of his *Journal* are, he admits, 'bien distinctes' (*JM*: ix)), reconfiguring it as a wider narrative project.

In the words of Cain, therefore, 'the "way" and the "what" of Marcel's thought are bound together' (1963: 14) — except that the scope of this observation is much greater than he, or indeed many other commentators have recognized. Troisfontaines, for example, states that, for Marcel, the question *que suis-je?* 'porte à la fois sur l'*être* et sur le *je*' (1953: I, 223). However, he fails to notice that if Marcel examines the general difficulties that arise (for any 'je') when pondering the question *que suis-je?*, these difficulties can also be explored on a second level with respect to Marcel's own 'je' — the double subject of his philosophical writings. Indeed, one might say — in keeping with his own philosophical language — that Marcel goes beyond a surface examination of Being that is based around *le problème*, in order to explore *le mystère de l'être* 'en quoi je suis moi-même engagé' (*EA*: I, 146). As he stresses in *En chemin* (1971), it is '*à partir de mon expérience* [que mes interrogations se développaient]' (*EC*: 13); 'j'ai eu vraiment conscience d'être engagé dans une recherche où tout l'essentiel de moi-même se trouvait concerné' (*EC*: 22). Thus, Marcel's autobiographical identity is inextricable from his philosophical identity.[55] This is what provides the experiential basis for his *approchès concrètes*,[56] and is also — as the third section now demonstrates — crucial to appreciate when considering the relation between Marcel's ontology and ethics.

Being-with-Others: Narrative Time, Ontology, and Ethics

One might protest, at this point, that the second section's Ricœurian reading of identity in Marcel seems primarily concerned with the question *qui suis-je?* — that is, with how I experience and understand my life as mine — whereas Marcel explicitly describes his philosophy in terms of a preoccupation with the question *que suis-je?*, suggesting that he seeks to establish a more general, ontological foundation for (the 'ruptured' experience of) temporal human identity. Indeed, as well as being instructive with respect to the *qui* of my identity, Marcel's notion of *profondeur* also relates to a (universal) *que*. In his diary notes from January 1938, for example, where he first reflects on the significance of *profondeur*, Marcel declares that 'cet ensemble de remarques [...] pointe [...] vers une métaphysique de l'essence' (*PI*: 44); and in *Les Hommes contre l'Humain* (1951) he then asserts:

> Il n'est de profondeur authentique que là où une communion peut être effectivement réalisée; elle ne le sera jamais ni entre les individus centrés sur eux-mêmes, et par conséquent sclérosés, ni au sein de la masse, de l'état de masse. La notion d'inter-subjectivité sur laquelle repose mon dernier ouvrage [*Le Mystère de l'être*] suppose une ouverture réciproque sans laquelle aucune spiritualité n'est concevable. (*HH*: 202)

Can a Ricœurian interpretation account for this ontological thesis as well?

Yes, is the answer, for as Dan Zahavi argues: 'the emphasis on narratives is not merely to be understood as an epistemological thesis. [...] narratives do not merely capture aspects of an already existing self, since there is no such thing as a pre-existing self, one that just awaits being portrayed in words' (2007: 180). Ricœur's narrative identity combines both an epistemological and an ontological thesis as it mediates between the phenomenological and the cosmological: 'la compréhension [narrative] de soi égale à la constitution de soi. L'acte ou l'activité de compréhension est de cette manière constitutive de son être', Carr explains (1991: 209). But, he continues, 'cette conception [...] ne veut pas dire que le sujet se crée *ex nihilo*'; 'je suis "enchevêtré" [...] dans les histoires déjà en cours, [et ...] c'est à travers ces histoires [communautaires] que je me comprends' (1991: 211). In addition to reconciling my identity with the otherness of myself, then, narrative identity emphasizes the connection between my identity and that of others. If one's identity can be likened to the protagonist in a story, the protagonist's identity equally intersects with the identity of other characters, and thus the protagonist's narrative becomes intertwined within a web of narratives, all of which mutually constitute and affirm one another.[57] One can therefore only make sense of oneself through one's involvement with others.[58] Ricœur, too, affirms Being as intersubjective; hence, narrative time continues to be apt as an interpretation of time in Marcel.

The narrative mode of Marcel's philosophical texts, observed at the end of the second section, makes this intersubjective foundation all the more explicit, owing to Marcel's own active (narrational) participation in his philosophy. Thus, continuing to draw parallels between Marcel and Ricœur, the final section of this chapter will re-engage with the form of Marcel's writings, in order to demonstrate how his entire philosophical methodology acts, through narrative, to affirm an underlying ontology of intersubjectivity. Narrative time therefore becomes pivotal to Marcel's philosophical *engagement*;[59] and as will be seen, this proves instructive with respect to understanding the ethical dimension of his philosophy as well.

As the second section established, Marcel's use of narrative is not just an idiomatic quirk; the coherence of his philosophy depends on it, for Marcel's totalizing interpretative narratives concerning his philosophical development are pivotal to shaping and unifying his philosophical position over time. In fact, Marcel uses narrative on a second, rhetorical level as well, where narrative acts as the 'logic' underwriting his investigations into the nature of human experience.[60] On this level, the narrative time of Marcel's texts provides a ground on which the narrator (Marcel) and the narratee (the reader or listener) can come together, drawing persuasive power from the possibility for identification this creates.[61] As he states in the introduction to his *Journal*, his aim is to 'inciter le lecteur à refaire avec moi le chemin que j'avais moi-même parcouru', so that the reader becomes his 'compagnon de route', and might 'live' the experiences to which he testifies (*JM*: ix).[62] Consequently, its structure can be said to conform to Bakhtin's (1895–1975) chronotope of the road. In general, the chronotope of an artistic work configures a relation between the text and the world, by means of a spatio-temporal structure of expression that it creates;[63] and 'of special importance' to the chronotope of the road, argues Bakhtin in *The Dialogic Imagination* (1975), is the close link it has with what he terms 'the motif of meeting' (1981: 98) — that is, its potential for encounter.

Indeed, out on the open road any variety of reader or listener can encounter the narrator. Time spent together then correlates with ground covered, as both parties go forth, explorer-like, toward whatever discoveries their journey might bring. Just as Bakhtin identified this with 'adventure' narratives in literature (1981: 98), Marcel also accorded an adventurous character to his philosophy,[64] drawing an analogy between the structure of his philosophy and the image of a road or path — a rather crucial analogy it seems, for in the introductory session to his 1949 Gifford Lectures he insisted upon asking his audience for

> la permission d'user d'une comparaison [pour marquer l'orientation générale de mon œuvre ...], car je partage la croyance que professait Henri Bergson en la valeur de certaines images en quelque sorte structurales. Celle qui s'impose à moi ici est l'image d'un chemin. Tout se passe, il me semble, comme si j'avais jusqu'à présent suivi des pistes à travers un pays qui m'apparaissait en grande partie comme inexploré [...]. (*ME I*: 10)

Through the use of this chronotopic first-person narration, therefore, Marcel seeks actively to create a spatio-temporal relation between his philosophy and the world — a relation where the philosophical and the literary in fact converge, demonstrating his rejection of systematic, presentifying modes of discourse.[65]

For this reason, Marcel's exploratory *approches concrètes* are more evocative than demonstrative. Nevertheless, Marcel still believes their indirect form to be of value.[66] Like Ricœur, Marcel refuses to separate truth from method, as if truth were something ready-made, an object that could simply be grasped.[67] For Ricœur, stories are not unreal or illusory; they project a possible world that can intersect with the world of the reader, proposing orientations toward others, objects, time, and space, with which readers can imaginatively experiment, and thereby explore their ontology. This then helps to inform and shape their relationship to the actual world, so their understanding is 'fictive' in a very real sense: 'c'est dans la fiction littéraire que la jointure entre l'action et son agent se laisse le mieux appréhender' (*SA*: 188), he writes in *Soi-même*. Marcel may not have made this argument in such explicit terms, but it is nevertheless implicit in his method, for stories provide a primary means for his ontological explorations. He frequently asks his readers or listeners to imagine a particular situation, such as losing one's watch,[68] coming across an unknown flower and wondering what it is,[69] or being confronted with a dilemma dramatized in one of his plays.[70] He then talks them through these examples — narrates these situations or experiences to his readers or listeners — so that they might ask themselves the same questions, encounter the same complexities, and subsequently identify with Marcel's thoughts.[71]

To give but one example, in the first of his 1950 Gifford Lectures ('Qu'est-ce que l'être?'), Marcel discussed the example of someone enquiring 'qu-est-ce que cette fleur?', while out on a walk with a friend. 'Si on me répond que c'est un orchis', Marcel observed, 'j'en conclurai qu'elle présente des caractères communs avec d'autres fleurs que j'ai déjà vues et que je suis capable de reconnaître' (*ME II*: 16), and I will be able to situate the flower within a classification system. Nevertheless, although this answer enables a certain amount of progress, it is not an exhaustive response, and is even, in a certain sense, evasive.[72] 'Je veux dire par là', Marcel explained, 'qu'elle laisse de côté la singularité même de cette fleur. Tout se passe [...] comme si ma question était interprétée de la manière suivante: à quoi *d'autre qu'elle* cette fleur elle-même peut-elle être réduite?' (*ME II*: 16). The ontological implications of this situation are then analysed: just as a botanical system can only classify the flower in accordance with characteristics possessed by flowers other than the one in question, any systematization of human Being is inevitably an abstraction, unable to express its singular experience. Furthermore, the question 'qu'est-ce que c'est?' can only apply to that which can be objectively designated, and implicit in such referentiality there is always a triadic order of relations: 'la question "Qu'est-ce que c'est?" porte toujours sur du désignable, et plus profondément [...] elle se réfère dans tous les cas à un ordre impliquant des relations triadiques' (*ME II*: 17). Indeed, the flower does not tell me its name; my botanist friend does.[73] The nominally objective relation between the flower and its name actually rests on the wider intersubjective context grounding all common understanding[74] — and thus the flower example becomes an argument for the intersubjective community that is primary to Marcel's ontology.[75]

If Marcel comes to any conclusions in this lecture, they are not reached through deductive analysis. Instead, their logic can be compared to that of a story. As Ricœur writes in *Temps et récit I*: '[La conclusion d'une histoire] n'est pas logiquement impliquée par quelques prémisses antérieures. [...] Comprendre l'histoire, c'est comprendre comment et pourquoi les épisodes successifs ont conduit à cette conclusion, laquelle, loin d'être prévisible, doit être finalement *acceptable*, comme congruente avec les épisodes rassemblés' (*TR I*: 130; my emphasis).[76] Similarly, to understand Marcel's philosophy is to identify and empathize with his personal reflections on the nature of experience; it is, in fact, already to legitimize his philosophical arguments, for to identify with Marcel's thoughts is to appreciate that all of these contribute to — or render 'acceptable' — the broader (ontological) conclusions that he draws. Narrative is the specific method that characterizes Marcel's phenomenology.

Furthermore, Marcel's narrative style itself can be interpreted as an active affirmation of the intersubjective ontology to which his phenomenological studies testify. Indeed, in the same lecture cited above Marcel describes intersubjectivity as 'le terrain sur lequel nous avons à nous établir pour poursuivre nos investigations' (*ME II*: 12). Accordingly, Marcel's philosophy is not closed in on itself, but instead, through the rhetoric of its narratives,[77] makes an outward appeal to others and their individual experience, requiring the presence of a reader or listener for the completion of its arguments.[78] The integral role played by the reader in the understanding (and subsequent impact) of a text is precisely what Ricœur wished to emphasize with respect to narrative (this is mimesis III). In his article, 'Life in Quest of Narrative' (published 1991), he declares:

> the process of composition, of configuration, is not completed in the text but in the reader and, under this condition, makes possible the reconfiguration of life by narrative. [...] more precisely: the sense or the significance of a narrative stems from the *intersection of the world of the text and the world of the reader*. The act of reading thus becomes the critical moment of the entire analysis. On it rests the narrative's capacity to transfigure the experience of the reader. (1991: 26)

This capacity was something Marcel wished to write into his philosophy too. 'Ce qui est vrai de la lecture d'un texte, s'applique presque exactement à la recherche philosophique telle que je l'entends' (*PI*: 24), he asserts in 'Mon propos fondamental' (1937), for as he explained in the 1973 paper 'De la recherche philosophique': 'j'ai toujours eu conscience d'un inachèvement, de prolongements non seulement possibles mais nécessaires et qu'il appartiendrait à d'autres de réaliser' (1976: 9). Indeed, 'the time of a narrative is public time', writes Ricœur in 'Narrative Time' — 'but not in the sense of ordinary time, indifferent to human beings, to their acting and their suffering. [...] Through its recitation, a story is incorporated into a community which it gathers together' (1980: 175–76). For this reason, Kevin Vanhoozer summarizes, 'narrative time, both in the text and outside it, is the time of *being-with-others*' (1991: 46). It is by means of this temporal mode that Marcel hopes his philosophy can make genuine intersubjective appeal, offering — and through a process of identification, also demonstrating — the possibility of an encounter with others.[79] The form of his philosophy is indissociable from its content.

Such an undertaking, however, is easier said than done, for while to narrate is always to suggest how a specific sequence of events should be considered meaningful, meaning is not something one can ever wholly control. Marcel is painfully aware of this, to the extent that in the introduction to the *Journal*, his philosophical meta-narrative warns of how his philosophical path will be 'rocailleux' and 'sinueux', because the reader, as his *compagnon de route*, will be 'associ[é ...] à mes hésitations, à mes incertitudes' (*JM*: ix). So, not only does the narrative form of Marcel's philosophy attempt to affirm the intersubjective foundation of Being through its creation of a shared temporal experience between the narrator and the narratee; it equally testifies to the difficulties of giving voice to human experience, relating the arduous and frustrating nature of Marcel's chosen philosophical approach, and lamenting especially the haziness, equivocity, or undesired nuances of words he committed to paper. 'J'emploie à regret cette terminologie inadéquate' (*JM*: 123), Marcel bemoans in a 1914 diary entry (7 May); 'tout cela est au fond horriblement difficile à penser clairement' (*EA*: I, 13), he writes in another on 29 February 1929.

But this only manifests more completely Ricœurian *attestation*, described in *Soi-même* as 'l'assurance d'être soi-même agissant et souffrant' (*SA*: 35),[80] and is in fact pivotal to Marcel's anti-dogmatism (anti-presentification). Indeed, for Marcel to formulate the experience of human subjectivity in any positive, universally valid manner is immediately to come into tension with his critique of objectivity, as he is forced to fix its nature in order to provide any positive characterization at all. As is visible in the meta-narrative of his narratives, Marcel struggles with this conflict of interest throughout his philosophical career, at once aware that he is grounding his reflections in his own singular experience and perspective, which he does not wish to impose on others, but also of a truth about subjectivity that he feels transcends his own particular situation.[81] One thus senses a certain loneliness in his philosophical writings, comparable to that which Alquié portrayed in a 1956 lecture, later published as *Qu'est-ce que comprendre un philosophe*. 'Il y a une solitude du philosophe, et c'est une solitude de l'universalité' (2005: 31), Alquié declares. 'Le drame du philosophe n'est pas de se découvrir le sujet d'états d'âme rares et non éprouvés par d'autres que lui. Le drame du philosophe, c'est celui d'un homme qui se sait porteur de vérités universelles, et qui découvre qu'il ne peut faire partager ces vérités aux autres' (2005: 27–28).

However, Marcel's decision to publish his diary and use of the narrative form in general manifest his determination to resist the barrier that universal solitude potentially represents,[82] his commitment to remain *disponible* to others and hope to 'socialize' his philosophical reflections.[83] 'On n'entreprend pas de philosopher exclusivement pour soi, c'est-à-dire pour sortir d'un état d'incertitude ou de trouble de façon à parvenir à un certain équilibre dont on puisse se satisfaire soi-même', he insisted in the late lecture 'La Responsabilité du philosophe dans le monde actuel' (published 1968).[84] 'Tout se passe bien plutôt comme si on entendait prendre en charge l'inquiétude ou l'angoisse d'autres êtres qu'on ne connaît pas individuellement, mais auxquels on se sent lié par un rapport fraternel' (*PST*: 40).[85] And this then helps us to understand the ethicometaphysical mix that Marcel's philosophy presents us with. For Marcel, 'il n'y a pas d'engagement purement gratuit [...]. Tout engagement est une réponse' (*EA*: I, 55; 7 November 1930). The mere configuration of Marcel's philosophy into narrative form is therefore insufficient; his narratives are not meaningful unless they succeed in communicating something to someone else. As such, the other is always already invoked. Each (narrative) philosophical assertion in the present is at once a commitment to engage with another in time; and thus,

Ricœur declares in a 1988 paper he presented on Marcel, 'éthique et ontologie se nouent' (1989: 158).

So a Ricœurian reading of Marcel has proved not only possible, but also fruitful, for it has helped to reveal the significance of narrative time in Marcel's philosophy itself, and has offered an explanation as to how the ethical dimension to his philosophy can be related to his ontology. The question that remains is whether such a philosophy of time still (necessarily?) condemns it to a metaphysics of presence, as Derrida has suggested (see Conclusion to Part I), or whether, from the perspective of this new interpretation, Marcel can be considered to progress in his project to reconceive metaphysics as something more concrete and faithful to lived experience. One might, *prima facie*, be tempted to reject Derrida's argument in the case of this hermeneutical recasting, since a notion of identity founded on a narrative understanding of time is conceived as something fragile and unstable, an identity '[qui] ne cesse de se faire et de se défaire' (*TR III*: 446) and thereby seems to challenge an essentializing foundation of presence. After all, concludes Ricœur in *Temps et récit III*, narrative identity remains as much 'le titre d'un problème' as it is a solution; '[elle] n'est pas une identité stable et sans faille' (*TR III*: 446). And yet such multiplicity is affirmed only, I would argue, to be renounced. Ricœur is clearly suspicious of the move to privilege eternity over the change, movement, and difference of time — as is manifest in his critique of Hegel and totalizing (historical) narratives,[86] and more generally when he insists, again in *Temps et récit III*, that: 'il ne sera pas dit que l'éloge du récit aura sournoisement redonné vie à la prétention du sujet constituant à maîtriser le sens' (*TN III*: 488). Nevertheless, the ontological horizon to which his notion of narrative identity ultimately relates — particularly following its development in *Soi-même* — 'presupposes', as Patrick Crowley remarks, 'the precedence of being over language and privileges the whole over the part' (2003: 4). No longer simply an account of an experiential interrelation between historical and fictional (re-)figurations of time, narrative identity, in *Soi-même*, becomes the model for personal identity itself, aiming to answer (or at least keep answering) the question as to who I 'am'. As such, Ricœur arguably succumbs to the very temptation of which he accuses Hegel, by assuming an overly 'philosophical' perspective (as he characterizes it) where 'tout le procès de temporalisation se sublime dans l'idée d'un "retour à soi"' (*TR III*: 360), and 'ce qui compte du passé [pour le philosophe], ce sont les signes [...] d'où rayonne une clarté suffisante sur l'essentiel' (*TR III*: 364).[87] It therefore seems, as Raphaël Baroni has observed, that 'l'équilibre de la position ricœurienne [...] débouche en fait sur une hiérarchisation, sur une inclusion de la discordance *dans* la concordance, ce qui revient à trahir le temps' (2009: 16).

For Marcel too, as illustrated by his notion of *profondeur*, the plural experience of the self that he intended to take seriously proves, in the end, only to be problematic for his account of identity, a problem that must be resolved — that is, returned to the stability of some kind of unity, albeit one that may be continually re-created and re-affirmed. Appearing to engage with the multiple nature of human identity, in 'Le Sens de "ma vie"' Marcel declared:

> il existe une multiplicité non dénombrable de présences qui interfèrent les unes avec les autres et qui entretiennent avec le moi agent des relations de types tout à fait variables [...]. Ceci pourrait se traduire d'abord par cette simple formule que je ne suis pas seulement moi ou même qu'à la rigueur il n'y a peut-être aucun sens à dire que je suis moi, car je suis aussi un autre [...]. Une véritable lutte pour l'existence peut s'engager entre celui que j'ai été hier et celui que je tends à être, que j'ai la velléité d'être aujourd'hui. (*ME I*: 200–01)

So not only do individuals enter into external, intersubjective relations with others; they also hold internal, intrasubjective relations with themselves: 'l'intersubjectivité affecte le sujet lui-même[; ...] le subjectif dans sa structure propre est déjà, est foncièrement intersubjectif' (*ME I*: 198).[88] As a result, and as Marcel acknowledged in the lecture, affirming any kind of personal identity becomes extremely difficult: 'n'est-ce pas là une voie infiniment dangereuse qui risque d'aboutir à la négation pure et simple de l'identité personnelle?' (*ME I*: 201). However, Marcel believes that it is still possible to 'transcender cette opposition', so that 'unité et pluralité se conjuguent au sein de l'être singulier que je suis' (*ME I*: 198). In other words, Marcel is unable to accept (discontinuous) plurality for itself. He feels the need to resolve this conflictual experience,[89] to re-assert his identity as singular and unified — the reality of which, it is important to notice, is never in question: 'je suis [un être singulier]', he affirms at the end of the sentence. This (re-)unification is what *la profondeur* achieves.

The critiques of Ricœur offered by Crowley and Baroni can thus also be applied to Marcel. As Crowley observes: 'Ricœur's methodology, though informed by hermeneutics and phenomenology, involves a constant effort to mediate between opposites, between change and constancy for example. Hence, Ricœur, when faced with an aporia, looks to find a new term or approach so as to get round the difficulty' (2003: 8–9). And such an observation seems equally pertinent in the case of Marcel. Though completion is constantly postponed, ultimately, Marcel and Ricœur both shy away from the threat of tension and uncertainty in favour of productive, teleological synthesis, as is manifest by the way in which narrative time smoothes out breaks in its continuity and seeks to understand them as necessary to the narrative as a whole.[90] Difference or otherness is therefore subordinated to the eternity of the narrative form — or, more specifically, to the eternal return to the present, in relation to which everything is understood. Narrative time both anticipates the future and recalls the past in relation to the present moment. The future and the past are thus only ever understood through the present; they are not significant in themselves. Hence, the movement and irreducible multiplicity of time, as well as the *ruptures* or *temps morts* that Marcel argued for (see Chapter 1) are betrayed. The eternal still has the last word and Derrida still appears justified. Such a *maîtrise* of discordance is precisely what Lévinas has reacted against; and Lévinas has in fact singled Marcel out by name, as a philosopher guilty of reducing otherness to a totalizing experience of presence. Does Marcel have the resources to respond to such a challenge? This is what Chapter 4 investigates.

Notes to Chapter 3

1. '[L]a spéculation sur le temps est une rumination inconclusive' (*TR I*: 24).
2. '[L]a temporalité ne se laisse pas dire dans le discours direct d'une phénoménologie, mais requiert le médiation du discours indirect de la narration' (*TR III*: 435).
3. Book XI of the *Confessions* contains Augustine's most systematic engagement with the nature of time. For consideration of time in his other works, see Sorabji (1983: 31–32), Leftow (1991: 73–111), and DeWeese (2004: 126).
4. See DeWeese (2004: 112).
5. This paradox was originally raised in Book IV of Aristotle's *Physics* (350 BCE). See Barnes (1984: I, 369–73).
6. See Augustine (1998: 231–33, XI. xv–xvi; 235–42, XI. xxi–xxvii).
7. Russell, for example, has referred to Augustine's theory of time as 'a better and clearer statement than Kant's of the subjective theory of time' (1991: 353).
8. As Sorabji notes, Augustine does not 'want to say that time is unreal: he finds it, with its distractions, all too real' (1983: 30). See also Wetzel (1995), Quinn (1999: 834 especially), and DeWeese (2004: 112, 126).
9. Newton established this understanding of time as dominant: 'Absolute, true, and mathematical time, of itself, and from its own nature flows equably without regard to anything external, and by another name is called duration: relative, apparent, and common time, is some sensible and external (whether accurate or unequable) measure of duration, by the means of motion, which is commonly used instead of true time; such as an hour, a day, a month, a year' (1729: I, 9). Previously, motion had actually been considered more fundamental than time. Augustine is thus one of the first thinkers to decisively separate the two. For discussion of how our modern conception of time developed, see Turetzky (1998: especially chaps 5–6).
10. It is interesting to note that both of these questions are present in Augustine's opening succession of questions about time. See above.
11. '[N]o body can be moved except in time' (Augustine 1998: 238, XI. xxiv.31).
12. *TR I*: 50–51.
13. '[I]l n'y a pas, chez Augustin, de phénoménologie pure du temps' (*TR I*: 23).
14. Where 'poetic' is understood in the sense of the Greek *poiesis*, which signifies the activity of creating or making.
15. Ricœur has suggested that Augustine's discussion of time might be read as moving toward a poetic resolution of its aporias in its fusion of philosophical argument and hymn — an interpretation which would equally privilege the importance of narrative in Augustine: 'En un sens, Augustin lui-même oriente vers une résolution de ce genre

[poétique]: la fusion de l'argument et de l'hymne dans la première partie du livre XI [...] laisse déjà entendre que seule une transfiguration poétique, non seulement de la solution, mais de la question elle-même, libère l'aporie du non-sens qu'elle côtoie' (*TR I*: 24). However, critics such as Protevi (1999) have objected to such a reading, contending that it misrepresents Augustine's purposes. The overall adequacy of a Ricœurian reading of both Augustine and Marcel will be evaluated further in Chapter 5.
16. *TR III*: 344–45.
17. Other scholars who argue for a narrative understanding of self include MacIntyre (1985), Taylor (1989), Dennett (1992), and Schechtman (1996; 2007).
18. *SA*: 12.
19. Schechtman (2007) argues similarly in her response to Strawson's (2005) critique of the narrative conception of self, which contends that we do not conceive of ourselves in accordance with one single, continuous narrative, but instead rely on a number of discrete, more temporary selves. Like Ricœur, Schechtman also considers it necessary to disentangle two different kinds of identity, and argues for a distinction to be made between 'persons' and 'selves'. In order to constitute oneself as a person, one only need conceive of one's temporality in terms of discrete micro-narratives, and thus it is not contradictory to say that someone is a different person from the person they used to be. In order to constitute oneself as a self, on the other hand, one's sense of personal continuity over time must be much stronger, and one must conceive of one's identity in terms of a whole narrative.
20. 'L'équivocité de l'identité concerne [...] la synonymie partielle, en français du moins, entre "même" et "identique". Dans ses acceptions variées, "même" est employé dans le cadre d'une *comparaison*; il a pour contraires: autre, contraire, distinct, divers, inégal, inverse. Le poids de cet usage comparatif du terme "même" m'a paru si grand que je tiendrai désormais la mêmeté pour synonyme de l'identité-*idem* et que je lui opposerai l'ipséité par référence à l'identité-*ipse*' (*SA*: 13).
21. 'L'attestation définit à nos yeux la sorte de certitude à laquelle peut prendre l'herméneutique. [...] C'est fondamentalement à la notion d'*épistèmè*, de science, prise au sens de savoir dernier et autofondateur, que l'attestation s'oppose. [...] L'attestation, en effet, se présente d'abord comme 'l'humiliation [de Descartes] chez Nietzsche et ses successeurs' (*SA*: 33).
22. Indeed, Ricœur is opposed to 'l'humiliation [de Descartes] chez Nietzsche et ses successeurs' (*SA*: 33); some form of certainty, he believes, can still be affirmed (*SA*: 34–35, 347).
23. 'La nature véritable de l'identité narrative ne se révèle, à mon avis, que dans la dialectique de l'ipséité et de la mêmeté' (*SA*: 167).
24. 'A narrative can link the past with the future by giving a sense of continuity to an ever-changing story of the self. [...] In Ricœur's words, there is a kind of "discordant concordance", which is conveyed through narrative, that in philosophical terms may be conceived as a "synthesis of the heterogeneous"' (Ramussen 2002: 62).
25. Discussions of personal identity in Marcel's philosophy can therefore be considered a precursor to Ricœur's theory of narrative identity.
26. See also *EC*: 15.
27. See also *JM*: 267.
28. As far as I am aware, there is no original French version of this article.
29. See also *TR I*: 291–92.
30. According to Busch, 'the greatest difficulty which besets Marcel's philosophy lies in explicating the "essentially positive" element in his definition of secondary reflection' (1995: 178). However, secondary reflection is actually characterized positively when Marcel introduces the notion of understanding one's life as a narrative (1995: 181); and as

a consequence, Busch suggests, secondary reflection is an act of interpretation and can be seen to prefigure Ricœur's philosophical project, even though this argument is not highly developed in Marcel's work (1995: 180). I am in agreement with Busch's overall argument here, but feel it is important to note that this is not how Marcel understands narrative. On the contrary, in fact: Marcel's explicit discussions of narrative present narrative form as reductive and essentially negative. Busch does not acknowledge this complexity.

31. '[C]omprendre objectivement, c'est user de catégories qui ne sont à personne' (*JM*: 227; 23 February 1920).
32. '[J]e ne peux pas séparer l'existant de la relation qu'il entretient avec lui-même, peut-être mieux vaudrait-il dire du fait qu'il est concerné par lui-même' (*PST*: 175).
33. Marcel identifies his position with Proust's here: 'ici bien entendu nous retrouvons l'expérience centrale autour de laquelle se déploie et s'organise l'œuvre de Proust' (*ME* I: 170–71). For a comparison of memory in Marcel and Proust, see Vigorito (1984: 396–403).
34. 'L'erreur de l'empirisme consiste [...] à méconnaître tout ce qu'une expérience authentique implique d'invention ou même d'initiative créatrice' (*RA*: 319).
35. Cf. Ricœur: 'L'œuvre narrative est une invitation à *voir* notre praxis *comme*...' (*TR* I: 155).
36. 'Without life-values and the temporal action of which they are part, personal identity could not and would not be narratively structured; with such life-values, it necessarily is' (Brockelman 1985: 65).
37. 'Si [l'activité narrative] les résout [les apories du temps], c'est en un sens poétique et non-théorique du terme' (*TR* I: 24).
38. Cf. also Bergson, who defines life as 'compénétration réciproque, création indéfiniment continuée' (*Œ*: 646); 'la réalité, imprégnée d'esprit, est création' (*Œ*: 1275).
39. Marcel first discusses *la profondeur* in his diary notes from January 1938 (*PI*: 37–45).
40. 'En réalité tout ce qui a été dit [...] au sujet du passé *que je suis* devrait être repris à la lumière de l'idée de profondeur' (*ME* I: 210).
41. '[J]e me trouve engagé *dans* et tendu *vers*. Engagé dans quoi [quel passé]? Tendu vers quoi [quel futur]? il [sic] n'est facile de répondre à ces questions que dans la mesure où je poursuis un certain travail créateur pour lequel je me sens à quelque degré indispensable' (*PI*: 47; 24 April 1939).
42. Marcel's understanding of the future is another difference between his understanding of time and that of Bergson. Marcel rejects Bergson's idea of the future as pure novelty, affirming instead a future (where it is authentically experienced) that is in some way recognizable as part of the essence of my individuality. In 'Le Sens de "ma vie"' he declared: 'Ce qui importe ici, c'est de voir que l'avenir ne peut pas dans cette perspective être conçu ou représenté comme simple innovation. L'avenir en tant que tout neuf peut être aussi attrayant que l'on voudra: il n'est sûrement pas éprouvé comme profond, justement parce qu'il est tout neuf' (*ME* I: 209). See also *HP*: 227 and *JM*: 194. Marcel's disagreement with Bergson concerning the future is discussed in relation to his metapsychical experiences as well (*EC*: 107; Monestier 1999: I, interview 3). The relation between the metapsychical and the future are discussed more generally in *EC*: 109–10 and *JM*: 263.
43. Cf. Bergson's *moi superficiel* and *moi profond*.
44. This act of emplotment is the second of three stages of mimesis that Ricœur identifies in the narrative act. Mimesis I refers to the pre-narrative structure (our pre-understanding) of the world: the network of human capacities (the ability to structure, temporalize, and symbolize action using signs, rules, and norms) and practical forms

of understanding (concepts such as agency, intention, circumstance, and consequence) which ground the possibility of narrative in the first place. Mimesis II is the creative configuration of events into an emplotted series — an imaginative composition or story; and mimesis III refers to the experiential impact of reading this story: by recognizing it as a representative form and integrating it (or at least a response to it) into one's understanding, the structure of the story is completed, and a connection is thus established between its text and the world (*TR I*: chap. 3).

45. Recall that, for Marcel, contingency is identified with temporality. Thus the necessity-contingency dialectic can be identified with the dialectic of time and eternity.
46. See also Ricœur (1980: 174).
47. '[L]'être qui est visé [...] doit être entendu comme verbe et non comme substantif' (*DH*: 107). See also *JM*: 180, *PST*: 84–85, and *RI*: 150.
48. Ricœur's use of the terms 'reckoning' and 'recollecting' in 'Narrative Time' is a reference to the Heideggerian notions *Rechnung* and *Gedenken*. However, Ricœur does also use the French 'recollection' in relation to mimesis III (e.g. *TR I*: 131).
49. *ME I*: 171–73.
50. For Blundell, '[the] mode of primary and secondary reflection is the single most important characteristic that Ricœur absorbed from his mentor [Gabriel Marcel], and it proceeded to shape the way he pursued his own philosophical research. It would not be quite accurate to call Ricœur a "Marcellian" philosopher, but it would be more accurate than ascribing a greater influence to anyone else' (2003: 92). On Marcel's influence on Ricœur, see Ihde (1971: 8–9), Spiegelberg (1994: 590–91), Ricœur (1995: 21, 35, 41–45), and Dosse (1997: chap. 2 especially). Other critics who consider Ricœur to be a disciple of Marcel include Bourgeois (1995) and Busch (1995).
51. In a diary entry dated 23 December 1932, Marcel declares that the question *que suis-je?* 'obsesses' him (*EA*: I, 146), and in his 1937 essay 'Mon propos fondamental' he writes: 'le seul problème métaphysique c'est: que suis-je?' (*PI*: 28–29). See also *DH*: 31–32. While many critics have commented on the primacy of this concern in the content of Marcel's philosophy, none has related it to the form of his writings.
52. '[T]his path, Marcel's thought, continually doubles back on itself, like switchbacks on a trail up the side of a mountain, for the path returns again and again to the same places [...]. Yet this repetition is not simply a mindless reiterating of the same, it is rather a deliberate return to the central regions of ontology and a continuing attempt to more deeply penetrate these regions' (Anderson 1989: 47).
53. E.g. *EC*: 80; *FP*: 5. Jeanne Hersch comments: 'Marcel, au fil des ans, devint de plus en plus, dans ses conférences, son propre biographe, comme si sa vie, désormais accomplie, le fascinait par sa manière d'être lui cessant de l'être' (Lévinas, Tilliette, and Ricœur 1976: 7).
54. Other philosophical shifts narrated by Marcel include a shift from a focus on 'le mystère de Dieu' to an emphasis on 'le mystère de l'homme' (Lhoste 1999: I, interview 3), a growing concern with justice and socio-political affairs (Lhoste 1999: II, interview 8), and a later reluctance to positively characterize his notion of *participation* (*PST*: 279).
55. 'When it comes to autobiography, *narrative* and *identity* are so intimately linked that each constantly and properly gravitates into the conceptual field of the other. Thus, narrative is not merely a literary form but a mode of phenomenological and cognitive self-experience, while self — the self of autobiographical discourse — does not necessarily precede its constitution in narrative' (Eakin 1999: 100).
56. 'Scarcely could Gabriel Marcel write a chapter of a book, an article or deliver a paper without becoming autobiographical. To many of his critics this evidences a

contamination with the singular which precludes his work from achieving the status of a philosophy. In the Marcelian context, however, it is a healthy sign since philosophical meanings are grasped not by abstraction from the singular but through a reflection sensitive to the singular' (Busch 1987: 1).

57. '[T]raitant de l'identité narrative, on a observé que c'est la vertu du récit de conjoindre agents et patients dans l'enchevêtrement de multiples histoires de vie' (SA: 370).

58. 'Being as [narrative] act/potency allows Ricœur to root [...] the self in relation to the other. The other is constitutive for the self. The self cannot be thought without the other. In its becoming, the self must encompass otherness, the dissimilar. That means Being includes passivity: all the experiences in which the self is forbidden to occupy the place of foundation. This is why the self can attest to itself only in a broken fashion [...]. The self must incorporate this other whether as one's own flesh, or as the intersubjective other, or as the "call" of one's own conscience. The self can no longer exist as an imperial self' (van den Hengel 2002: 86). See SA: 368–69.

59. '[O]n ne peut philosopher authentiquement qu'avec tout soi-même' (Marcel 1953: 14).

60. The autobiographical narrative in Descartes's *Discours* (1637) and the diary-like style of his *Méditations* (1641; 1647), on the other hand, seem to jar with his philosophical quest for certainty. Although, in the *Discours,* Descartes explicitly states that he wishes the reader to come to their own conclusions about his reflections, Descartes's *Méditations* take the *Discours*'s method very seriously, building on it in order to expound a universal — and allegedly irrefutable — philosophical method for truth, and yet still presenting his thought in the most subjective of ways. This renders questionable both the opening rhetoric of the *Discours*, which presents its text as nothing more than 'une histoire' or 'une fable' (1951: 31), and, at the same time, the mode of philosophizing that underpins the allegedly universal discoveries in the *Méditations*, which Rée refers to as 'narrative content [...] masquerading as the antithesis of narrative' (1991: 77). Judovitz (1988) provides an in-depth examination of the paradoxical relation between the autobiographical and the philosophical in Descartes's writings.

61. 'La philosophie, c'est bien une certaine façon pour l'expérience de se reconnaître, de s'appréhender' (RI: 25).

62. 'Je souhaiterais qu'on *vît* dans ce Journal l'effort conjugué de certaines puissances de l'esprit apparemment distinctes, d'abord pour désorienter notre regard intérieur, et ensuite pour l'accommoder à des champs nouveaux d'expérience' (JM: ix; my emphasis).

63. It is defined as 'the intrinsic connectedness of temporal and spatial relationships that are artistically expressed in literature' (Bakhtin 1981: 84).

64. 'À mesure que mes investigations se poursuivaient, je prenais toujours plus nettement conscience non seulement de la répugnance que m'inspirait en soi un procédé d'exposition dogmatique, mais encore du caractère hypothétique, aventureux d'un très grand nombre des thèmes autour desquels ma réflexion gravitait' (JM: ix).

65. As O'Malley remarks: 'The philosopher's task is envisaged by Gabriel Marcel in terms of a personal research in which the ontological status of the person undertaking it is intimately involved. At the same time, the concept of person itself furnishes a focus of integration for a philosophy that is admittedly impatient of systematic presentation. These two features [...] are in fact reciprocal' (1966: 7).

66. Bourgeois has therefore described Marcel as a 'corrective force to postmodern deconstruction' (2003: 193): 'It is this very possibility of philosophical thinking that postmodernity has recently challenged' (2003: 198), he argues; but 'the philosophy of Marcel reestablishes some faith in philosophical analysis as having something worthwhile to say, especially when it interprets existence and language' (2003: 205).

Freeman (2003) also believes that narratives are epistemologically significant. He argues for a need to rethink our conceptions of the fictive, the real, and the true, and thinks that Ricœur's analyses of narrative are particularly valuable for such a reconsideration. In fact, Freeman cites Marcel's brief commentary on telling a story to one's friends (*ME I*: 171) in support of his opposition to the equation of reality with the immediate (2003: 123). 'What I would like to do is try to think beyond this perspective', he writes; 'and the key to doing this, I believe, lies in moving beyond clock time and seeing in narrative time [...] a possible inroad into rethinking the problem of truth' (2003: 124).

67. 'Reality, argues Ricœur, is larger than the positivists' conception of it' (Vanhoozer 1991: 49).
68. E.g. *JM*: 140; *ME I*: 92.
69. *ME II*: 16.
70. Plays commonly cited include *Le Palais de sable* (1913), *L'Iconoclaste* (1923), *Un homme de Dieu* (1925), *Le Monde cassé* (1933), *Le Dard* (1936), and *L'Émissaire* (1945).
71. Hocking comments: 'Exploration beyond the limits assigned to conceptual thinking will naturally take the form of autobiography' (1954: 441).
72. 'Il y a ainsi une possibilité de progrès dans la réponse de la question: "Qu'est-ce que c'est que cette fleur?" Nous voyons cependant aussitôt que même cette réponse plus scientifique qui me permet de situer la fleur dans une classification n'est pas exhaustive, que même en un certain sens elle est une évasion' (*ME II*: 16).
73. Marcel does consider the possibility of a dyadic relation: 'Ne pourrait-on se demander [...] si [...] la question posée ne situait pas sur un plan non triadique mais dyadique, comme si j'avais demandé à la fleur: "Qui es-tu?"'. However, such an analysis is inadequate: 'Seulement cette question se dénature inévitablement; ce n'est pas la fleur qui me dit son nom par le truchement du botaniste; je serai amené à voir clairement que ce nom est une convention' (*ME II*: 17).
74. '[N]ous avons déjà dû précédemment dépasser le plan des relations pures. Ce que nous appelons communément des relations, ce n'est en somme qu'une figuration abstraite de ce qui, dans le registre qui nous intéresse ici, doit être reconnu comme communication vivante' (*ME II*: 18).
75. '[S]i j'ai tant insisté sur l'inter-subjectivité, c'est justement pour mettre l'accent sur la présence d'un tréfonds senti, d'une communauté profondément enracinée dans l'ontologique sans laquelle les liens humains réels seraient regardés comme exclusivement mythiques. [...] je ne me soucie de l'être que pour autant que je prends conscience plus ou moins distinctement de l'unité sous-jacente qui me relie à d'autres êtres dont je pressens la réalité [sic]' (*ME II*: 20).
76. See also Ricœur (1980: 174); *TR I*: 267–68.
77. 'The choice of format, stylistic and rhetorical devices plays a crucial role in shaping and conveying the ideational content of philosophical discourse' (Judovitz 1988: 5).
78. '[L]'expérience philosophique, si elle commence nécessairement comme un solo instrumental, tend dans son développement à devenir concertante' (*PST*: 24).
79. The significance of the form of Marcel's philosophy is also affirmed (indirectly) in his lecture 'Que peut-on attendre de la philosophie?' (published 1968). In this lecture Marcel notes that 'en art, la subjectivité tend à se transmuer en une inter-subjectivité toute différente de l'objectivité telle que la conçoit la science, mais qui déborde cependant absolument les limites de la conscience individuelle réduite à elle-même' (*PST*: 20–21). 'Cette valeur', Marcel explains, 'n'apparaît qu'avec les structures qui se constituent à travers ce que nous appelons le processus créateur et qui viennent se proposer à l'appréciation, non seulement du sujet, c'est-à-dire ici de l'artiste, mais

d'autres contemplateurs ou auditeurs possibles' (*PST*: 19–20). The 'philosophical experience', he then argues, ought to be conceived of in an analogous way: 'Or des considérations à certains égards analogues peuvent être présentées en ce qui concerne ce que nous avons le droit d'appeler l'expérience philosophique' (*PST*: 21).

80. Cf. Marcel: 'On peut dire, tout à fait en gros, que [l'exigence ontologique est] une exigence de cohésion et de plénitude. Il faut ajouter aussitôt que cette exigence ne prend sa signification, sa valeur d'aspiration que par rapport à un être souffrant et écartelé' (*PST*: 86).
81. 'J'avoue d'ailleurs bien volontiers que cette métaphysique me cause à moi-même un sentiment de malaise; est-ce que je ne risque pas de confondre le pensé et le senti?' (*JM*: 250; 27 October 1920).
82. '[Le philosophe] est tenu de cheminer sur une arête, [... et] du même coup, il est voué à une certaine solitude. De cette solitude, je pense qu'il n'a point à s'enorgueillir. C'est même là une autre tentation à laquelle il lui faut résister' (*PST*: 49).
83. See especially *PI*: 32–34.
84. Here a parallel may be drawn between Marcel's metaphysical diary, and Roquentin's diary in Sartre's *La Nausée* (1938). As was the case for Marcel, Raoul describes how Roquentin 'writes initially [...] for himself alone and is therefore his own narratee and potential reader' (1983: 705). However, Raoul continues, 'the desire for a reader other than himself [...] leads Roquentin finally to opt for the novel rather than the diary: "Un roman. Et il y aurait des gens qui liraient ce roman et qui diraient: c'est Antoine Roquentin qui l'a écrit"' (1983: 707; for the exact context of the citation, see Sartre 1938: 250). Similarly, it is the desire for a reader other than himself that leads Marcel to publish his diary as a philosophical work in itself. Not only this: as Jefferson observes, in Roquentin's account of the jazz song 'Some of these Days', which is presented as an ideal work of art and is what inspires him to write a novel, Roquentin 'seems to require [biographical information] as a necessary counterpart to the song[, ...] as an indispensable element in his understanding of the creative enterprise' (2005: 191–92). As has been seen, biographical context is an equally essential accompaniment to Marcel's philosophical reflections. Consequently, both Marcel and Roquentin are, as Jefferson writes with respect to Roquentin's lingering fear of bad faith, 'placed in a kind of double bind whereby, having understood that a certain kind of art has been made possible by a certain use of biography, the art in question is then permanently vitiated by this knowledge' (2005: 191). That is, the art becomes self-reflexively aware of its paradoxical status as both personal and universal — just as Marcel's meta-narrative on his experientially grounded reflections illustrates.
85. See also *DH*: 32.
86. *TR III*: 349–72.
87. For John Caputo too, Ricœur, like Hegel, 'balances', both thinkers being 'driven by a deep sense of unity, a metaphysics of unity which will always be unshaken by diversity' (Marsh 1992: 160). Interestingly, as Chapter 2 noted, Marcel's position also suggests itself to be more Hegelian than his (earlier) critiques might lead one to expect.
88. Although Marcel uses the term 'intersubjectif' (sometimes hyphenated, sometimes not) for both internal and external personal relations, for reasons of clarity it is useful to make a distinction between (internal) intrasubjective relations, and (external) intersubjective relations.
89. It is telling that, when reflecting on his own life, Marcel writes: 'quand je réfléchis sur ce que ce mot "terme" [i.e. d'une vie] désigne, je me vois forcé de constater que, quelque aspect que ce terme doive présenter en fait, celui de l'accident out de la maladie, il n'y a pas la moindre chance pour qu'il affecte le caractère d'une *résolution*, au sens

harmonique de ce mot. Et pourtant, à mes yeux, c'est la résolution seule qui importe, celle qui mettrait fin aux dissonances, à la cacophanie qui me blesse jour après jour' (*EC*: 285).

90. 'Ricœur does not problematise the role of language, preferring instead to see it as a tool at the service of self-knowledge rather than as something that constitutes consciousness and is open to constant misreadings' (Crowley 2003: 8).

CHAPTER 4

~

Marcel's Theatre: An-Other Time

As with Marcel and Ricœur, in Lévinas's philosophy the Self's (*le Moi*) relation to time equally serves as a model for its relation to the Other (*l'Autre/Autrui*);[1] subjectivity and intersubjectivity are inextricably linked. Unlike Marcel and Ricœur, however, (authentic) Lévinassian time is not conceived in terms of presence, but is absolutely Other. Rather than justifying a totalizing return to the Self, Lévinas believes that the paradoxes of (temporal) identity bear witness to a time that simply cannot — and therefore should not — be resolved. Only if such *ruptures* are accepted for what they are, he contends, can any relation with the Other be possible, a relation where the Other is not simply reduced to the eternal return to the Same (*le Même*)[2] — that is, not simply identified with, and therefore appropriated by, the Self.[3] But philosophy cannot help but appropriate the Other if it takes ontology to be its fundamental question, argues Lévinas, because of its desire to obtain a panoramic view of all things and relate existents to some common ontological condition or situation. For Lévinas, the whole of Western philosophy is therefore guilty of neglecting the Other's otherness, because it has always taken the Self's being and freedom as primary (as indeed Marcel and Ricœur seem to do), and sought to understand Being as a whole from this starting point. Thus, he maintains, the Other's being has only ever been understood in terms of the Self, reduced to the same ontological ground so that it has never been specific to the Other as a singular individual, but has remained a Self-centred construction.[4]

Given that recognition of intersubjectivity and individual singularity is fundamental to Marcel's project, Lévinas's critique of Western philosophy has the potential to dismantle his entire project — although, it must be said, Lévinas's judgement of Marcel in particular is more complex than his view on Western philosophy in general. Lévinas has in fact praised Marcel's work, acknowledging that, in many respects, Marcel (and also Buber, whom Lévinas frequently discusses alongside Marcel as a fellow 'philosophe du dialogue') anticipates his own intellectual contribution.[5] He is credited, for example, with offering 'une nouvelle sagesse' or 'une rationalité nouvelle' (*EN*: 73), that is, '[une] réponse nouvelle [au privilège traditionnel de l'ontologie] où la source et le modèle du sensé se cherchent dans les rapports interhumains' (*HS*: 33). More specifically, writes Lévinas in *Totalité et infini* (1961): 'Grâce à un courant d'idées qui s'est manifesté indépendamment dans le *Journal Métaphysique* de Gabriel Marcel et dans le Je-Tu de Buber la relation avec Autrui comme irréductible à la connaissance objective a perdu son caractère insolite' (*TI*: 64), thus giving credence to '[une] *socialité* irréductible au savoir et à la vérité' (*HS*: 36), where 'le moi de l'homme n'est plus le point central, ni initial, ni final' (*HS*: 42). In the final analysis however, for Lévinas, Marcel remains 'profondément enraciné, malgré tous les bouleversements que la notion du Toi y introduit, dans l'ontologie' (*HS*: 37): 'et Buber et Marcel caractérisent la relation Je-Tu en termes d'*être* [...]: la co-présence, le co-esse', so that, 'à suivre la lettre des textes, l'être, la présence demeurent l'ultime référence du sens' (*HS*: 36–37).[6]

By 'textes', here, Lévinas seems to refer only to Marcel's formal philosophical writings. Later in his career, however, Marcel claims that to read his philosophy alone is in fact to misread his philosophy, for no understanding of his philosophy is complete without an equal appreciation of the works he produced for the theatre.[7] What implications, then, does this have for interpreting his thought? If Marcel's dramatic works are considered alongside his philosophical writings, is it still true to say that his theory of (intersubjective) time is bound to a metaphysics of presence? This is what Chapter 4 aims to determine. The first of its three sections observes how Marcel himself was not entirely comfortable with (Self-centred?) narrative form, to the extent that he later privileges the non-narrative temporality of his theatre over and above his philosophy. The second section then analyses the different time of Marcel's theatre in detail, and explores how it challenges the discussion of time found in his theoretical writings, thereby complexifying the presentation of time in his work as a whole. Finally, the third section will ask how Marcel's philosophy and theatre might be thought together — and, more specifically, whether any continuity can be found between the two which might offer grounds for a defence of his theory of time.

Philosophy and Theatre, Narrative and Ethics

If Lévinas contests the very foundation of Western metaphysics, his critique of the tradition is also ethical. In *Totalité et infini* he writes: 'La philosophie occidentale a été le plus souvent une ontologie: une réduction de l'Autre au Même'. And 'l'ontologie qui ramène l'Autre au Même', he continues, 'promeut la liberté qui est l'identification du Même, qui ne se laisse *pas* aliéner par l'Autre' (*TI*: 33; my emphasis). Western philosophy's ontologizing is thus unethical, for, in effect, what it affirms is the arrogant appropriating power of the Self, which is allowed to dominate everything else: 'Philosophie du pouvoir, l'ontologie, comme philosophie première qui ne met pas en question le Même, est une philosophie de l'injustice' (*TI*: 38).[8] Not only, then, does this call into question the success of Marcel's and Ricœur's efforts to accommodate otherness and time in their (ontologically oriented) philosophies; it also raises the further question as to whether the precedence they both give to the ontological renders their narrative conceptions of (temporal) identity unethical.

In this respect it is interesting to note that while, through his narrative acts of philosophy, Marcel seems to testify personally to the intersubjective ontology he wishes to affirm, the narrative act itself also troubles him, leading him to call it into question. In *En chemin* (1971), for example, Marcel describes how

> il m'est [...] arrivé de ressentir une déception assez amère lorsque l'autre ne semblait pas apprécier suffisamment ce *bien* auquel je l'avais cru digne de participer avec moi. [...] c'était pour moi comme si l'être même de ce que j'avais voulu communiquer se trouvait profané, je dirais presque souillé par l'incompréhension à laquelle je me suis heurté. (*EC*: 23)

Not only was the initial formulation of narrative difficult for Marcel, then; experiences he believed to be common and thus 'shareable' were not always recognized as such by others. This perhaps explains why, later in his career, Marcel began to speak of 'une volonté d'intersubjectivité', rather than any definitive affirmation of intersubjectivity as ontological reality (e.g. 1969: 257; *EC*: 133). Indeed, immediately following the above remarks concerning his disappointment, Marcel asks himself whether his appeals to others had not been motivated, above all else, by his own need for reassurance:

> En référence à ces mêmes expériences, je note aussi le besoin, que j'ai toujours ressenti lorsque j'avais fait une découverte, d'y associer l'un de mes proches, bien loin de vouloir garder pour moi tout seul ou d'enfouir en moi ce trésor. Je m'interroge aujourd'hui sur la nature véritable de ce besoin. N'était-ce pas, au moins dans certains cas, comme s'il me fallait trouver une confirmation dans l'assentiment que je comptais bien rencontrer chez celui ou celle à qui je confiais ma découverte. Dès lors, si cet assentiment m'était refusé, je me voyais comme rejeté dans un doute que m'était insupportable, au moins jusqu'au moment où une meilleure occasion se présenterait pour moi, de m'en dégager. (*EC*: 24)

This might be read as a personal confession of Marcel's inability to tolerate uncertainty, and of the evasive way he has reacted to such experiences, seeking, through acts of narrative, to remove doubt and regain security above all else. 'J'ai toujours, oui toujours, depuis la prime enfance, aspiré à me sentir en consonance avec l'autre', he writes, wondering whether this is not 'l'essentiel' — the very crux of the matter when it comes to interpreting his philosophy (*EC*: 25).[9] The fact that Marcel has not always achieved such *consonance*,[10] seems, then, to have made him realize that he has possibly been too eager to affirm harmonious, intersubjective (and indeed intrasubjective) identification.

In keeping with this self-evaluation, Marcel also — again in the latter part of his career, and especially in the 1961 William James Lectures[11] — begins to rethink the status of his philosophy, and this leads him to emphasize the importance, if not precedence, of his theatre as compared with his theoretical writings.[12] Marcel thus comes to feel that his theatrical works in some sense 'complete' his *œuvre*;[13] and because of this, many critics attempt to consider Marcel's philosophy and theatre together. Yet, as Hilda Lazaron remarks in the preface to her 1978 study, 'whilst stressing the importance of Marcel as a dramatist, [the works that have been written about Marcel's dramatic works during the last quarter century] fail to consider the plays solely as drama[, ... seeking only] to find analogies between the plays and the philosophy' (1978: 6). This continues to hold true for scholarship to date (including Hanley (1987), in addition to earlier works such as Chenu's (1948)), which tends merely to transpose themes from Marcel's theory onto his dramatic works, so that his plays (automatically) provide illustrations of (or even, in Hanley's case, serve as a tribute to) the ideas discussed in Marcel's philosophy — especially his affirmation of intersubjectivity. Lazaron, on the other hand, offers a very different interpretation. Only concerned with Marcel's plays as plays — that is, as dramatic, performative works as opposed to philosophical mouthpieces — she identifies loneliness as the predominant theme, observing that man appears 'a stranger to himself and incomprehensible to others' (1978: 25).[14] Indeed, the very capacity Marcel's characters seem to lack is the ability to narrate experiences, in a manner that enables identification and understanding.[15] Furthermore, the marked absence — suggested impossibility, even — of narrated time in his theatre seems deliberate: '[Le romancier] sera toujours tenté d'expliquer, et par conséquent d'intervenir', writes Marcel in his 1946 article 'Finalité essentielle de l'œuvre dramatique'; 'le dramaturge, au contraire, ne peut pas intervenir, à moins qu'il n'ait la malencontreuse idée de se munir d'un porte-parole, mais il est alors aussitôt dépisté' (1946: 286).[16] If one wishes to take the later Marcel's (philosophical) preference for his theatre seriously, this difference must be taken into account.[17]

Marcel, for one, certainly considered his theatre's difference to be significant — to the extent that he presented his later preference for his plays as a kind of philosophical corrective. In a 1970 interview with Monestier, for example, he speaks of his theatre's 'valeur de rectification [...] au nom d'une certaine vérité de la vie' (Monestier 1999: II, interview 4),[18] and in *En chemin* he declares: 'c'est dans mon théâtre avant tout, que se manifeste le caractère existentiel de mon œuvre' (*EC*: 230). More specifically, he explained to Boutang in 1970, 'la priorité, en un sens très profond, appartient au théâtre', because his theatre maintained a prospective (dynamic, non-presentifying) outlook 'que souvent la philosophie dégage pour [...] fixer, pour [...] cristalliser [ses idées]' (*GM*: 39). 'Le théâtre montre mieux [...] ce que j'appellerai le cheminement de la pensée pensante, par opposition à la pensée pensée' (Lhoste 1999: I, interview 4), he reiterated to Lhoste in 1973.[19]

More interestingly still, Marcel's preference for his theatre also appears ethical in motivation. In response to World War II in particular,[20] Marcel became increasingly preoccupied with ethics. 'Le souci de la justice commande de plus en plus mes attitudes en présence des événements' (*EC*: 243), he writes in *En chemin*.[21] From the early 1930s onward he was gripped by a premonition of the war to come,[22] causing him to become ever more anxious about the European situation,[23] and prompting greater socio-political participation on his part.[24] However, as Marcel writes in *En chemin*, his concern with justice did not come to its height until the years immediately following the war:

> toutes [les] circonstances [au cours des années qui suivirent immédiatement la Seconde Guerre mondiale] aiguisèrent en moi une exigence de justice qui avait toujours été mienne,[25] mais qui se présenta à partir de cette époque avec un caractère de rigueur qu'elle n'avait peut-être pas affecté jusque-là au même degré. [...] depuis ce temps-là le problème de la justice s'est posé pour moi au tout premier plan.[26] (*EC*: 226)

'La découverte, à la fin de la [deuxième] guerre, de ce qu'avaient été les camps d'extermination l'a bouleversé',[27] explains René Rémond (1989: 37), and as a result, Marcel dares to speak out forcefully against the *épuration*,[28] and the conditions surrounding the trials of Pétain (although Marcel was opposed to the Vichy régime)[29] and Maurras (despite being a *dreyfusard*).[30] These may appear strange reactions, but some elucidation is offered in *En chemin*: 'le souci de justice de plus en plus exigeant qui m'habite se conjugue de façon parfois bien embarrassante avec une conscience toujours plus distincte des difficultés, voire des contradictions auxquelles se heurte l'institutionnalisation de la justice' (*EC*: 244). Thus as René Poirier notes: 'il eût signé deux pétitions en sens inverse, pourvu qu'elles fussent inspirées par une pensée généreuse, plutôt que de profiter de leur antithèse pour n'en signer aucune. Je l'ai entendu, et non moi seul, défendre la peine de mort, alors qu'il venait de signer un manifeste pour son abolition' (1989: 43). For Marcel, who was acutely aware of ambivalences et paradoxes, justice and individual circumstance took precedence over consistent judgement. Hence, Rémond concludes: 'Si [...] on cherche quel peut être le dénominateur commun de tant de prises de positions, de tant d'attitudes, dont certaines surprenantes ou contradictoires, la clé me semble être dans la détestation de l'intolérance et une constante exigence de justice' (1989: 37).[31]

Yet Marcel did not feel that this concern was adequately communicated in his philosophical writings, and in his 'Testament philosophique' (Vienna, 1968) he reprimanded himself for not sufficiently engaging with ethical questions in this context: 'Je serais porté aujourd'hui, à l'âge avancé qui est le mien, à me reprocher de n'avoir peut-être pas poussé aussi loin qu'il aurait fallu, la recherche éthique fondamentale portant sur la nature de ces essences [de l'être que j'ai discutées]' (1969: 258).[32] However, his theatre, Marcel notices, does address these issues: 'au fond, les problèmes d'éthique se sont posés à moi surtout sous la forme dramatique', he explained to Boutang — and hence, 'c'est surtout dans mon théâtre beaucoup plus que dans mes essais philosophiques qu'on trouverait des éclaircissements sur ce que j'ai pensé, et, surtout, sur ce que j'ai contesté' (*GM*: 94). 'A partir d'un certain moment, et en tout cas à partir du *Dard*, et inclusivement, mes pièces ne sont pas du tout désactualisées', insisté Marcel at a 1973 Cerisy-la-Salle colloquium; 'le rapport à l'actuel fondamental, dans *Le Dard* [1937], le sera également dans [...] *L'Émissaire* [1949] et *Le Signe de la Croix* [1949][33] [..., et] aussi dans *Rome n'est plus dans Rome* [1951], pour ne rien dire de *Croissez et multipliez* [1955]' (*EAGM*: 124).[34] Indeed, all of these plays explore socio-political questions much more overtly and in greater depth than his philosophical writings and previous dramatic works (which tend to focus on interpersonal relations and questions of fidelity more generally),[35] addressing such ethical issues as anti-Semitism and genocide (*Le Dard*, *L'Émissaire*, *Le Signe de la croix*); Jewish response to persecution (*Le Dard*, *Le Signe de la croix*); treachery, collaboration, and resistance during the French Occupation (*L'Émissaire*); political engagement and patriotism in post-World War II France — including questions concerning the *épuration* and the rise of Communism (*Rome n'est plus dans Rome*); and (a little different, but expressing similar social engagement) the ethico-personal implications of Catholic dogma for the daily lives of Church adherents — especially Catholic non-acceptance of birth control (*Croissez et multipliez*).[36]

Furthermore, in his paper 'De la recherche philosophique' (also presented at Cerisy, 1973) Marcel remarked, after identifying a rejection of '[tout] formalisme dans le domaine moral' in his early works and onward: 'On serait assurément tenté de dire que c'est le dramaturge, en moi, qui émettait cette protestation, et je crois qu'effectivement il en était bien ainsi' (1976: 18).[37] More than simply offering an increased engagement with the ethical, then, one might say that for the later Marcel, the form of his theatre actually represents the ethical, as is equally suggested in the preface to *La Dignité humaine* where Marcel writes: 'l'expression dramatique est existentielle par excellence, parce que l'être est ici traité comme sujet et éventuellement comme décidant de soi' (*DH*: 9) — that is, because here individual human subjects are treated with (respectful) openness; free from the confines of an overarching narrative, characters' identities are approached and presented in an ethical manner.[38] The privilege Marcel later gives to his theatre's mode of expression over his philosophy, I therefore suggest, might be understood as setting up a debate, internal to his *œuvre*, concerning not only the legitimacy, but — furthermore — the ethics of narrative time; and on a broader scale, this shift in the presentation of his position could be interpreted as a kind of Lévinassian move away from traditional, totalizing metaphysics, in order to engage ethically with the lived, intersubjective time of reality. However, the extent to which Lévinassian time — with its emphasis on non-recuperability and an ethical relation to the Other — can be read into the specifics of Marcel's theatre still remains to be shown. This is my task in the next section.

The Lévinassian Time of Marcel's Theatre

'L'étrangeté d'Autrui — son irréductibilité à Moi — [...] s'accomplit précisément comme une mise en question de ma spontanéité, comme éthique' (*TI*: 33), writes Lévinas in *Totalité et infini*. And this *mise en question* happens during what Lévinas terms 'le face-à-face', where the Self is confronted with the absolute (or infinite) otherness of the Other and realizes that, because it cannot appropriate such radical difference, its free — that is, Self-sufficient — attempts to understand the world are arbitrary: 'La conscience morale accueille autrui. C'est la révélation d'une résistance à mes pouvoirs, qui ne les met pas, comme force plus grande, en échec, mais qui met en question le *droit* naïf de mes pouvoirs, ma glorieuse spontanéité de vivant. La morale commence lorsque la liberté, au lieu de se justifier par elle-même, se sent arbitraire et violente' (*TI*: 83; my emphasis).[39] Thus, 'il s'agit d'inverser les termes de la conception qui fait reposer la vérité sur la liberté' (*TI*: 338), Lévinas continues. Metaphysical truth does not rest on freedom. Rather, metaphysics and truth are to be seen as anterior to freedom, as this ethical relation with the Other — an event that is not yet representation, nor objectification, narration or a return to the Self, but the pure affectivity of the Other which cannot be grasped, and therefore precedes all ontological theorization. This, for Lévinas, is metaphysics. Ethics is 'la philosophie première' (*TI*: 340).

Consequently, contends Lévinas in *Le Temps et l'Autre* (a compilation of four lectures presented 1946–47), situating the Self in the present — as narrative time does — 'n'est pas encore introduire le temps dans l'être' (*TA*: 32), for, he observes in *Autrement qu'être* (1974): 'Comme le temps narré se fait, dans le récit, et dans l'écrit, temps réversible, tout phénomène est dit: simultanéité' (*AE*: 64). But, he proceeds to ask: 'La temporalisation ne saurait-elle pas signifier autrement qu'en se laissant entendre dans le Dit où sa diachronie s'expose à la synchronisation?' (*AE*: 67). Yes, is the answer for Lévinas. True metaphysical time is a relation to the Other. Thus, while appreciating that 'le rapport avec l'autre est généralement recherché comme une fusion [ou bien, une présence]', Lévinas explains in *Le Temps et l'Autre*: 'J'ai voulu précisément contester que la relation avec l'autre soit fusion. La relation avec autrui, c'est l'absence de l'autre; non pas absence pure et simple, non pas absence de pur néant, mais absence dans un horizon d'avenir, une absence qui est le temps' (*TA*: 83–84). In *Totalité et infini* he therefore insists: 'L'œuvre du temps va au-delà de [...] la continuité de la durée. Il faut une rupture de la continuité et continuation à travers la rupture. L'essentiel du temps consiste à être un drame' (*TI*: 316–17) — namely, the dramatic confrontation of the *face-à-face*, which fractures the Self's totalizing present, opening it out onto something Other that cannot be contained, cannot be presentified.

It should be noted that the importance of time for Lévinas's account of Self–Other relations is not ubiquitously emphasized. Though in early works — especially *Le Temps et l'Autre* — alterity is described in temporal terms, these are replaced with what are predominantly spatial metaphors in his most famous work, *Totalité et infini*. Indeed, the subtitle to *Totalité et infini* is 'essai sur l'extériorité', and accordingly, interiority and exteriority dominate its discussions: the Self and the Other share no 'frontière' (*TI*: 222) and are separated by '[une] distance plus radicale que toute distance au monde' (*TI*: 230); the intersubjective relation is described in terms of '[une] courbure de l'espace' (e.g. *TI*: 323, 324); and 'l'homme en tant qu'Autrui nous arrive du dehors' (*TI*: 324). Lévinas's next major work, *Autrement qu'être*, acts as corrective, however. Explicitly a reaction against the failings of *Totalité et infini*, and most notably its 'ontological' language,[40] *Autrement qu'être* erases the language of symmetry and asymmetry, interiority and exteriority — which only make sense in relation to the present.[41] The Self–Other relation is temporal once again, described in terms of diachrony[42] rather than synchrony,[43] so that the Other is 'plus tôt' (e.g. *AE*: 47) rather than 'plus haut [que le Soi]'.[44] Lévinas is thus able to write, in his preface to *Le Temps et l'Autre* (not published until 1979): 'Si nous avons [...] accepté l'idée de [l'édition de ce texte] en volume et avons renoncer à le rajeunir, c'est que nous tenons encore au projet principal dont il est — au milieu de mouvements divers de la pensée — la naissance et la formulation première' (*TA*: 8) — namely, that time is 'la relation même du sujet avec autrui' (*TA*: 17).

Marcel's theatre coincides most closely with *Autrement qu'être*'s analyses, though moments in his plays also chime with Lévinas's earlier writings on time and Otherness — aspects which do not change substantially as Lévinas's thinking evolves to recognize, increasingly, the complexity of interpersonal relations, and to establish time as central to the ethical *face-à-face*. The following discussion now offers a reading of Marcel's theatre which traces the evolution of Lévinas's philosophy of time. In so doing, it will demonstrate how a Lévinassian conception of time can be seen to resonate with the time of Marcel's plays.

Lévinas formulates his philosophy of time primarily in opposition to Heidegger, who, in *Being and Time* (1927), famously explored human Being as defined by temporality, taking time as the ultimate horizon for its understanding. The specific Heideggerian notion Lévinas attacks is that of Being-toward-death. If Being is time in Heidegger, Being is also finite, for time comes to an end with our death. Living authentically therefore requires that we confront this horizon, that our worldly projects recognize this inevitability and seek a way to assume this fate, so that death becomes meaningful and can be made our own. In so doing a future is opened up, and this, for Heidegger, constitutes genuine human time. As Lévinas contended in a 1975 lecture, however: 'La relation à mon propre mourir n'a pas le sens de savoir ou d'expérience [...]. On ne sait pas, on ne peut assister à son anéantissement' (*DMT*: 28). How, then, can one 'be-toward-death', make death one's own? It is impossible, Lévinas argues: 'Le maintenant, c'est le fait que je suis maître'; but, he points out in *Le Temps et l'Autre*, 'la mort n'est jamais maintenant' (*TA*: 59). He thus proposes, as he summarized in 1976, to 'penser la mort à partir du temps — et non pas, comme chez Heidegger, le temps à partir de la mort' (*DMT*: 122).

Contrary to any horizon for my being, then, death in Lévinas is radically Other. He writes in *Totalité et infini*:

> Je ne peux absolument pas saisir l'instant de la mort — 'surpassant notre porté', comme dirait Montaigne. [...] Ma mort vient d'un instant sur lequel, sous aucune forme, je ne peux exercer mon pouvoir. [...] La mort est une menace qui s'approche de moi comme un mystère [...], pure menace [...] qui me vient d'une absolue altérité. (*TI*: 261)

When confronting the reality of death the Self is therefore rendered passive, for its objectifying projects and pretentions to certitude and control are called into question, as they suddenly become powerless. Interestingly, Gérard's imminent death in Marcel's 1911 play *La Grâce* seems to exemplify precisely such a relation. Up until the point at which he becomes conscious of death as an approaching reality, Gérard is a character who seeks certainty and stability in religious belief, to the point of rejecting life in this world and fixing his sights on the beyond: 'je découvre des semences de religion dans toutes les folies de mon passé', he declares; 'je ne pourrais plus accepter maintenant les distractions perpétuelles de cette vie' (Marcel 1914: 89). When he senses that death is close however, everything changes. Gérard is no longer certain; anxiety overwhelms him and he feels powerless in the face of his impending fate. Another character, Olivier, who has admired Gérard's faith, tries to reignite his former conviction; but it is never enough. Death is too Other for any *maîtrise*:

> OLIVIER: [...] rien au monde puisse être plus réel que ta croyance... qui fait graviter le monde autour de soi. La force de la croyance est sûrement la mesure de l'être. *(Sentant toujours le regard qui pèse sur lui.)* Cela ne suffit pas? ta foi est réelle à mes yeux comme le rêve et comme la vie.
> FRANÇOISE [Gérard's wife]: Tu l'épuises.

GÉRARD: Non, non... encore...
OLIVIER: Elle est plus qu'une vérité; elle est un acte et une création; elle est l'idée vivante qui réalise et qui transforme... davantage? Je sens toujours peser sur moi ton anxiété.
GÉRARD, *indistinctement*: Et Lui?
OLIVIER: Il est l'esprit qui affirme son unité, il est la foi qui se dépasse et se projette... davantage encore? (Marcel 1914: 207–08)

Importantly though, just as Lévinas will argue with respect to the human Other, my relation to death, though unknowable, is still a relation. As he stated in 1975: 'Ma relation avec ma mort est non-savoir sur le mourir même — non-savoir qui n'est pas cependant absence de relation' (*DMT*: 28). However, in the face of this infinite Otherness, one might ask, how is it possible for the Self to remain a subject, to remain in some form of personal relation with the Other and not simply have its subjectivity negated? One cannot start with the autonomous Self if one wants to recognize the true otherness of the Other, for if everything rests on the Self's individual free action, the Self's totalizing identification will dominate and reduce the Other to the Same. Lévinas himself acknowledges the legitimacy of this question when, in *Le Temps et l'Autre*, he asks: 'Y a-t-il dans l'homme une autre maîtrise que cette virilité [...]?' (*TA*: 73). 'Si nous la trouvons', he then proceeds to respond, 'c'est en elle, en cette relation, que consistera le lieu même du temps. [...] cette relation, c'est la relation avec autrui' (*TA*: 73).

The latter part of *Totalité et infini* (in anticipation of *Autrement qu'être*) addresses this issue in more detail than *Le Temps et l'Autre*. Here, to 'have time' — in a non-possessive, non-dominating way — is neither to continue to act for one's Self (*pour soi*), by making time one's own, nor to assume pure passivity (in the sense of inertia or apathy) by resigning oneself to fate (ultimately death). Instead, it is to recognize possibilities for-the-Other (*pour autrui*). 'Un être à la fois indépendant de l'autre et cependant offert à lui — est un être temporel', states Lévinas;

> à la violence inévitable de la mort il oppose son temps qui est l'ajournement même. [...] Le temps est précisément le fait que toute l'existence de l'être mortel — offert à la violence [de la mort] — n'est pas l'être pour la mort [soit sa propre mort (passivité), soit la mort de l'Autre (virilité, autarcie, impérialisme, égoïsme)],[45] mais le 'pas encore' qui est une façon d'être contre la mort, [...] au sein même de son approche inexorable. (*TI*: 247)

To 'have time', therefore, is to have a time directed away from a concern with the perpetuation of my own powers — which might, in the extreme, drive me to murder, in defence of my autocracy[46] — and turned toward my responsibility for the Other, who in the *face-à-face* commands: 'tu ne commettras pas de meurtre' (*TI*: 217). So, not only is the *face-à-face* experienced negatively, as the ineffability of the Other's transcendence; the *face-à-face* is also experienced positively, as this responsibility I feel after my Self-centred action has been called into question. The Other 'm'oppose [...] non pas une force plus grande [...], mais avec quelque chose d'absolument *Autre*: [...] la résistance éthique' (*TI*: 217);[47] and recognition of this opens up a more meaningful, responsible time, which Lévinas refers to as 'volonté' (as opposed to *liberté*) and defines as follows:

> La volonté, déjà trahison et aliénation de soi [parce que rejetant le *pour soi*], mais qui ajourne cette trahison, allant vers la mort, mais toujours future, qui s'y expose, mais pas *tout de suite*, a le temps d'être pour Autrui et de retrouver ainsi un sens malgré la mort. Cette existence pour Autrui [...] n'en conserve pas moins un caractère personnel [donc je suis toujours sujet, mais pas le sujet autonome d'une autarcie]. [...] Le Désir où se dissout la volonté menacée, ne défend plus les pouvoirs d'une volonté, mais a son centre hors d'elle-même, comme la bonté à laquelle la mort ne peut enlever son sens. (*TI*: 263)

This new orientation requires not a concept of finite time (à la Heidegger), but one of infinite time: 'against the congealing of *fini*-tude [...] the will turns [...] to a time that cannot be collected into a totality even after its death, to an infinite time', writes Jeffrey Dudiak (2001: 276) — infinite, because this time, though it is related to me (and thus maintains my subjectivity), is not simply my own. Rather, it frees me from my ego by opening up a relation to the Other, a relation for the Other that, as such, cannot be contained within my present, but opens up a future.[48] This, Lévinas describes as 'la fécondité':

> [Dans la vraie temporalité,] il ne s'agit pas de se complaire dans un je ne sais quel romantisme des possibles, mais d'échapper à [...] l'existence qui vire en destin [qui revendique un retour à Moi et mes possibles], et de se reprendre à l'aventure de l'existence pour être à l'infini. [...] Un être capable d'un autre destin que le sien est un être fécond. (*TI*: 314)

One of the few genuinely positive characters in Marcel's theatre — Werner, from the 1936 play *Le Dard* — exemplifies precisely such a way of assuming time. A musician and singer, he has left Germany because his Jewish friend and accompanist Rudolf was forced to flee persecution. However, at the end of the play he decides he must return to Germany, in spite of the danger such a decision entails (Werner himself is Aryan, but having associated with and assisted Jews such as Rudolf, his status is compromised). Béatrice, the wife of the friend with whom Werner stays in France, protests that his decision is suicide; but Werner feels a responsibility toward those suffering in Germany, which he wishes to pursue in spite of the threat of death he knows he faces:

> BÉATRICE: Ce retour, c'est un suicide.
> WERNER: Absolument pas, Béatrice, vous vous trompez. [...] Je me mets simplement à la disposition.
> BÉATRICE: De qui, Werner? de la cause? de la révolution?

> WERNER: La cause ne m'intéresse pas; les hommes m'intéressent [...].
> (Marcel 1967b: 150)

That death might encourage a sense of responsibility toward others is also suggested in *L'Horizon* (1945). In this play, the protagonist Germain, after his death has been predicted by a celebrated clairvoyant at a séance, begins to worry about what will become of his wife and two children after he is gone, and tries to arrange certain things in advance — most notably a second husband for Thérèse (of whom he approves). However here, it is suggested, Germain is acting in bad faith, for his actions seem more symptomatic of his allergy to change than of a genuine concern for his family's well-being.[49] Indeed, when Thérèse learns of the prophecy (which, incidentally, is not fulfilled; it is Germain's friend Bernard who dies) and broaches the subject with her husband, Germain vehemently denies that it has provoked any change in him. Thus, if Germain has been confronted with the Otherness of the future and death, and experienced a call to responsibility as a result, his response is ultimately a refusal of the Other. As Marcel Belay comments:

> Germain ne peut supporter un futur qui soit vraiment à venir [...]. On comprend dès lors l'importance qu'à ses yeux revêt la fidélité: elle est, [...] telle qu'il la comprend, l'antidote par excellence à ce malaise engendré par l'idée d'imprévisible. [...] Il veut convaincre Thérèse de l'efficacité de ce pouvoir sur soi-même dont dispose l'homme pour demeurer inchangé, pour se garder de toutes les variations de l'âme, que les vicissitudes de la vie pourraient susciter. (Marcel 1973b: 389-90)

This only damages his relationship with Thérèse, and others close to him.

Such a negative argument for action *pour autrui* is equally manifest in the time of characters such as Jeanne, in *Le Mort de demain* (written 1919; published 1931), and Aline in *La Chapelle ardente* (1925), both of whom eternalize the death of a man very dear to them and, by living only for a static, constructed present, not only alienate and cause suffering to those around them, but damage these individuals' relationships with others as well.[50] Jeanne has resigned herself in advance to the death of her husband Noël, who is fighting on the frontline (World War I). Before the war, their married life was successful and happy. However, when Noël returns home on an unexpected visit, Jeanne's determination to live as though her husband were dead causes her to receive him as a stranger, and Noël returns to the front bereft of all his courage and faith in victory, convinced of his inevitable death.[51] Jeanne's repudiation of 'un autre destin que le sien' (her *indisponibilité*, in Marcellian terms) has closed off Noël's previously positive, open future — and this, it is suggested, is likely to impact negatively on (the time of) Noël's fellow comrades as well.

Aline's timeless existence, on the other hand, is a response to the death of her son, Raymond, who was killed in battle during World War I. Preserving his memory has become an obsession (Raymond's little nephew Jacques, for example, is forbidden to play with his old toys),[52] and no one else, Aline frequently insinuates, has the slightest respect for his absence.[53] Seeking control in the face of the change that Raymond's death threatens, Aline tries to influence the life of Raymond's fiancée, Mireille, a well-meaning girl who only wishes to please Raymond's grief-stricken mother. Aline persuades Mireille to marry her sickly nephew André instead of Robert, a boy who has shown interest in her and to whom she is attracted. Octave, Aline's husband, is horrified by this behaviour and leaves Aline. Mireille, on the other hand, is unable to stand up for herself, and becomes ever more unhappy as she is denied an identity or a future on her own terms.

So for *La Chapelle ardente*, just as for the two previous plays, an understanding of characters' relations to time proves crucial in determining their *disponibilité* toward others; and, more specifically, the time of *disponibilité* emerges as a time that affirms the alterity of death and the Other, for if these are denied, so too is any genuine future. Paradoxically, these plays therefore seem to suggest, it is only by accepting the uncertain alterity of death that my solitude can be broken, and I can enter into genuine relation with an Other. This echoes Lévinas's refutation of Heidegger's view: 'La mort, c'est l'impossibilité d'avoir un projet. Cette approche de la mort indique que nous sommes en relation avec quelque chose qui est absolument autre [...]. Ma solitude ainsi n'est pas confirmée par la mort, mais brisée par la mort' (*TA*: 62–63), he states in *Le Temps et l'Autre*.

The temporal structure of *fécondité* also grounds the possibility of forgiveness in Lévinas's philosophy. 'Le temps discontinu de la fécondité rend possible [...] un recommencement, tout en laissant au recommencement une relation avec le passé [...]. Ce recommencement de l'instant, ce triomphe du temps de la fécondité sur le devenir de l'être mortel et vieillissant, est un pardon, l'œuvre même du temps' (*TI*: 315), he writes in *Totalité et infini*. The structure of the pardon is equally explored in Marcel's theatre, and again, it is the self's relation with time that proves crucial in either creating or denying the possibility of forgiveness. *Un homme de Dieu* (1922) for example, explores, as Jacques Francis notes, 'la contre-épreuve [du pardon]' (1989: 242). Edmée, the wife of a pastor, Claude, has had an affair with another man twenty years previously, during the early days of their marriage. Claude has forgiven her, but, it appears, this was more out of his professional desire to embody the saintly figure that befits a pastor, for though Edmée is aware that Claude went through a difficult period early in their marriage, she learns from his mother that this was actually before finding out about the affair. Afterwards, Claude's letters to his mother reveal, he suddenly feels at peace (should he not instead have been scandalized?); and this is because, Edmée believes, Claude's supposed act of forgiveness bore no relation to her, but served merely to bolster his opinion of himself as virtuous. Edmée therefore rejects the pardon.[54] It was not a relational act and was not therefore genuine; it had nothing to do with her.[55] Claude was trying to appropriate time, single-handedly, to secure a certain future. However, this was not a real future; only a propagation of the eternal Good he wished to bestow upon himself, which turned a blind eye to Edmée's and Claude's past and thereby refused the Other, refused time.

Marcel's 1933 play *Le Monde cassé*, on the other hand, shows the positive possibility of a pardon, the temporality of which is equally Lévinassian. The action also revolves around an unhappy married couple (Christiane and Laurent), each of whom lives only for him- or herself, shutting the other person out. Christiane is a social butterfly who surrounds herself with (superficial) friends and interests in an attempt to find satisfaction: she is an actress and a novelist; she travels, involves herself in artists' projects, and is interested in musical composition. However, with none of these ventures does she appear genuinely engaged. Christiane is merely keeping up appearances to impress others.[56]

Her husband Laurent is no better. He gains his pleasure from the fact that he is not like his wife, nurturing pride and a sense of superiority in response to her and her friends' every action, while playing the irreproachable husband by giving Christiane absolute freedom to do whatever she pleases: 'Tu feras exactement comme tu voudras [...]. Tu sais très bien que je ne demande jamais rien' (*CPM*: 133). Yet Christiane does not want this freedom. 'Je ne sais rien de plus pesant que cette façon de me laisser tout à fait libre', she complains. 'Si la vie ne te plaît pas, rien ne t'empêche de la modifier', retorts Laurent; but, Christiane replies: 'Si par hasard j'avais besoin d'une volonté?' (*CPM*: 135) — *une volonté*, this play will suggest, that is Lévinassian, one of responsibility for the Other.

At the end of *Le Monde cassé*, the audience learns of the past that Christiane has buried beneath her frenzied social agenda: her love for a man — Jacques — who left her to become a monk. So Christiane married Laurent instead, not because of love, but because she could think of no reason not to. However, when she is visited by Jacques's sister, who, after Jacques's recent death, has stumbled upon his diary and discovered confessions of his love for Christiane, Christiane is finally able to have closure and begin to live for the future. She is therefore able to speak for the first time to her husband, and to recognize her responsibility for their unhappiness together. She has been haunted 'par nostalgie de l'amour' (*CPM*: 215), she confesses. 'J'ai honte pas seulement pour moi. Pour nous deux. [...] Ta faute, c'est ma faute; ta faiblesse, c'est la mienne' (*CPM*: 216). However, by asking for Laurent's forgiveness, the closing lines of play suggest, the possibility of a more genuine future is finally opened up:

> CHRISTIANE: Je te jure que je n'appartiens plus qu'à toi... je suis délivrée.
> LAURENT: Ah! c'est comme si tu m'étais rendue après ta mort...

CHRISTIANE, *humblement*: Ce mot-là, je vais maintenant tâcher de le mériter. (*CPM*: 216)

This future is necessarily uncertain ('je vais [...] tâcher de le mériter'); there is no guarantee that Christiane and Laurent will not simply lapse back into their old ways. But this is precisely because it is now relational, because it depends on both of them. 'L'avenir me vient non pas d'un grouillement de possibles indiscernables qui afflueraient vers mon présent et que je saisirais', writes Lévinas in *Totalité et infini*; 'il me vient à travers un intervalle absolu dont Autrui absolument autre [...] est seul capable de jalonner l'autre rive et d'y renouer avec le passé' (*TI*: 316).

So, for Lévinas, and also — it seems — for Marcel's theatre, a true encounter with the Other is at once a true encounter with time. Furthermore, such an event is not simply experienced negatively as uncertainty or as my inability to grasp the otherness of the Other. It is also experienced positively, as my responsibility for the Other. What Lévinas does not succeed in doing in *Totalité et infini*, however, is establishing the ethical authority of the Other. As Michael Newman argues: 'the subject cannot *only* be constituted as ethical in relation to a singular, personal Other (*autrui*)[; ...] a further condition is required to maintain the *otherness* of this Other, and to prevent its collapse into the identity of the alter ego' (2000: 104).[57] Lévinas himself recognizes this at the end of *Totalité et infini*, when he acknowledges that infinite time cannot guarantee the truth it promises: 'le temps infini est aussi la remise en question de la vérité qu'elle promet' (*TI*: 318). The Other is still at risk from a reduction to the Same. The indefinite, open future which infinite (or 'fecund') time proposes — if it is not simply closed off again by Self-appropriation — therefore seems, at best, to postpone the ethical relation of responsibility for the Other, for in the moment I experience a hint of a possible future with the Other, I still have yet to be responsible.

Autrement qu'être offers a solution. It restructures time to recognize the Other's relation to my past, as well as (possibly) to my future, so that, because I am already marked by the Other's trace, I am always already responsible for the Other. The ethical relation is thus not postponed by an indefinite, open-ended future, and it does not depend on the Self's choice. It is already a part of the Self's past, a past that cannot be recuperated or appropriated (true time is radically Other), and which therefore forces the Self to respond to the Other — because it cannot do otherwise. It is in this 'diachrony', as Lévinas refers to it, that the ethical imperative is reinforced: 'La positivité de [...] cette diachronie, [...] c'est ma responsabilité pour les autres. [...] Le paradoxe de cette responsabilité consiste en ce que je suis obligé sans que cette obligation ait commencé en moi' (*AE*: 28).

Indeed, in *Totalité et infini* the Self and the Other were still essentially independent, as expressed by its dualistic terminology which discussed the ethical relation in terms of a *Moi–Autre* binary. As such, Lévinas continued to give primacy to the Self, for as the Other was not already there, affecting the Self, its otherness depended on the Self's agreement to relinquish the (apparent) security of its eternal present, and to let the Other in.[58] In *Autrement qu'être*, however: 'La liberté d'autrui n'aura jamais pu commencer dans la mienne' (*AE*: 24). There is no longer a connection between the Other and my present (to be in my present is precisely not to be Other), and whereas before, I could conceivably have avoided the Other completely, *Autrement qu'être*'s diachronic relation to the Other means that the Other is no longer entirely separate. If anything, the Other is too close:[59] even if I choose to refuse the Other (and I can),[60] I am nevertheless still responding to the Other (irresponsibly).[61] Accordingly, Lévinas begins to describe the ethical in terms of 'proximité',[62] and introduces a third notion into the ethical relation: 'il n'est plus question du Moi, mais de *moi*. Le sujet [...] n'est pas un sujet en général [c'est-à-dire généralisable, comme le Moi; ...]. L'identité du sujet tient ici [...] à l'impossibilité de se dérober à la responsabilité' (*AE*: 29; my emphasis). The Other is now already in me (*moi*),[63] inseparable from my past; the Other is inextricably bound to me, and the responsibility I have for the Other shapes my identity before any conscious, intentional *Moi* considerations even arise (in the present).[64] Dudiak summarizes:

> From out of a time without beginning, the other calls me to responsibility in his regard, exceeding every present and *preceding* every present. Preceding in that, before the subject finds its feet as a constituting ego [and indeed, as a constitutor of time], it finds itself already in relation to that which calls to it from a profound past, [...] a past to which it is already in relation (in the ethical relation that Levinas refers to as proximity), and to which it is always already a response — before choosing to be so or not to be so — a response that *is* its subjectivity. (2001: 286)

Thus, Lévinas affirmed in a 1975 lecture: 'L'identité intérieure signifie tout juste l'impossibilité de se tenir en repos. Elle est d'emblée éthique' (*DMT*: 127).[65]

Indeed, although I can contest or refuse responsibility for the Other in any particular situation, reject any ethical codification into a moral norm, as Diane Perpich observes: 'the *moment* of normativity — *that the other makes a claim on me to which I cannot be entirely indifferent* — is incontestable' (2008: 147–48). This, Perpich refers to as 'normativity without norms' (2008: 126), and is now, in *Autrement qu'être*, a moment of normativity that infinitely recurs.[66] For now that the Other is also in my past, I am already vulnerable, always already exposed to the Other's otherness.[67] And this is something that continually impacts on, and will continue to impact on, my present, while at the same time being irre-present-able (or 'immémorial',[68] as it is often referred to) — unlike in *Totalité et infini*, where a connection between the Other and my present still remains. Lévinas therefore describes how the Other 'obsesses' and 'persecutes' me,[69] how I am the Other's 'hostage'[70] to the point of 'substitution'[71] — because I do not choose this election to responsibility, and because I cannot escape this unplaceable past, prior to any and all pasts to which I can knowingly relate myself, in the same way as I could the infinite time of the Other as pure future. 'Dès le départ, Autrui nous affecte malgré nous' (*AE*: 205), asserts Lévinas in *Autrement qu'être*, for 'le coup de l'affection fait impact [...] dans un passé plus profond que tout ce que je suis à même de *rassembler* par la mémoire, par l'historiographie, de dominer par *l'a priori*: dans un temps d'avant le commencement' (*AE*: 140).

The past is hugely significant in Marcel's theatre. Though the examples I have given above hint at possible (or impossible) futures, in many of these plays the past is actually the driving force:[72] Jewish suffering from Werner's past calls him to responsibility in Germany; Raymond's recent death haunts Aline; Edmée's past affair and Claude's past pardon return to plague them, calling them to take responsibility for what they have never faced up to; and Christiane's past love of Jacques and irresponsible motivation for marrying Laurent gnaw at her married life, creating a *malaise* that Christiane will only belatedly confront. In fact, as Anne Mary notes: 'l'action principale, dans la plupart des drames marcelliens, a eu lieu avant le commencement de la pièce' (2008: 97), so that the action on the stage continually refers back to something that precedes its beginning, a key event that happened before the play even started, the significance of which, however, is always equivocal or enigmatic, so that characters spend the duration of the play itself confused and preoccupied. The explanation or resolution they seek never seems achievable: implications or consequences of the event are always left uncertain; additional questions and the realization of further-reaching responsibilities continue to destabilize the present.

Characters' responses to each other are also frequently delayed. There is a temporal gap (*intervalle*) separating the Self's response to the Other's demand; time is the otherness preventing intersubjective comprehension or unity. Sometimes this lag extends until after the Other's death, where the encounter with death not only brings a character's own actions into question, but makes them aware that the Other has always already been there and incites a response[73] — only, this response is too late; the Self and Other can never be in synchrony.[74] The best example is perhaps that of *Le Fanal* (1936), where the recent death of Madame Parmentier (which precedes the play's opening) prompts her son, Raymond, to question the significance of his relationship with her, and as a result, aspects of his life in general — because, he realizes, the way he had been responding to other people was in fact often just a response to her. Most notably, Raymond re-evaluates the motivation behind his engagement to his fiancée Sabine, which, he now believes, was more a rebellion against his mother that took advantage of her waning health, than it was a positive affirmation of any relation with Sabine. 'Ce n'est pas toi, c'est nous que je déteste', he says to Sabine:

> Moi surtout, je me méprise... [...] voilà, moi qui ne lui [ma mère] avais jamais menti, j'ai pu la tromper... parce que je savais qu'elle allait mourir. La pensée de sa mort était là comme un fanal. Elle partie, je toucherais au port. Et toi aussi, tu le regardais, ce fanal. Tu évaluais, tu minutais... (Marcel 1936: 38-39)

Now that his mother has died, however, Raymond realizes that he has not been able to 'toucher au port'. A Self-centred ambition, founded on antagonistic egoism, it brings him no satisfaction. His encounter with his mother's actual death then makes him aware of the injustice of his project, and this 'contact' with an Other in (and yet beyond his own configuration of) his past compels him to change his ways, opening up a future. Although — indeed, because — this remains uncertain (just like his past), Raymond has now been confronted with the Other, and with time. As such, Madame Parmentier, though dead and anterior to all the action on the stage, is arguably the main character,[75] the Other whose call is prior to, and assumes precedence over Raymond's Self. Raymond is still in relation with her,[76] but their relation is one of diachrony rather than synchrony.

In plays where a greater number of characters are involved, and therefore where it is less possible to isolate one-to-one Self-Other relations, Marcel's plays continue to conform to Lévinassian time by presenting characters who are in the situation of needing to respond already: to take a stance regarding the French Resistance during World War II, for example (*L'Émissaire*), to respond to unexpected visits where the Other impacts on the self prior to any decision or anticipation (*Le Mort de demain, L'Émissaire*), to act in support of something during a war despite not wanting to support the war itself (*Un juste*) — even to finish a sentence (the line 'Finis ta phrase!' (or equivalent) is in almost every play).[77] 'Je n'ai pas le temps de faire face', writes Lévinas in *Autrement qu'être*. 'L'adéquation est impossible. Les obligations sont disproportionnées à tout engagement pris ou à prendre ou à tenir dans un présent. Rien, en un sens, n'est plus encombrant que le prochain. [...] Le prochain qui ne saurait me laisser indifférent' (*AE*: 140). Indeed, characters in Marcel's theatre question the legitimacy of one another's actions incessantly — frequently with the challenge 'De quel droit?'[78] — and they experience recurring calls to normativity while faced with an absence of, or contradictions or ambiguities between, norms. In Marcel's later, more overtly ethical plays especially, characters are called to responsibility on more than one count. Not only are they required to choose, they must also justify their (difficult) choices — something that tends to paralyse characters, for as Lévinas emphasizes, the face of the Other is always equivocal and enigmatic; there can be no certainty as to what constitutes the right response.[79] On the contrary, according to both Lévinas and Marcel's theatre, I can never be responsible enough.[80] This is well illustrated in *Rome n'est plus dans Rome* (1951), where the protagonist Pascal, a political journalist, feels that he must defend his political beliefs in response to threats from opposing parties, and yet is also aware of his concurrent (and in this case seemingly conflicting) responsibility to his family and their safety. He chooses his family, and in fact decides to move to another country (Brazil); but this does not eradicate family tensions, and he continues to be plagued by the question of his loyalty to France (though, as the play also suggests, it is difficult to pinpoint exactly what it might mean to be faithful to one's country). So here again, Marcel's theatre presents us with normativity without norms; and it is with respect to the infinite reach of this normativity — I can always do more: for one Other, for many Others — that human finitude is presented, not with respect to my death as Heidegger argued.

The interpersonal relations dramatized in Marcel's theatre, then, asymmetrical and ineffable as they are, suggest that characters are still very much in relation. However, as Mary observes: 'les personnages marcelliens ne se battent pas pour de l'argent, des honneurs, une femme. Il ne s'opposent pas par intérêt, mais plutôt par manque d'intersubjectivité. Ils se comprennent mal' (2008: 99). As in Lévinas, the Self–Other relation in Marcel's theatre is, instead, one of discourse, which reflects an ethical (rather than a physical, power- or lust-driven) resistance. 'La relation du Même et de l'Autre [...] se joue originellement comme discours' (*TI*: 29), states Lévinas in *Totalité et infini*; and as Marcel's plays seem to demonstrate, the essence of discourse is not first and foremost the exchange of information — characters' positions and identities are generally too uncertain for there to be anything determinate to defend or disclose. Rather, discourse is (the act of) expression in the face of the Other,[81] whose absolute alterity breaks the monopoly of meaning held by the Same, and incites a response. Discourse in fact requires the absolutely Other: 'l'absolument étranger seul peut nous instruire' (*TI*: 70–71), writes Lévinas in *Totalité et infini*; 'le langage suppose des interlocuteurs, une pluralité',' and 'leur commerce [...] est éthique' (*TI*: 70).[82]

In this sense, one might say that the form of Marcel's theatre is, in itself, 'ethical'. Its non-narrativity favours a plurality of individual consciousnesses that are distinct, but which, through discourse, are in relation. And rather than understanding the ethical in terms of something secondary, which is merely added to a pre-established acting subject, it suggests that ethics *is*, already — an ambiguous, inter-relational web of interpellation, non-indifference, and responsibility in which my subjectivity is always already implicated, but which, in turn, is always in the making, as emphasized by the transition made between *le discours* and the explicitly diachronic notion of 'le Dire' (contrasted with 'le Dit') in *Autrement qu'être*.[83] If Marcel's characters learn things about themselves and, in so doing, open up the possibility of a more positive future with others (e.g. Christiane), this is not by virtue of a definitive resolution, or struggle being brought to an end. Rather, it results from entering into discourse with Others, and, feeling the weight of their address, being alerted to one's (continual) responsibility *pour autrui*. The non-presence of characters to one another not only makes linguistic 'contact' necessary; it is what makes such a relation meaningful and able to open up a time that transcends the present — unlike characters such as Aline, who, failing to engage in any discourse, are trapped in a present that is dislocated from all Other times.

Furthermore, this seems to be how Marcel wishes his theatre to be understood in general: as a 'saying' rather than a 'said'. 'L'art dramatique n'est rien s'il n'est pas libérateur, s'il n'est pas animé par le souci d'une justice supérieure' (1946: 286), he declares in 'Finalité essentielle de l'œuvre dramatique'; and this *justice supérieure* is, for Marcel, linked to the 'éclairage interrogatif' (*GM*: 47) which, as he explained to Boutang, the spectator themselves should (ideally) experience as they also become the subject of interpellation. 'Il avait à traiter ce spectateur comme une conscience qu'il s'agit d'éveiller [...]. Il ne saurait être question de lui administrer je ne sais quelle vérité toute faite qu'il aurait à absorber comme un calmant ou un tranquillisant' (*EC*: 169), states Marcel in *En chemin*.[84] As he proclaims in an interview with Lhoste: 'la justification véritable [que se fait une pièce d'elle-même], c'est une espèce de vue post mortem, c'est-à-dire [...] le fait qu'une pièce qu'on a vue continue à vous habiter, continue à vous occuper, continue à vous troubler' (Lhoste 1999: II, interview 6). According to Marcel, therefore, a true dramatic work is never 'dit', but instead confronts the spectator with (unknowable, ambiguous) Others who infiltrate his/her present, and, in virtue of a diachronic relation that is created between them, continue to 'haunt' the spectator,[85] calling him/her into question. 'L'auditeur [ou le spectateur] ne doit pas se retrouver dans le héros, mais se découvrir', writes Mary. 'Et l'importance de cette découverte se mesure à la faculté de survie des personnages dans son esprit, ainsi qu'au questionnement que la pièce fait surgir en lui' (2008: 57).

Thus if time internal to Marcel's theatre appears Lévinassian, a similar notion seems to shape his dramaturgy, reinforcing the difference that, this chapter has proposed, sets Marcel's plays apart from his philosophical writings. The question that remains, therefore, is whether Marcel's theatre really does present a time absolutely Other to that underlying his philosophy (in which case his later preference for his theatre would represent a definitive break with his earlier position), or whether Chapter 3's Ricœurian interpretation in fact obscures certain aspects of Marcel's thought, which would suggest there to be more continuity between the two than is immediately evident.

Thinking Marcel's Philosophy and Theatre Together

If the primacy of intersubjectivity is in any sense suggested by Marcel's theatre, it is in the Lévinassian sense of *proximité* — an uneasy, ethically charged relation which, as primary, precedes all (present) knowledge and initiative, and in which the subject's inauguration into temporality itself coincides with an ineffable, yet indubitable feeling of non-indifference toward Others. As Marcel's plays seem to demonstrate, the Other's proximity always already marks the self's subjectivity. However, ungraspable in its absolute singularity, this cannot be accessed or comprehended through an ontological investigation of (the timeless, universal characteristics of) human existence. Rather, and as we see in Marcel's drama, such a diachronic relation is approached through an exploration of ethical horizons and exigencies, which appear both beyond, and prior to, any thematization. Even positive endings (e.g. *Le Monde cassé*) are only positive by virtue of a promise of fidelity, which, as a temporally extended notion, remains incomplete in the present, incomplete as a conclusion. Implicating not only the future but also the past — as a commitment that recognizes an Other who has already been calling me to responsibility, from a time before any that I could ever configure — such fidelity at best remains in the subjunctive. Interestingly, Marcel himself has also described proximity, as presented in his plays, in terms that explicitly defy presentification: 'ce qui est le plus lointain, c'est aussi, c'est essentiellement ce qui est le plus proche', he writes in the 1950 article 'Théâtre de l'âme en exil'; 'seulement cette proximité-là n'est pas celle des objets vers lesquels il nous est donné de tendre la main et dont nous pouvons nous saisir' (1950b: 10). One thus wonders: if Marcel's engagement with interpersonal relations in his plays seems so different from in his philosophy, and if Marcel has acknowledged this (different) sense of proximity himself (though he does not relate it back to his philosophy), might his theoretical discussions of intersubjectivity actually overstate a position which, at root, is more akin to Lévinas's?

Closer examination of Marcel's texts reveals a surprising number of statements that could be seen to challenge Lévinas's interpretation of him — that is, as suggesting genuine intersubjectivity to involve a reciprocity where self and other are unified in a present of love and mutual understanding. In 'La Fidélité créatrice' (1940), for example, Marcel defines 'l'exigence de l'*être*' as 'la hantise des êtres saisis dans leur singularité et en même temps atteints dans les mystérieux rapports qui les lient' (*RI*: 192–93); and such heterogeneity is all the more emphasized in Marcel's discussions of equality, which forcefully dismiss such a relation as self-centred and rigidly totalizing, favouring instead the (asymmetrical) notion of fraternity.[86] 'Tous les hommes font partie intégrante d'un même tout; mais on ne dépassera pas par là un jugement d'homogénéité' (1940a: 167), asserts Marcel in the 1940 article 'Considérations sur l'égalité'. Thus, he declares:

> le mot qui exprime le plus fidèlement cette interdépendance active et même créatrice n'est pas le mot égalité, mais bien le mot fraternité. Or entre frères on ne se demande pas si l'on est entre égaux; cette question ne se pose pas, elle est sans intérêt, et, ajouterai-je, sans signification. (1940a: 168)

I may recognize myself to be inferior or superior to my brother in certain respects, Marcel observes, but this does not change the 'qualité fraternelle qui nous unit' (1940a: 168). 'Une communauté n'est possible qu'à partir du moment où des êtres se reconnaissent mutuellement comme différents, comme existant ensemble dans leur différence même' (*RI*: 14), he writes in the introduction to *Du refus à l'invocation* (1940); 'nous sommes frères à travers toutes nos dissemblances' (*DH*: 173), he affirmed in his William James Lecture 'Dignité humaine'. Moreover, for Marcel, thinking in terms of equality actually distorts our understanding of human relations — crucially, because of its atemporality. Marcel writes: 'L'intemporel où s'établit le philosophe de l'égalité est un intemporel de mensonge qui dissimule aux regards [...] un processus de dégradation, ou plus exactement de dévalorisation' (1940a: 169) — a devaluation of the other's otherness, that is, for as Marcel declares to Monestier: 'l'égalité prend son centre dans le moi';[87] only 'la fraternité est centré sur l'autre' (Monestier 1999: I, interview 1).[88] In these discussions, therefore, time (and not an eternal present) emerges as pivotal to any true relation with another, the otherness of whom, Marcel argues, will simply be denied if one's understanding of intersubjectivity is restricted to egalitarian identification with the self.[89]

Furthermore, Marcel's later works in general seem more inclined to treat the self and other as distinct, while at the same time affirming their relation. Attempting to define his philosophical ambition in the William James Lecture 'L'Intégrité menacée', for example, Marcel identified the personal experience of the universal with solitude as opposed to shared presence or communion, and quoted several verses from Claudel's *Grandes Odes* (1910) in order to illustrate his position: 'Faites que je sois entre les hommes comme une personne sans visage et ma / Parole sur eux sans aucun son comme un semeur de silence', he cites; for, 'des mots tels que *semeur de silence* ou *semeur de solitude* répondent bien à ce que j'aurai tenté d'être' (*DH*: 201). Though this may appear in contradiction with his affirmation of intersubjectivity or fraternity, Marcel quickly corrected such an interpretation:

> La contradiction n'est, je crois, qu'apparente, ou plus exactement elle

implique une confusion que je crois indispensable de dénoncer. Le mot solitude est ambigu; en réalité il ne signifie pas l'isolement; celui-ci est un manque, une privation, au lieu que la solitude est un plein. [...] la solitude est essentielle à la fraternité comme le silence l'est à la musique. (*DH*: 203–04)

And thus, he declared in another William James Lecture ('Participation'): 'c'est la référence à ce type d'universel senti dans l'individuel qu'il faut se référer si l'on veut seulement entrevoir la signification de toute [ma] recherche' (*DH*, 45). Marcel's references to the universal should perhaps, therefore, be understood apart from totalization[90] — though the two notions can of course easily be conflated, as Marcel's constant gravitation back toward ontological language demonstrates.

The late lectures published in 1968, in *Pour une sagesse tragique et son au-delà*, also accentuate what may be described as Lévinassian themes, including the impossibility of Being-toward-death and one's vocation *hors-soi*, in addition to the lectures' general tendency to foreground uncertainty — sometimes even in conjunction with time. 'Les mots "une fois pour toutes" sont sans doute applicables au seul domaine de la rationalité, ou de la technique en laquelle celle-ci prend corps, mais non pas de la vie comme telle', states Marcel in the volume's last and eponymous lecture, before observing, in language not dissimilar to Lévinas's, that experience itself 'comporte toujours des reprises, des récurrences, de nostalgiques appels surgissant du passé' (*PST*: 295). Regarding death, in 'L'Humanisme authentique et ses pré-supposés existentiels' Marcel comments: 'lorsque j'exerce ma réflexion sur le fait que ma mort à venir peut exercer sur moi une action pétrifiante, je suis amené à reconnaître que cette action n'est possible que par la collusion de ma liberté. Ma mort ne peut rien contre moi que par la collusion d'une liberté qui se trahit elle-même' (*PST*: 73-74) — a statement which echoes the above discussion of Lévinas and death, and which is restated, along with an assertion of the priority of the other, in the volume's final lecture: 'chacun participe à l'être-contre-la-mort, non pas seulement en vertu [...] d'un instinct de conservation, mais bien plus profondément et intimement contre la mort de ce qu'il aime, et qui, pour lui, compte infiniment plus que le soi, au point qu'il est, non pas de sa nature, mais de sa vocation, d'être décentré ou polycentré' (*PST*: 309). Finally, the link between Marcel, and Lévinassian time and the Other, is made decisive when this same lecture concludes: 'la véritable sagesse consiste à s'aventurer, prudemment certes, mais avec une sorte de frémissement heureux, sur les chemins qui conduisent, je ne dis pas hors du temps, mais hors de *notre temps*' (*PST*: 309–10). If Marcel has sometimes appeared to advocate the transcendence of time's otherness, he now seems to be correcting this overstatement. In so doing, Marcel's position emerges as (intentionally, at least) very much akin to Lévinas's — as a philosophy which reacts against the egoism of Self-centred time, in order to open up the possibility of a genuine encounter with the Other, and a true encounter with time.

I consequently find it difficult to agree with Treanor, who states that there is an impasse between Marcel and Lévinas which 'can, without oversimplification, be addressed in terms of the way in which each thinker characterizes otherness' (2006a: 151), as it is not at all clear to me that Marcel and Lévinas always interpret otherness so differently. It should be noted that Treanor states his understanding of Lévinas to rest primarily on *Totalité et infini*, so the substantial evolution that the Self–Other encounter undergoes between *Totalité et infini* and *Autrement qu'être* is excluded.[91] 'Levinas insists the other is absolutely other, while Marcel would clearly reject such a black and white distinction', reports Treanor (2006a: 121). But I feel that the (more mature) Lévinas of *Autrement qu'être* would equally reject such a black and white distinction. The strength of the contrast Treanor draws between Marcel and Lévinas is thus undermined: both wish the other to be appreciated on his or her own terms, beyond the appropriating grasp of the self; and for neither does this entail radical separation, understood in the sense of a non-relation — only a separation that presentifying knowledge cannot bridge, a temporal separation this chapter has proposed.

Treanor has another argument, however, which declares the opposition between Marcel's and Lévinas's interpretation of otherness to be consolidated by the different way in which each prioritizes love and justice: 'Levinas characterizes the relation of the same to the other primarily in terms of justice and responsibility, while Marcel describes *disponibilité* primarily in terms of love and hope' (2006a: 153). Again, though, I would hesitate to make such a straightforward juxtaposition, because I do not believe their use of 'love' is directly comparable. Treanor does acknowledge that 'each thinker addresses both justice and love as modes of relating to the other person, and the attention paid to the less-emphasized mode of relation — in Marcel's case, justice, and in Levinas's case, love — appears to moderate or otherwise introduce ambiguity into each philosopher's principal emphasis'. Treanor contends that justice ultimately overrides love in Lévinas, whereas the primary sense of justice in Marcel is really an aspect of love — so 'what might appear to be moderation or ambiguity', he concludes, 'is in fact further entrenchment of the principal position through a reinterpretation of either love or justice' (2006a: 154). Now, if 'love' meant the same thing in Marcel and Lévinas, Treanor's argument would be valid. However, while Lévinas refuses to give love priority because he tends to understand it in the sense of the Greek *eros* (passionate, sexual love) — indeed, Lévinas often uses this term, 'eros', rather than simply 'amour'[92] — discussions of 'love' (*l'amour*) in Marcel are never concerned with the passions, nor do they offer phenomenological analyses of experiences such as the caress, as Lévinas does. Love, in Marcel, is more akin to the Greek *agape* (Christian self-sacrificial love, often explicitly opposed to *eros*).⁹³ 'Si mon amour peut exercer une action sur l'être aimé, c'est seulement en tant que cet amour n'est pas un désir; dans le désir en effet, je tends à subordonner l'être aimé à mes fins propres, je le convertis au fond en objet' (*JM*: 218), writes Marcel on 2 December 1919, for example.⁹⁴ Instead, love, as a mode of *être* rather than *avoir*, is to be identified with respect,⁹⁵ and it is a relation that is both beyond and prior to knowledge,⁹⁶ which 'implique [...] la libération du *je* [...]. L'amour surgit comme invocation' (*JM*: 217; 1 December 1919). Thus, Marcel's treatment of love cannot be directly compared to Lévinas's. What is comparable, however, is their shared suspicion of a love grounded in desire and self-satisfaction — and for this reason (contrary to Treanor's analysis) Marcel's emphasis on love and Lévinas's emphasis on justice represent a very similar argument: both advocate ethical respect and responsibility for the other.

One cannot of course deny that in many places, Marcel's portrayal of intersubjectivity appears to be one of totalization, ostensibly opposed to Lévinas's position because of its determination to grasp and ontologize the relation with the Other for all time. Presence is undoubtedly the most common image used to describe this relation, as Chapter 2 observed, and in his *Journal* Marcel even goes so far as to say: 'l'être que j'aime n'a pas pour moi des qualités; je le saisis comme totalité' (*JM*: 157; 18 December 1918). Nonetheless, it is useful to recall that Marcel's affirmation of intersubjectivity is motivated by his struggle to establish an individual philosophical position apart from idealism, and from Bergson's somewhat solipsistic theories of the temporal self — a struggle that has a certain urgency, since Marcel's First World War experiences are a rather rude awakening to the gap between theory and experience, which he then becomes determined to close. Could the intensity of these concerns perhaps be the reason why Marcel emphasizes a connection between the self and the other to such an extent, and why his philosophical arguments at times descend into affirmational hyperbole and polemic?⁹⁷ After all, Marcel's war experiences not only confronted him with a contrast between the theoretical and the real, but also between a worrying contemporary situation where intersubjectivity was equally absent, and the hope for a more positive future to come. In this light, Ricœur's conclusion that 'éthique et ontologie se nouent' (1989: 158) in Marcel's work seems especially well observed; and I wonder whether, in spite of Marcel's undeniably ontological language (the only 'philosophical' language he knows?), the ethical does not actually take precedence.⁹⁸

Here one must concede, as Treanor remarks, that 'the topic of justice is more often than not implicit rather than overt in Marcel's thought' (2006a: 89–90). Nevertheless, in retrospect Marcel affirms that ethics has always been important to him, telling Ricœur, for example: 'il m'est apparu de plus en plus clairement que le problème de la justice était premier' (*EPR*: 102).⁹⁹ And the fact that such emphasis on the significance of the ethical is concurrent with Marcel's concerted effort to attribute philosophical significance to his dramatic works — which, crucially, do engage with the temporality of intersubjectivity, and the temporal grounding of ethical exigency and aspiration — suggests that Marcel might be attempting to counteract the excesses of his earlier, timeless presentations of intersubjectivity.[100] Interpreted in this way, the difference of Marcel's theatre would not undermine Chapter 3's analysis of time in his philosophical writings, but rather depend on it, for Marcel's privileging of his plays could be understood as a reaction against a presentifying, narrative structure of time, which, as Wood phrases it, 'stresses the intelligible organization of events at the expense of [...] the ethical moment' (1991: 4). Marcel's explicit critique of narrative, referenced in Chapter 3, could still, then, be taken seriously, with his later preference for his theatre reinforcing this all the more so, as an implicit realization that he himself had been overly inclined to narrate. In fact, one could compare Marcel's ontological language to that of Lévinas in *Totalité et infini*: though Marcel and Lévinas are desperate to find solid grounds on which to found an ethical relation to the Other, and are therefore tempted by the stability of ontology, both realize that the uncertainty and ambiguity of temporality cannot be escaped, for these are what actually define human experiences of normativity and intersubjectivity. A metaphysics of presence is not the answer. Lévinas thus writes *Autrement qu'être*, and, one might suggest, Marcel subordinates his philosophy to his theatre.

Unfortunately, this is not the end of the story. Although Marcel's plays do not themselves offer any totalizing narratives, which reduce their significance to an intemporal present, such narratives are constructed in Marcel's theoretical texts, where, having stressed the philosophical importance of his theatre, he appropriates it for eternalizing purposes. To give but one example, Werner's positive characteristics (in *Le Dard*) are reified, so that they become emblematic of the authenticity Marcel wishes to argue for: 'Werner Schnee est au sens le plus profond du mot, un être disponible et consacré' (*DH*: 163), he declared in the William James Lecture 'Dignité humaine'; 'et cette sympathie qu'il inspire à tous est sûrement liée très directement au fait qu'il existe aussi peu que possible pour lui-même' (*DH*: 164). Such objectification of Werner's character surely undermines Marcel's theory of theatre, which insists that the dramatist maintain '[un] respect absolu de ses personnages et de leur liberté' (Marcel 1951d: 151).[101] Werner may appear more exemplary than other characters in this particular play, but his decision to return to Germany could still be viewed as controversial, since the other dynamic of the plot concerns Werner's and Béatrice's growing love for one another: if Werner leaves, he leaves Béatrice behind, to live with a husband who does not seem to appreciate her and has been having an affair with another woman. Marcel's definitive characterization of Werner dissuades his audience from considering other such possible perspectives, however. Moreover, as the play ends before we can see the consequences of Werner's decision, Marcel's conclusions are simply premature. Thus if, as Mary notes, 'le théâtre de Gabriel Marcel se veut avant tout ouverture' (2008: 339), it is not allowed to maintain this openness, for Marcel homogenizes its significance when he tells us 'ce que j'ai cherché à montrer' (*DH*: 165).

So while Marcel might in one sense be said to make a Lévinassian move, when he begins to favour his dramatic works over his formal philosophical writings, this is followed by a Ricœurian return that arguably reverses the correction, because he fails to integrate the (retrospective) instruction that he claims to receive from his theatre. The bi-polarity of his work is denied; it defaults on the philosophical — because Marcel seems unable to escape the lure of eternalizing narrative, unable to face up to the unsettling instability of time. Although an Other conception of time may ostensibly be offered by his theatre, therefore, the theory of time offered in his *œuvre* as a whole remains at best inconsistent, and at worst, representative of a failure to confront time's lived reality.

Notes to Chapter 4

1. When referring to Lévinassian conceptions of the Self and the Other (person — things are not absolutely Other for Lévinas, because they can be mastered by the Self), capital letters will be used, following Lévinas's own practice. Not only does this affirm the Self and Other as absolutely distinct, but also reflects the transcendental character of his phenomenology, which, seeking to ascertain the conditions for the very possibility of subjectivity (and all aspects of its experience), entails a more abstract discussion. As will be seen, Lévinas later speaks of *le soi*, in addition to *le Moi*. This development remains intimately bound up with his thought on time, and thus engagement with the evolution of his philosophy will be important for this chapter.
2. To be understood as loosely synonymous with the 'Self' (*le Moi*), with particular emphasis on the totalizing, presentifying nature of its comprehension (of its own being and actions, as well as its relation to others and the world).
3. 'L'Autre métaphysique est autre d'une altérité [...] antérieure à toute initiative, à tout impérialisme du Même' (*TI*: 28).
4. 'La relation avec l'être, qui se joue comme ontologie, consiste à neutraliser l'étant pour le comprendre ou pour le saisir. Elle n'est donc pas une relation avec l'autre comme tel, mais la réduction de l'Autre au Même. Telle est la définition de la liberté: se maintenir contre l'autre, malgré toute relation avec l'autre, assurer l'autarcie d'un moi' (*TI*: 36–37).
5. For discussion of Marcel in Lévinas, see especially *HS*: 33–55 and *EN*: 72–74. See also *DMT*: 164, *HS*: 10, and *TI*: i.
6. Marcel never (directly) replies to Lévinas's critique, but this, in all likelihood, is simply due to a lack of acquaintance with Lévinas's many writings. The sole reference Marcel makes to Lévinas appears in the seventh William James Lecture ('Dignité humaine'; 1961), where Marcel praises *Totalité et infini*'s 'grande pénétration' and, more generally, Lévinas's non-Hegelian interpretation of alterity. He appears surprised that Lévinas does not employ the term 'prochain' to refer to 'l'autre en tant que l'autre, mais en

même temps tel qu'il se présente à moi pour être, non pas seulement affronté, mais accueilli', for, recognizing such a relation is, in Marcel's eyes, the only way to avoid atomistic pluralism (*DH*: 170). In *Autrement qu'être* (1974), however, Lévinas does use this term. Indeed, though Lévinas insists on the Other's absolute alterity, this is not to be understood as arguing for the absence of a relation between the Self and Other.

7. '[C]hez moi le lien entre philosophie et théâtre est le plus étroit, le plus intime qui soit. [...] ma philosophie est existentielle dans la mesure même où elle est en même temps théâtre' (*EPR*: 52).

8. In *Totalité et infini* the term 'justice' is often treated as synonymous with the ethical ('l'éthique'). Later, however, a distinction is drawn between the two, where ethics is identified with the Self–Other encounter, and justice with the institution of moral codes in an effort to command respect for the Other in action. In a 1986 interview Lévinas states: 'now "justice" is for me something which is a calculation, which is knowledge, and which supposes politics; it is inseparable from the political. It is something which I distinguish from ethics, which is primary. However, in *Totality and Infinity*, the word "ethical" and the word "just" are the same word, the same question, the same language' (Bernasconi and Wood 1988: 171).

9. See also *PI*: 19.

10. See also *EC*: 254.

11. But see also *EC*: 82, 230–31, 247–48; *EPR*: 52–55, 56–58, 62–69; *GM*: 39, 54, 72, 93–94; Lhoste (1999: I, interview 4); Marcel (1976: 9–10); Monestier (1999: II, interview 4).

12. 'Je crois qu'on voit beaucoup plus profondément en moi [mon œuvre] à travers mon théâtre qu'à travers mes écrits philosophiques' (Monestier 1999: II, interview 4), states Marcel to Monestier in 1970; in his 1973 paper 'De la recherche philosophique' he declares: 'je tends de plus en plus à préférer mes pièces à mes écrits philosophiques' (1976: 9); and in an interview with Lhoste in 1973, Marcel also states that he would prefer to be judged by his theatre rather than his philosophy (Lhoste 1999: I, interview 4).

13. '[M]on œuvre dramatique, bien loin de constituer une sorte de compartiment étanche de ma vie, complète indissolublement mes écrits philosophiques' (*DH*: 18).

14. As Lazaron does not reflect on the philosophical implications of the aesthetic and performative dimensions with which she engages, however, her work fails to further enquiries into the relation between these two aspects of Marcel's *œuvre*. And although Mary's recent doctoral thesis on Marcel's theatre aims to shed light on what his plays bring to his philosophy, her analysis, despite being thorough and well observed, is more descriptive than it is critical, and merely concludes that Marcel's theatre acts as 'un laboratoire de la pensée' (2008: 435). A significant gap thus remains in Marcellian studies, in reaction to which this chapter hopes to sketch out a more analytical response.

15. Theatrical form in itself does not in itself preclude narrative (see for example Richardson (2007)), but Marcel's characters find narrative problematic.

16. 'La non-intervention est fondamentale dans mon œuvre dramatique' (*EAGM*: 105). See also Marcel (1949b: 236; 1951d: 151, 154, 177), and *PI*: 34.

17. I therefore cannot agree with Chenu when he declares: 'Drame et métaphysique sont [...] deux moments de la même élucidation de l'existence. [...] Dans cette union [...], c'est toute l'unité de pensée de G. Marcel qui se révèle à nous' (1948: 178).

18. See also *GM*: 54, and Marcel (1976: 16).

19. See also *EPR*: 68, *GM*: 39, 94, Marcel (1967b: 11–12), and *RA*: 296.

20. The First World War was, of course, influential; but as Marcel writes in *En chemin*, although 'la pensée des ruines laissées par la [première] guerre ne me quittait pas' (*EC*: 19), 'l'avenir n'éveillait pas en nous d'inquiétude particulière. [...] Les nuages qui devaient si vite s'amonceler à l'horizon politique, n'étaient pas encore clairement

discernables' (*EC*: 18–19). Loubet del Bayle remarks similarly on the contrast between the decade immediately following World War I in France, and the 1930s: 'Pour la France, les années 1930–1932 furent celles d'un cruel réveil qui dissipa les rêves de paix et de prospérité qu'elle avait cultivés depuis 1918' (1969: 11).

21. Marcel also emphasizes this in an interview with Ricœur (*EPR*: 96), Lhoste (1999: II, interview 8), and Boutang (*GM*: 94–95). When Boutang enquires as to how such a concern has impacted on his work, it is interesting that Marcel cites an example from his theatre rather than his philosophy.
22. *EC*: 146–47.
23. The title of Marcel's 1933 play, *Le Monde cassé*, testifies to this directly. Rémond reports: '[Dans les années 1930, Marcel] redoute la résurgence de l'esprit nationaliste outre-Rhin. Il déclare avoir éprouvé dès 1932, soit donc avant même l'accession de Hitler à la Chancellerie, un sentiment d'angoisse qui ne le quittera plus. [...] Il a conscience de vivre dans "un monde cassé"; le titre d'une de ses pièces les plus connues lui a été suggéré par le spectacle de l'Europe' (1989: 35).
24. In 1934, he signed Maritain's and de Gandillac's manifesto, 'Pour le bien commun', which emphasized the responsibility of Christians at that historical moment, and rallied those who opposed the Conservative Right (Rémond 1989: 36). In 1937, he signed — along with Maritain, Mauriac, and Mounier — a manifesto in defence of the Basque people, following the bombing of Guernica and killing of non-combatants (ibid.; see also *EC*: 241–42). 25 February 1938 saw the publication of an article urging France to assume its responsibilities, so as not to betray the values it stood for (Marcel (1938)). Marcel declared France's violation of its promise to Czechoslovakia (the 'Little Entente', which pledged to provide assistance if its territory was violated; breached in 1938 when Hitler invaded the Sudetenland) to be a scandal (*EC*: 171); and in November 1940, he published an article in (Mounier's review) *Esprit*, condemning the air of 'mea culpisme' surrounding the French defeat, which, he warned, 'peut [...] affecter le caractère morbide d'une complaisance à rebours' (1940c: 17). On Marcel's socio-political action and reaction, see especially Marcel (1983), but also *EC*: 98–99, 146–50, 163–66, 171–72, 177–79, 181–83, 187–89, 191, 193–97, 222–28, 241–46, 248, 259–64. For commentary on Marcel's *engagement*, see Boutang (1989), Poirier (1989), Rémond (1989), and Volkoff (1989) (all presentations at the 1988 Marcel conference, organized by the *Bibliothèque nationale* and *Présence de Gabriel Marcel*).
25. Especially since the Dreyfus affair, which, as for so many writers and intellectuals, had a significant impact on Marcel (*EC*: 40–41; *EPR*: 56, 97). Interestingly, Marcel states that what this event alerted him to in particular, was 'l'aspect foncièrement dramatique de la vie humaine', and that this was exactly what his theatrical works were concerned to convey (*EC*: 40–41). So again, Marcel's interest in dramatic, ethically charged situations can be related to theatrical expression.
26. See also *EPR*: 99.
27. See *EC*: 197.
28. See especially Marcel (1983: 77–103) (reprint of 'La Philosophie de l'épuration I & II', published in the Canadian journal *La Nouvelle Relève*, 1946), but also *EC*: 223–24, 226, and *GM*: 94–95.
29. 'Le procès intenté au maréchal Pétain et les conditions inhumaines dans lesquelles s'accomplit sa détention à l'île de Yeu, restent pour moi un objet de scandale. Non que j'éprouve pour la victime une sympathie ou même une indulgence sans contrepartie' (*EC*: 225). See also *EC*: 178.
30. At the first public meeting concerning the trial, Marcel announces: 'Plus que je me sens en opposition sur des points essentiels avec la personne de Maurras, avec ce que

j'appellerai globalement son nationalisme, mieux je me sens armé pour déclarer qu'il a été victime d'un déni de justice. [...] se montrer inique envers lui, ce sera l'imiter, donc le justifier' (1949a: 5). See also Maurice de Gandillac's comment, during the discussion following Boutang's presentation at the 1988 Marcel conference (Sacquin 1989: 89).

31. 'Il est vrai [...] que nous écrivains qui avons été si souvent sollicités, nous avons probablement trop prodigué notre signature [...]. Mais, dans bien des cas, je ne me suis pas reconnu le droit de refuser la mienne et j'ai parfois même tenté une démarche pour réparer une injustice flagrante. En ce sens très général, je me déclare volontiers *engagé* — mais à la condition expresse que cet engagement n'affecte pas un caractère partisan [...]. J'ai eu surtout horreur des protestations à sens unique comme si l'injustice n'était pas l'injustice quel que soit le camp où elle se manifeste' (*EC*: 227–28).
32. See also *EPR*: 102–03.
33. Republished with an epilogue in 1953.
34. Here Marcel is reacting against Joseph Chenu's suggestion (in the paper he presents here) that Marcel's theatre, despite, in one sense, becoming more focused on the contemporary, largely transcends actuality (*EAGM*: 117).
35. If the First World War is referenced (e.g *La Chapelle ardente, Le Mort de demain, Le Regard neuf*), it tends to be much more of a background context than the later World War II plays. The only exception is *Un juste*, which was written in 1918 and asks how a non-supporter of the war should respond to his fellow countrymen who are fighting on the frontline. However, *Un juste* was not published until 1968 (along with a speech Marcel made when receiving a peace prize in Frankfurt, 1964). In this respect, it can still be seen as testifying to a later preoccupation with ethics.
36. Accordingly, Boutang includes some discussion of Marcel's plays in his 1988 paper on Marcel's politics (1989: 84).
37. See also *EC*: 198.
38. Though of course, Marcel adds: 'Ceci n'est [...] vrai que dans la mesure où le dramaturge est fidèle à sa mission de dramaturge, et ne se comporte pas comme un simple manipulateur ou comme un montreur de marionnettes qui se content de tirer des ficelles sans rien pouvoir changer à ses poupées qui sont là avec leur caractère invariable' (*DH*: 9).
39. Lévinas describes this situation in terms of the Self being in the accusative (e.g. *AE*: 31, 91, 135, 177, 233, 239) or the vocative case (e.g *DVI*: 129, 131, 156; *EN*: 19). While useful images, both are problematic. Reference to the accusative case, helpful because of its emphasis on the Self's *mise en question*, also carries the undesirable implication that the Self is an object — something that Lévinas wanted to avoid; a genuine encounter with the Other, despite not being freely chosen, is non-violent and non-objectifying (e.g. *TI*: 215–18). The advantage of the vocative is the reverse: it emphasizes the non-violence of the *face-à-face*, but does not connote the Other's interrogation of the Self and demand for justification.
40. In the preface to the German edition of *Totalité et infini* (January 1987), Lévinas declares: '*Autrement qu'être: ou au delà de l'essence* évite déjà le langage ontologique [...] auquel *Totalité et infini* ne cesse de recourir' (*TI*: i–ii).
41. In some instances, *Totalité et infini* implicitly acknowledges the inadequacy of its spatial presentism. Although its analyses of the Self's 'dwelling' (*la demeure*: the hospitable 'place' into which the Other is welcomed before the *face-à-face*, thus allowing for the possibility of a non-violent encounter with the Other) very much privilege presence, for example, time shows its face sporadically, as if to recognize that speaking of the 'presence' of the Other is not quite the way to express things (e.g. *TI*: 178–80). Time also seems to become crucial in Part III ('Visage et extériorité'). Here Lévinas returns to the

question of freedom, and asks how we can understand this in the context of beings who are 'at war' (where beings refuse totalization by affirming themselves as distinct — but in a world where ontology takes precedence, so that asserting themselves as separate from the whole only reinforces totalization, because it remains Self-centred, rigidly refusing the Other), while still being in some kind of relation. Not in causal terms, he argues, for if freedom is understood as finite because it is subject to exterior forces it cannot control, it is not clear in what sense it is still free. The solution to the problem, Lévinas in fact concludes, is time: 'Ce n'est pas la liberté finie qui rend intelligible la notion du temps; c'est le temps qui donne un sens à la notion de liberté finie. [...] La prise qu'a la violence sur [un] être — la mortalité de cet être — est le fait originel. La liberté elle-même n'en est que l'ajournement par le temps' (*TI*: 247–48).

42. 'Le présent — c'est l'essence qui commence et qui finit, commencement et fin rassemblés, en conjonction thématisable [...]. La diachronie, c'est le refus de la conjonction, le non-totalisable' (*AE*: 26).
43. '[C]ette impossibilité d'être ensemble est la trace de la diachronie de l'un-pour-l'autre' (*AE*: 127).
44. On the Other's 'hauteur' in *Totalité et infini*, see for example *TI*: 73, 166.
45. All of these terms are used in *Totalité et infini* to describe the Self's violent acts of appropriation and totalization.
46. 'Tuer n'est pas dominer mais anéantir, renoncer absolument à la compréhension. Le meurtre exerce un pouvoir sur ce qui échappe au pouvoir. [...] Je ne peux vouloir tuer qu'un étant absolument indépendant, celui qui dépasse infiniment mes pouvoirs et qui par là ne s'y oppose pas, mais paralyse le pouvoir même de pouvoir. Autrui est le seul être que je peux vouloir tuer' (*TI*: 216).
47. See also *TI*: 215, 223.
48. In Butler's terms: 'To be undone by another is [...] an anguish, to be sure, but also a chance — to be addressed, claimed, bound to what is not me, but also to be moved, to be prompted to act, to address myself elsewhere, and so to vacate the self-sufficient "I"' (2005: 136).
49. Thérèse criticizes him for his intransigence on numerous occasions. She makes reference to 'cet amour qui ne bouge pas [... et] ressemble à un prison' (Marcel 1973b: 233), for example, and exclaims in frustration: 'Il faut que tu voies devant toi. C'est comme en voyage. Ton itinéraire est fixé le jour de ton départ; rien ne t'amènera à le modifier. Tu es rentré avant d'être parti. Ah! si seulement il y avait des agences pour la conduite de la vie!' (1973b: 287).
50. Marcel in fact compares Germain to Jeanne: 'Germain Lestrade, comme Jeanne Framont, est amené à se rendre coupable d'une infraction précise à une loi profondément inscrite au cœur même de notre condition. Cette loi est ce qu'on pourrait appeler le devoir de non anticipation; à la base de cette infraction, comment ne pas discerner une impatience, une intolérance radicale en présence de l'incertitude à laquelle nous condamne notre existence d'êtres engagés dans le temps?' (1973b: 369).
51. 'Il est entré ici vivant; il en sortira... mort' (Marcel 1931: 181), comments Noël's dismayed brother, Antoine, to Jeanne at the end of the play.
52. Aline's daughter Yvonne protests: 'je trouve que ça n'a pas le sens commun de ne pas utiliser ce qu'on a: tu aimes mieux que tous ces jouets moisissent au grenier sans servir à rien?' (Marcel 1950a: 17); and Raymond's fiancée Mireille remarks: 'ces jouets, Raymond les aurait sûrement donnés à son neveu' (1950a: 19).
53. 'Je ne te répondrai qu'un mot: quand on a été capable d'aller au bal trois mois après la mort de son frère, on n'a pas qualité...' (Marcel 1950a: 18), snaps Aline at Yvonne, for example.

54. 'Si tu ne m'as pas pardonné parce que tu m'aimais, qu'est-ce que tu veux que j'en fasse de ton pardon?' (*CPM*: 37).
55. 'Longtemps les philosophes ont placé tout le poids du pardon sur le sujet moral et l'autonomie de sa volonté', observes Francis; 'peut-être oublie-t-on trop vite que les problèmes moraux sont des affaires qui intéressent une personne *à l'égard* d'une autre personne' (1989: 238).
56. This is an excellent illustration of Lévinas's notion of *jouissance* — 'satisfaction et égoïsme du moi' (*TI*: 96) — where the Self tries to lose itself in objects, to prevent the eternal return to the Self to which it is condemned by its egoistic projects. As Lévinas argues in *Le Temps et l'Autre*: 'l'identité n'est pas seulement un départ de soi; elle est aussi un retour à soi. [...] L'identité n'est pas une inoffensive relation avec soi, mais un enchaînement à soi; c'est la nécessité de s'occuper de soi. [...] Sa liberté est immédiatement limitée par sa responsabilité. C'est son grand paradoxe: un être libre n'est déjà plus libre parce qu'il est responsable de lui-même' (*TA*: 36).
57. See also Perpich (2008: chap. 3 especially).
58. For this reason, Derrida is right to argue that the Self–Other relation in *Totalité et infini* cannot escape a certain violence. 'Levinas est très proche de Hegel, beaucoup plus proche qu'il ne le voudrait lui-même' (Derrida 1967b: 147), he remarks. This is necessitated by *Totalité et infini*'s dualistic and overly spatial (ontological) language, which only understands the Self in terms of *le Moi/le Même*, so that any relation between the Self and the Other — and Lévinas wants to argue for a relation — forcibly relates the Other to the Self (as Ego), which cannot understand the Other's asymmetry without recourse to its own present: 'ou bien il n'y a que le même et il ne peut même plus apparaître et être dit, ni même exercer la violence [...]; ou bien il y a le même *et* l'autre, et alors l'autre ne peut être l'autre — du même — qu'en étant le même (que soi: ego) et le même ne peut être le même (que soi: ego) qu'en étant l'autre de l'autre: alter ego' (Derrida 1967b: 188).
59. Ricœur's critique of Lévinas, which contends that Lévinas has overly insularized and isolated the self (*SA*: 221, 236, 387–93, 408–09), seems a legitimate response to *Totalité et infini*, but is less justified with respect to *Autrement qu'être* (to which Ricœur makes some reference, but appears to misunderstand; see Cohen (2002)). As Newman writes, in *Autrement qu'être* 'there is no longer any need for an independent analysis of the phenomenon of the Other' (2000: 94), for 'from the start — indeed before the start, before any beginning — [the subject is] in a relation with the Other' (2000: 93).
60. 'The face is not a force. It is an authority. Authority is often without force. [... The] idea that God commands and demands[, that he] is extremely powerful[, and that if] you try not doing what he tells you, he will punish you[, ...] is a very recent notion. On the contrary, the first form, the unforgettable form, in my opinion, is that, in the last analysis, he can not do anything at all. He is not a force but an authority' (Lévinas 1988: 169).
61. As Waldenfels explains, Lévinas's novelty is his '*retraduction* of "responsibility" into a kind of responding' (1995: 39): 'the respondent [...] is not a subject in the traditional sense, [...] but a respondent [...] who in a certain sense remains unknown to him- or herself. If we want to continue calling him a "subject", then we do so in the sense of his "subjection" to the demands of the other' (1995: 42).
62. 'La proximité — suppression de la distance que comporte la "conscience de"... — ouvre la distance de la dia-chronie sans *présent commun* où la différence est passé non rattrapable, un avenir inimaginable, le non-représentable du prochain sur lequel je suis en retard — obsédé par le prochain — mais où cette différence est ma non-indifférence à l'Autre. La proximité est dérangement du temps remémorable' (*AE*: 142).

63. Also referred to as 'soi', 'soi-même', or 'le *se*': 'Malgré moi, pour-un-autre — voilà [...] le sens du soi-même, du *se*' (*AE*: 26), explains Lévinas; 'l'obsession par l'autre [...] ramène le Moi à *soi* en deçà de mon identité, *plus tôt* que toute conscience de soi' (*AE*: 147). For Lévinas, therefore: 'C'est à partir de la subjectivité comme soi — [...] où le Moi ne s'apparaît pas, mais s'immole — que la relation avec l'autre peut être communication et transcendance et non pas toujours une autre façon de rechercher la certitude, ou la coïncidence avec soi' (*AE*: 188).

64. As a consequence, Lévinas makes reference to the self's passivity: 'Au passé diachronique, irrécupérable par la représentation [...], c'est-à-dire incommensurable avec le présent, correspond ou répond la passivité inassumable du soi. [...] La réponse qui est responsabilité — responsabilité incombant pour le prochain — résonne dans cette passivité' (*AE*: 30–31).

65. '[T]his Levinassian disruption not only challenges a successive or uninterrupted time, but also questions the understanding of the subject, for the conception of time as duration attributes a continuous identity to the subject whereas the disjunctive time of the relation to the other inaugurates the subject as "constituted in response to the arrival of alterity"' (Hodge 2007: 107).

66. Lévinas uses this term, 'récurrence', to describe — as Dudiak writes — 'my being driven back, ever and again, upon myself and beyond myself out of an ever resuming present in response to an unrepresentable past that demands my response in the form of responsibility, demands my passive undergoing for the sake of the other as patience, signifies as the very temporalization of time' (2001: 289). See especially *AE*: 162–73.

67. This, Lévinas refers to as 'sensibilité': 'La sensibilité est exposition à l'autre. [...] Avoir-été-offert-sans-retenue où l'infinitif souligne le non-présent, le non-commencement, la non-initiative de la sensibilité — non-initiative qui, plus ancienne que tout présent, est non pas une passivité contemporaine et contre-partie d'un acte, mais l'en deçà du libre et du non-libre qu'est l'an-archie du Bien [...]: la vulnérabilité même' (*AE*: 120). 'En partant de la sensibilité interprétée non pas comme savoir mais comme proximité', Lévinas explains, 'nous avons essayé de décrire la subjectivité comme irréductible à la conscience et à la thématisation. La proximité apparaît comme la relation avec Autrui, qui ne peut se résoudre en "images" ni s'exposer en thème' (*AE*: 157).

68. '[L]e laps de temps, c'est aussi de l'irrécupérable, du réfractaire à la simultanéité du présent, de l'irreprésentable, de l'immémorial, du pré-historique. [...] L'immémorial n'est pas l'effet d'une faiblesse de mémoire, d'une incapacité de franchir les grands intervalles du temps, de ressusciter de trop profonds passés. C'est l'impossibilité pour la dispersion du temps de se rassembler en présent — la diachronie insurmontable du temps' (*AE*: 66).

69. 'Nous avons appelé obsession cette relation [avec l'Autre] irréductible à la conscience. [...] sous les espèces d'un Moi incapable de penser ce qui le "touche" l'emprise de l'Autre s'exerce sur le Même au point de l'interrompre [...]: [cette] anarchie est persécution. L'obsession est persécution' (*AE*: 159–60). Importantly, Butler explains, 'Levinas is not saying that primary relations are abusive or terrible; he is simply saying that at the most primary level we are acted upon by others in ways over which we have no say, and that this passivity, susceptibility, and condition of *being impinged upon* inaugurate who we are' (2005: 90).

70. '[Ê]tre soi — condition d'otage — c'est toujours avoir un degré de responsabilité de plus' (*AE*: 185–86).

71. '[L]'Autre dans le Même est ma substitution à l'autre selon la *responsabilité*, pour laquelle, *irremplaçable*, je suis assigné. [...] Dans cette substitution où l'identité s'invertit, dans cette passivité plus passive que la passivité conjointe de l'acte, [...] le soi

72. The prominence of the past in *Autrement qu'être* does not undermine Lévinas's reflections on the future, in his writings on *fécondité*. Rather, it recognizes how the future cannot, in fact, be so clearly separated from the past. Thus, it is still possible to discuss Marcel's plays in conjunction with both *Totalité et infini* and *Autrement qu'être* — indeed, Cohen argues that the two works 'are distinguished less by a difference in content than by a difference in emphasis' (2001: 147).
73. As Mary observes, death is often a 'moteur de l'action'; 'cependant, la mort n'intervient que rarement comme conclusion: les disparus ne viennent pas clore la pièce, mais au contraire ils resurfissent et enclenchent l'action' (2008: 103). Death, along with sickness and suffering, 'cristallisent les interrogations et déclenchent les rares actions des pièces' (2008: 106).
74. 'Ma présence ne répond pas à l'extrême urgence de l'assignation. Je suis accusé d'avoir tardé' (*AE*: 141).
75. Lazaron agrees: 'The principal character is a woman who has recently died, the mother of Raymond Chavière' (1978: 91).
76. As further suggested when the housekeeper describes Raymond as 'le portrait de la pauvre Madame' (Marcel 1936: 10).
77. E.g. *CPM*: 201, 253; Marcel (1925: 73, 126, 138; 1950a: 39; 1951d: 31).
78. E.g. *CPM*: 229; Marcel (1949b: 64; 1951d: 133).
79. 'Cette diachronie est elle-même énigme' (*AE*: 37).
80. '[D]ans la responsabilité que nous avons l'un de l'autre, *moi* j'aie toujours une réponse de plus à tenir' (*AE*: 134). The use of the subjunctive here reflects the diachrony of the relation, which has its roots in an 'immemorial' past, and implicates me in an unimaginable future.
81. 'L'essence originelle de l'expression et du discours ne réside pas dans l'information qu'ils fourniraient sur un monde intérieur et caché. Dans l'expression un être se présente lui-même' (*TI*: 218).
82. See also *TI*: 212. As Butler writes of Lévinas, there is 'an ethical valence to my unknowingness' (2005: 84).
83. 'La responsabilité pour autrui, c'est précisément un Dire d'avant tout Dit' (*AE*: 75). Not conjugated to signify in a particular tense, the infinitive form of *le Dire* manifests how the relation spans all time. Ungraspable in itself, it transcends any particular moment — unlike the opposing notion *le Dit*, which is rigidly tied to the present alone.
84. Cf. Lévinas: 'Il s'agit d'apercevoir la fonction du langage non pas comme subordonnée à la *conscience* qu'on prend de la présence d'autrui [...], mais comme condition de cette "prise de conscience"' (*EN*: 17).
85. '[I]l y a d'abord le dénouement qui permet de décider si un ouvrage se tient, s'il réalise un minimum d'existence dramatique; il y a, en second lieu, une mystérieuse survie qui est en réalité, au sens métapsychique du mot, une *hantise*. L'œuvre et le personnage qui ne continuent pas à nous habiter en quelque manière après la fin du spectacle se révèlent par là même dépourvus de réalité dramatique profonde, et ce terme de hantise est encore de ceux qui font ressortir ce caractère existentiel du drame sur lequel j'ai tenté d'attirer l'attention' (Marcel 1946: 294-95).
86. Lévinas also uses the term 'fraternité', which, again, is contrasted with collective similarity, to describe intersubjective relations (e.g. *AE*: 138; *TI*: 235).
87. '[L]'égalité est essentiellement revendicative, elle est au sens le plus fort du mot égocentrique' (*DH*: 172).
88. See also Lhoste (1999: II, interview 8).
89. See also *HH*: 28, 119-20.

90. '[L]'intersubjectivité elle-même ne peut être en aucune façon traitée comme une structure comparable à celle qui tombe sous les prises de la connaissance objective' (*ME II*: 109).
91. The difference between the two is not merely a question of focus, as Treanor suggests: 'Although arguing the same general thesis, *Totality and Infinity* concentrates on the epiphany of the other's face (alterity) — which is our guiding question and concern — while *Otherwise than Being* concentrates on ethical subjectivity and the selfhood of the subject' (2006a: 274). Lévinas's general concern does indeed remain constant, but *Totalité et infini*'s failures cannot be overlooked. The epiphany of the Other's face itself has changed by the time Lévinas writes *Autrement qu'être*, and this shift must be taken into account.
92. In *Le Temps et l'Autre* for example, and especially *Totalité et infini*. Lévinas's (increasing) references to love in later works, on the other hand, are decidedly distanced from *eros* (indeed, another striking difference between *Totalité et infini* and *Autrement qu'être* is the fact that Lévinas abandons discussion of the erotic relation as a genuine approach to the Other); they are more akin to *agape*.
93. See for example *HH*: 24, 141, and *ME II*: 110.
94. On the opposition between desire and love see also *EA*: I, 190, 210.
95. E.g. *EA*: I, 56.
96. See especially Marcel (1954b), but also *EA*: I, 149, and *JM*: 63-64, 226-27.
97. For Gillman, for example, 'to claim that when I "love" someone, the lines of our distinctive personalities are blurred so that we "participate" in each other's being is [...] to indulge in hyperbole' (1980: 165-66).
98. Sweetman might agree: 'Marcel did little direct writing either on moral philosophy in general, or on specific ethical problems. Nevertheless his work is deeply ethical [...]; indeed, from one point of view, the whole of his thought is a sustained discussion on the issue of how to live ethically in a world that is making it increasingly difficult to do so' (2002: 271).
99. Later, in *Soi-même*, Ricœur in fact identifies Marcel with Lévinas (and also Jean Nabert) on the basis of their shared concern with ethics (*SA*: 198).
100. As Chapter 2 observed, love and intersubjectivity are frequently discussed in relation to fidelity, the temporality of which Marcel does seem to want to take seriously. However, because Marcel's conclusions then tend to speak of intersubjectivity in terms of presence, its temporality is then eclipsed — perhaps even forgotten.
101. This particular reference is to a criticism made of Camus's theatre, in a 1951 lecture concerning Marcel's play *Rome n'est plus dans Rome*. Marcel states: '[Camus] ne me semble pas authentiquement dramaturge, je ne vois pas qu'il ait évité nulle part l'écueil de la pièce à thèse; je ne trouve pas chez lui ce respect absolu de ses personnages et de leur liberté qui doit apposer son sceau à une œuvre dramatique'.

CONCLUSION TO PART II

~

Between Ricœur and Lévinas

According to Richard Cohen, 'what Levinas wants to account for [...] is not the relation *between* self and other, but the encounter with alterity as transcendence [itself], as the outside, the other' (2000: 141). While Marcel's theatre — and indeed some of his philosophical writings — have been shown to engage with this aspect of alterity, his philosophical project cannot simply be confined to this because, as Marcel's narrations of his plays reinforce, he is also concerned with the self in action — that is, with the basis on which the self might make decisions, and in what light ('authentic' or otherwise) these decisions might be viewed. It is for this reason that Marcel can still be read through a Ricœurian lens, for, as Muldoon informs us, Ricœur 'devoted his philosophical career to unraveling the web of relations that must be determined to understand how we come to view practical actions as potentially ethical ones' (2005: 64). Thus the (different) ethical interests of Ricœur and Lévinas seem to intersect in Marcel's œuvre,[1] which, as Part II has shown, is also deeply — if not first and foremost — ethical in motivation. Marcel's philosophical texts have emerged as primarily Ricœurian, and Marcel's theatrical works as Lévinassian; and interestingly, a different theory of time also appears to underwrite each of these two poles — namely, time as narrative, and time as Other.

As far as Marcel's own ethical project is concerned, if any ethical 'message' is provided, this, by Marcel's own admission, is not developed in his philosophy, nor can it be identified with the content of his theatre — for as has been seen, Marcel's plays dramatize only conflict and ambiguity. Rather, a response from the spectator or reader is required: a response that is non-totalizing, and lies somewhere between the 'limits' of what is presented in Marcel's philosophy and in his theatre. Thus while the time of Marcel's dramatic works suggests Marcel to be (at least) open to the idea that the call to responsibility originates in the Other, as opposed to the innate capacities and reflexivity of the self (as Ricœur proposes), this insistence on transcending his theatre's content also indicates dissatisfaction with a Lévinassian position in itself.

Indeed, Lévinas has been criticized for not adequately addressing the question as to what, in practice, an ethical response to the Other involves,[2] or how one might decide which Others to respond to, since one's finitude consists in an impossibility to assume responsibility for all. Kearney, for example, points to the problem of discernment that arises if philosophy only approaches the Other as one who surpasses interpretation and representation. 'Radical undecidability [...] needs to be addressed by a critical hermeneutics of self-and-other', he argues, and 'this [...] calls for a practice of narrative interpretation capable of tracing interconnections between the poles of sameness and strangeness'. Only if we move beyond the 'postmodern fixation on inaccessible alterity, [and] build paths between the worlds of *autos* and *heteros*[, ... can philosophy] help us to discover the other in our self and our self in the other — without abjuring either' (2003: 10). Such a limitation is equally suggested by Marcel's theatre, in which action is markedly absent. Characters tend to remain silent or paralysed by indecision, and this is frequently the result of their inability to judge, or to enter into a critical relation with the other. Hence, Raymond, in *Un juste*, is left mute in his intermediate position as a non-supporter of the causes for which World War I has come to stand for, and as a close friend to men who are fighting in the trenches and are in need of moral support.[3] And the identities of characters such as Jérôme and Violette, in *Le Chemin de crête* (1935), are eroded by an other (Ariane) so enigmatic that they cannot determine if she is a devil or saint.[4] Even Werner and Christiane are not proven to be successful: if they are presented in a more positive light, this is only because the possibility of a relation with the Other is not ruled out; it is not by virtue of any actual relation being established. Thus, it seems that there might still be a place for narrative time. As Ricœur maintains when he responds to Lévinas in *Soi-même*:

> Il faut bien accorder au soi une capacité d'accueil qui résulte d'une structure réflexive. [...] Bien plus, ne faut-il pas joindre à cette capacité d'accueil une capacité de discernement et de reconnaissance [...]? Et que dire de l'Autre, quand il est le bourreau? Et qui donc distinguera le maître du bourreau? (*SA*: 391)

One might criticize Ricœur for taking the other's otherness for granted, by starting with hermeneutics rather than with an investigation of what makes the hermeneutical situation possible as such; but this is because Ricœur is predominantly concerned with *praxis*, and with the interpretative framework that enables the self to conceive of its actions and interactions teleologically, as it strives to live a 'good' life. This is not to turn a blind eye to conflict or the other, Ricœur would therefore argue, for such teleology seeks not to resolve, but rather to ground itself in, the lived tensions of experience. It is for this reason that Ricœur introduces, as an ethical development of his narrative conception of identity, the notion of 'tragic wisdom' (*la sagesse tragique*),[5] a wisdom that gives a certain privilege to Greek tragedy and its teachings about reason's scope and limits, and which shares striking similarities with Marcel's work — both his theory of theatre, and his conception of *le tragique*, the consciousness of which, he writes in *Homo viator* (1944), 'est liée à un sentiment aigu de la pluralité humaine, c'est-à-dire tout à la fois de la communication et du conflit, mais avant tout de l'irréductible que nul accord rationnel ne peut faire disparaître' (*HV*: 192).

Discussing practical wisdom in *Soi-même*, Ricœur interrupts his reflections to include a commentary on *Antigone*, 'afin de [...] faire entendre une autre voix que la philosophie' (*SA*: 281). 'Si [...] j'ai choisi *Antigone*', he explains, 'c'est parce que cette tragédie dit quelque chose d'unique concernant le caractère inéluctable du conflit dans la vie morale' (*SA*: 283). The kind of wisdom it offers, however, is 'non point un enseignement au sens le plus didactique du terme, mais une conversion du regard' (*SA*: 286). Just as Ricœur sees an illustration of such a conversion in the chorus of *Antigone*, which at no point offers straightforward advice as to how the conflict between Creon and Antigone might be resolved, Marcel's plays deliberately refuse to offer totalizing messages. For both (to use Ricœur's wording): 'la tragédie, après avoir désorienté le regard, condamne l'homme de la *praxis* à réorienter l'action, à ses propres risques et frais, dans le sens d'une sagesse pratique en situation qui *réponde* à la sagesse tragique' (*SA*: 288). Yet again, though, if Ricœur speaks of ethical conflict, uncertainty, and risk, it is not clear that he is willing to accept such temporal discordance for what it is. Instead, as Crowley has forcefully argued, Ricœur's overriding desire for stability continues to motivate this circumscription of narrative identity into a theory of ethical action (2003: 4). 'It is the word of praxis that finally stabilizes meaning', Crowley writes, for the definition of identity that ensues from the self's ethical conviction (having decided how to respond to *le tragique*) 'involves not only an action but a promise'. And in making a promise of fidelity to another, 'one ties speech to action and defines selfhood'; 'by holding to one's word both language and the subject [seek to] resist change over time' (2003: 8).

So while Lévinas might be accused of neglecting the question of response, Ricœur appears to affirm a response too soon. He may speak of *le tragique* and of the non-philosophical, but ultimately these discussions remain grounded in the theoretical, the intention of which is to synthesize (or totalize) above all else. Thus, here again, a parallel with Marcel can be maintained. Marcel, too, slides between references to *le tragique* as drama and *le tragique* as philosophical principle. In the preface to his first volume of plays, *Le Seuil invisible* (1914), for instance, he discusses dramatic forms of tragedy in direct conjunction with 'le tragique de pensée' that he considers his theatre to express.[6] And Marcel, too, chooses (eternalizing) philosophical principle over an engagement with time: in the same way as Ricœur's shift to ethics serves to write stability into his (ostensibly temporal) theory of narrative identity, the non-narrative time of Marcel's theatre is re-narrated in his philosophical texts, allowing him to draw conclusions about how the self can conceive of itself ethically, in a unified way that resolves conflict. Although, then, it has been suggested that there might still be a place for narrative time, a Ricœurian conception — even when ethically situated — does not appear to be the solution. Our understanding of narrative itself, it seems, must first be reconceived;[7] and thus, the challenge Lévinas's philosophy poses to Western philosophy's metaphysical tradition continues to be a justified critique of positions such as Marcel's and Ricœur's.

Part II's Ricœurian reading of Marcel has ultimately failed, therefore, to reconcile the tensions between time and eternity in his work. Despite repeated efforts to engage with time, Marcel chooses eternity in the end. One question still remains to be answered, however: why is it that Marcel always subordinates time to eternity? Part I suggested that Marcel might have failed to escape idealism, but another explanation is also possible — namely, Marcel's religious belief. Is this where it might be helpful to think of his philosophy as religious, or, more specifically, as Christian? This is what the fifth and final chapter endeavours to ascertain.

Notes to the Conclusion

1. For more specific discussion regarding their individual projects, see Bourgeois (2002) and Cohen (2002).
2. See Vanni (2004).
3. 'Il n'y a que les idées confuses qui paralysent' (Marcel 1965: 158), Raymond exclaims.
4. Violette says to Ariane, for example: 'Et vous... je ne vous comprends pas... tantôt je vous admire plus que personne, tantôt... [...] Il n'y a pas de mot. Je perds pied, oui, c'est comme si je tombais. C'est affreux' (*CPM*: 284).
5. See *SA*: chap. 9.
6. '[Les drames dans ce volume] sont essentiellement des drames d'idées, ils se meuvent dans la sphère de la pensée métaphysique, et pourtant ce ne sont à aucun degré des dialogues philosophiques; ils portent sur les antinomies [...] que l'esprit découvre en réfléchissant sur lui-même [...]. Je me suis efforcé [...] de montrer le tragique de pensée se réalisant [...] dans la conscience' (Marcel 1914: 1). See also Marcel (1921; 1924; 1926).
7. This is exactly what Baroni (2009) aims to do, in specific response to the inadequacies of a Ricœurian conception of narrative. See also Kearney (2003).

PART III

Time and Eternity

CHAPTER 5

~

Time and God

If Chapter 3 suggested that a Ricœurian reading of Marcel might be possible, this was also inspired by similarities identified between Marcel and Augustine, whose philosophy of time Ricœur drew upon to support his argument that phenomenological and cosmological time are incommensurable. If Augustine's concern with time is philosophical,[1] however, it is also linked to his religious outlook — just as, this chapter will suggest, might be argued for Marcel. However, this theological aspect is rather neglected in Ricœur's interpretation of Augustine. For Ricœur, God's eternity in Book XI of the *Confessions* merely functions as time's Other, intensifying and deepening our understanding of time. Yet eternity is much more than this for Augustine: not simply a philosophical concept that serves as a useful counterpoint when reflecting on the nature of time, it is generally considered to correspond to a form of being — the being of God. If Ricœur's hermeneutical reading fails to take this into account, could it be that Part II's interpretation of Marcel overlooks something equally fundamental?

Ricœur is not unaware of eternity's importance for Augustine's reflections on time. 'Je n'ignore pas que l'analyse du temps est enchâssée dans une méditation sur les rapports entre l'éternité et le temps, suscitée par le premier verset de la *Genèse: In principio fecit Deus...*', he writes in *Temps et récit I*. Nevertheless, although — as Ricœur acknowledges — 'isoler l'analyse du temps de cette méditation, c'est faire au texte une certaine violence', he feels that his approach is justified by Augustine's own interpretative focus: 'Toutefois, cette violence trouve quelque justification dans l'argument même d'Augustin qui, traitant du temps, ne se réfère plus à l'éternité que pour marquer plus fortement la déficience ontologique caractéristique du temps humain' (*TR I*: 22). I would argue, however, that Augustine's lack of specific engagement with the eternal in this context is not a sufficient justification for Ricœur's inattention — at least, not if one is interested in what time means for Augustine; and given that Augustine's conclusions about time are ultimately unclear, it seems rather crucial to consider the wider context in which Augustine is writing.[2]

The fact that Augustine feels the need to introduce a notion of eternity when philosophizing about time is, in itself, significant. Moreover, as a consequence of drawing this relation, what Augustine means by 'eternity' becomes pivotal to understanding what he means when he speaks of time.

Ricœur may insist that he does not intend to 'enfermer l'éternité selon Augustin dans la fonction kantienne d'une idée-limite' (*TR I*: 57), but he nevertheless treats eternity in Augustine as a purely negative conception,[3] where eternity is considered to be just as mysterious as time.[4] And this misrepresents Augustine's position, for if, in response to time's paradoxes, Augustine is willing to rethink its fundamental nature, the same does not apply for eternity. Eternity, for Augustine, is God; and in spite of all that Augustine is unable to understand about God, His reality and nature are not in question, but are further affirmed as what is most truly real. When Augustine writes on time, what he really seems to be struggling to understand is the relation between (temporal) human beings and (eternal) God, the unshakeable foundation of his metaphysics. Thus, Augustine's meditations on time must be considered in conjunction with eternity; and from this perspective one should arguably conclude — as John Protevi does, and contrary to Ricœur — that 'the *distentio animi* has [in fact] been grounded', in 'a metaphysical order revealed in God's light' (1999: 86), the unity of which is asserted by Augustine's expressions of faith.

The question this chapter asks of Marcel, therefore, is whether his reflections on time are also an expression of a wider concern with the relation between humans and God; for if they were found to be bound up in such a religious context, Part II's Ricœurian interpretation of time in Marcel at least risks inadequacy. Marcel's own need to introduce a notion of eternity when he philosophizes about time, combined with the privilege he consistently gives to the eternal and the fact that he himself was a religious believer, certainly suggest such a reading to be possible. Indeed, as the first section will propose through its comparison of Marcel and Augustine, there is evidence to indicate that the framework within which Marcel understands the structure of time is Augustinian, and that God might be the eternity Marcel is favouring. The second section then compares Marcel's position with that of Plotinus before the Christianization of his Neoplatonic ideas, and asks whether Marcel might not also contest this Augustinian understanding of time. To what extent is it useful — or indeed legitimate — therefore, to situate his project within a specifically Christian context? What is the significance of 'God' in Marcel's philosophy? And what implications do Marcel's references to God have for interpreting and placing his discussions of time? These are the questions that guide the third section, in its attempt to make sense of the two equally possible, but seemingly contradictory, readings presented in the first and second sections.

Time and Eternity: Marcel and Augustine

As DeWeese informs us: 'Augustine profoundly influenced both theology and philosophy in the West for centuries. [...] his theological views of God's nature and the nature of eternity were unquestioned for nearly a millennium' (2004: 112).[5] According to Augustine, God's eternal being, as immutable and indivisible into parts (e.g. of time), is atemporal, and as such is (absolutely) beyond the existence of worldly beings. Given that Marcel also seems to draw a sharp contrast between temporal human worldliness, and the eternity of an 'au-delà' which in places is related to God, might he, too, be positioned within the Augustinian tradition?

Both Augustine and Marcel are alike in their presentation of time as ontologically subordinate to, and an epistemological distraction from, authentic eternal Being — which may be identified with God. Owing to the fact that the past is no longer, the future not yet, and the present only an ephemeral 'now', for Augustine temporal beings tend toward non-existence, and are fallible because of the way in which their understanding is scattered over these temporal fragments. Book XI of the *Confessions* therefore describes the misunderstanding that underpins various people's attempts 'to taste eternity when their heart is still flitting about in the realm where things change and have a past and future; it is still "vain" (Ps. 5: 10)'. 'They do not yet understand how things were made which came to be through you [God] and in you' (1998: 228, XI. xi.13), Augustine observes. As he explains in *The City of God* (413–26 CE), 'the world was not created *in* time but *with* time' (2003: 436, XI. 6). Time thus shapes the ontological and epistemological capacities of human beings, the reality of which is dependent on their (atemporal) eternal creator,[6] who, existing 'in the sublimity of an eternity which is always in the present' (1998: 230, XI. xiii.16), is ontologically and epistemologically constant, and therefore most fully real.[7] Humans on the other hand, as created beings, are unable completely to escape time's imperfection and finite constraints. They can, however, through devoted contemplation, resist the distraction of temporal succession and, in so doing, partake of eternal life insofar as their limited nature allows. It is this that Augustine's three-fold, synoptic present aims to encourage, for it recognizes the self as essentially unified, despite the perpetual risk of distraction by (the non-being of) the past or future. If the self can focus its attention and wholly dedicate itself to God, with God's grace it is possible to experience hints of Being's (eternal) plenitude.[8]

For Marcel, too, time — as temporal succession — tends toward non-existence, and as such constitutes an epistemological distraction. Thus Marcel also argues for a conversion of attention: away from a conception of Being that reduces the self to a linear chronology (*réflexion primaire*), toward a fuller, more recuperative engagement with Being (*réflexion seconde*) which affirms the self as unified and grounded in eternal presence. On 7 March 1929 Marcel writes: 'on commet une grave erreur en traitant le temps comme mode d'appréhension (car on est alors contraint de le considérer aussi comme l'ordre selon lequel le sujet s'appréhende lui-même, et ceci n'est possible qu'à condition que le sujet se distraie, si j'ose dire, de soi [...])'. Time, he therefore affirms, is 'la forme même de l'épreuve' (*EA*: I, 19), and he encourages us — in the 1943 lecture 'Valeur et immortalité' for example — to 'axer [...] notre vie sur l'au-delà' (*HV*: 213). Accordingly, when speaking of the *pente de l'existence* (see Chapter 2), Marcel describes the scope of human Being as ranging from object-like existence, which hardly 'is' at all,[9] to 'l'être dans son authenticité' (*ME II*: 30), which can never be objectified but where the affirmation 'je suis' may be

> murmurée sur un ton qui est à la fois celui de l'humilité, de la crainte et de l'émerveillement. De l'humilité, dis-je, car après tout cet être [...] ne peut que nous être accordé, c'est une grossière illusion de croire que je puis me le conférer à moi-même; de la crainte, car je ne puis même pas être tout à fait sûr qu'il ne soit pas, hélas, en mon pouvoir de me rendre indigne de ce don au point d'être condamné à le perdre si la grâce ne vient pas à mon aide; de l'émerveillement enfin, parce que ce don porte avec soi la lumière, parce qu'il *est* lumière. (*ME II*: 34)

Save the absence of an explicit reference to God, this account of Being seems to map almost exactly onto Augustine's, with its implication that Being has been bestowed upon the self by a greater power, and its reference to the grace and light that allow the self to become aware of the nature of its individual existence. On other occasions, Marcel does actually refer to human beings as 'créatures';[10] and crucially, in certain instances Marcel also suggests that authentic being depends on, and is a concurrent discovery of, God's transcendence. 'Le rapport à Dieu, la position de la transcendance divine permettent seuls de penser l'individualité' (*JM*: 86), writes Marcel in his diary on 20 February 1914; 'sans doute suis-je d'autant plus que Dieu est davantage pour moi' (*JM*: 206), he states on 20 October 1917; and in the closing remarks of his 1941 lecture 'Moi et autrui', Marcel cites the following from a text by Gustave Thibon, which he believes to 'traduire admirablement cette exigence d'incarnation à laquelle la personne ne peut se dérober sans trahir sa mission véritable':

> Tu te sens à l'étroit. Tu rêves d'évasion. Mais prends garde aux mirages. Pour t'évader, ne cours pas, ne te fuis pas: creuse plutôt cette place étroite qui t'est donnée: tu y trouveras Dieu et tout. Dieu ne flotte pas sur ton horizon, il dort dans ton épaisseur. La vanité court, l'amour creuse. Si tu fuis hors de toi-même, ta prison courra avec toi et se rétrécira au vent de ta course: si tu t'enfonces en toi-même, elle s'évasera en paradis. (*HV*: 35)

Furthermore, many of Marcel's references to the *exigence ontologique* or *assurance existentielle* he so often speaks of, support an Augustinian reading of his philosophy. As Erwin Straus and Michael Machado note:

> There is a close parallel between the movement of conversion, by which reflective thought passes from the objective order of existence to the transcendent order of being, and that by which the self opens into communion out of its prior community with the world. [...] Marcel discovered that the quest for being contained an implicit awareness of the exigence for God. Following the Augustinian tradition, perhaps unconsciously, Marcel viewed us 'as bound to God by a sort of fundamental and radical pre-awareness of God as the source of [our] being [...]'.[11] (1984: 139)

An analogous notion of personal identity thus appears to permeate the philosophies of Augustine and Marcel, as a result of very similar, first-person methods of reflection. Fumbling and always interrogative and self-critical, Marcel's and Augustine's autobiographically grounded meditations both come to recognize glimpses of a mysterious, yet indubitable source of (inner) knowledge,[12] which, at the deepest level of self-consciousness, is said to reveal a yearning oriented (outward and upward) toward fulfilment in God's plenitude.[13] 'Through my soul itself I will ascend to Him' (1998: 185, X. vii.11), pronounces Augustine in Book X of the *Confessions*. 'Nous n'appartenons pas entièrement à ce monde des choses auquel on entend nous assimiler' (*HH*: 23), writes Marcel in *Les Hommes contre l'Humain* (1951), who then affirmed, in his second 1961 William James Lecture ('Participation'), 'l'espèce d'assurance existentielle [...] qui me portait à reconnaître à la religion comme un mystérieux primat' (*DH*: 43). Further still, argues Marcel in 'Ébauche d'une philosophie concrète' (1940), 'une philosophie concrète ne peut pas ne pas être *aimantée* [...] par les données chrétiennes', for, he contends, 'le philosophe qui s'astreint à ne penser qu'en tant que philosophe, se place en deçà de l'expérience'. According to Marcel, however, 'la philosophie est une surélévation de l'expérience, elle n'est pas une castration' (*RI*: 109). Philosophy must engage with experience rather than treat it as irrelevant; and to engage properly with experience is to realize that 'chacun de nous, dans une part considérable de sa vie ou de son être, est encore inéveillé' (*HV*: 28) — that is, to recognize the reality of the *exigence ontologique*, which incites the self to transcend this dormant state and 'be' more authentically, by following its yearning for God.

As John Quinn reminds us, though, Augustine's references to time are in fact equivocal, meaning 'both psychic and moral distension', where 'the disintegration proper to moral time is the upshot of sin' (1999: 836).[14] Indeed, according to orthodox Christian belief, human beings are not only finite, as spatio-temporal beings, but also 'fallen', as a result of Original Sin.[15] This conception of human Being appears to be shared by Marcel, who writes in his diary (14 March 1943) that 'il n'y a peut-être pas de sens à parler de temps avant la chute. Le temps est relatif au monde — et peut-être n'y a-t-il un *monde* que par et après la chute' (*PI*: 118). Notions of responsibility and individual salvation thus emerge as simultaneous with (temporal) personal identity in Augustine and Marcel, for both present teleological conceptions of Being,[16] arguing for what Marcel terms 'une transcendance enveloppée dans le sentiment du péché' (*PI*: 110; 6 March 1943). As beings endowed with freedom, humans are not causally determined (hence the two philosophers' opposition to thinking human temporality in terms of linear succession), but are instead free to develop and better themselves. However, having inherited a propensity to sin, time and all its worldliness becomes, for us, '[une forme] de la tentation' (*EA*: I, 27; 23 March 1929),[17] and personal spiritual progress oriented toward God becomes a difficult, yet vital responsibility.

Indeed, as Desmond comments, 'Marcel was profoundly disturbed at the godlessness of western modernity' (1994: 116); and for Marcel this godlessness was, at least in part, at the root of the dehumanization he saw around him: 'avant tout, ce qui manque à ce monde, c'est la conscience de Dieu' (1955b: 40), he declares in a 1955 survey conducted by *L'Âge nouveau*.[18] Accordingly, it therefore seems, Marcel will often emphasize Christian aspects of his thought when differentiating his position from other philosophies, presenting non-religious conclusions as, in some sense, inadequate — especially when ethical questions are at stake. This is most evident with respect to Sartre, in response to whom Marcel has positively embraced the Christian label attributed to him in Sartre's 'L'Existentialisme est-il un humanisme?' (before later rejecting it), using this as a basis on which to question the authenticity of Sartre's own (atheistic) brand of existentialism. 'Il n'y a aucun sens à considérer Sartre, je ne dis pas comme fondateur — lui-même serait le premier à en convenir — mais seulement comme le chef de fil de l'existentialisme contemporain', states Marcel in the 1947 article 'Existentialisme et pensée chrétienne'. 'Sa position est en réalité marginale, et il serait facile de montrer que la métaphysique qu'il nous présente est presque irrésistiblement attirée par des conceptions qui se situent aux antipodes de l'existentialisme proprement dit' (1947a: 157). 'Pour une raison connexe', Marcel continues, 'il est absurde de s'étonner qu'un existentialisme puisse être chrétien'.[19] On the contrary, he argues: 'je suis convaincu qu'on serait à peu près dans la vérité en disant que l'existentialisme est en soi d'essence chrétienne, et qu'il ne peut devenir athée que par accident et en se méprenant sur sa propre nature' (1947a: 158). As Marcel states in *L'Existence et la liberté humaine chez Jean-Paul Sartre* (written 1946), in the case of existentialism proper — that is, 'si je m'interroge sincèrement et sans me référer à une philosophie préconçue' — 'je m'apparais non pas du tout comme choisissant mes valeurs, mais comme les reconnaissant' (1981: 86). For philosophers such as Sartre, however, Marcel explains in the late lecture 'Vérité et liberté' (published 1968), 'la liberté se confond avec ce qu'ils appellent le choix. Ils estiment d'ailleurs que notre condition nous oblige perpétuellement à choisir' (*PST*: 127). To define freedom only in relation to the self's projects, however, is, in Marcel's eyes, arrogant and can only encourage *indisponibilité*. Moreover, it actually fails to engage with concrete existence, because it ignores the *exigence ontologique* that Marcel feels lies at its very depth, and which communicates to us our metaphysical dependency on God.[20] 'Cette liberté que nous avons à défendre *in extremis*, [...] ce n'est pas la liberté d'un être qui serait ou qui prétendrait être *par soi*', insists Marcel in *Les Hommes contre l'Humain*. 'Je ne me suis pas lassé de répéter depuis des années, la liberté n'est rien [...] si, dans un esprit d'humilité absolue, elle ne reconnaît pas qu'elle est articulée à la grâce, et quand je dis la grâce, je ne prends pas ce mot dans je ne sais quelle acception abstraite et laïcisée, il s'agit bien de la grâce du Dieu vivant' (*HH*: 187). For Marcel, it therefore appears, situating human temporality in relation to God's eternity is crucial to establishing genuinely free individuals capable of engaging in positive human relations, the possibility of which Marcel feels Sartre's atheistic existentialism denies by centring freedom and values on the human self alone,[21] and refusing to acknowledge the broader community in which we all exist as created beings[22] — a community which, we must be humble enough to acknowledge, is ultimately dependent on God.

It is in this light that we can understand Marcel's recommendation, in his lecture 'Valeur et immortalité', that a Christian might 'venir en aide spirituellement à l'incroyant', in order to 'éveiller chez la conscience de ce qu'il est, [...] de sa filiation divine, lui apprendre à se reconnaître comme enfant de Dieu à travers l'amour qui lui est témoigné' (*HV*: 223). Such existential humility, as a requirement for intersubjective *disponibilité*, is necessary in order to 'be', in its most authentic sense. Furthermore — and as Marcel's last remark above hints — it is necessary to understand one's (temporal) existence in relation to God's eternity because, without such a relation to the eternal, love and fidelity to others would not be possible in the first place. Time's flux and finitude are unable to ground the permanent, absolute, or unconditional; we need, as it were, God's help. In an undated diary entry (c. 1930) Marcel writes:

> Il faut [...] admettre que la relation [d'une fidélité vouée à un être] implique quelque chose d'inaltérable; d'où partir pour le saisir? Nécessité de partir de l'*être même* — de l'engagement envers Dieu.
> Acte de transcendance avec contrepartie ontologique qui est la prise de Dieu sur moi. Et c'est par rapport à cette prise que ma liberté même s'ordonne et se définit. (*EA I*: 66)

Hence values, for Marcel, become (or at least have to be able to become) ideals — otherwise Marcel fears that they will simply collapse, without anything to ground the possibility of their unconditionality. 'Does an absolute commitment demand that one be committed to an Absolute Personal Being — to God?', Keen asks. As this diary entry suggests, and as Keen is himself led to conclude, 'Marcel's answer is that ultimately it does', for the truest love or fidelity demands unconditional loyalty, and such a vow 'can be fully articulated only where there is [also] religious faith' (1984: 113). Accordingly therefore, in the case of hope, Marcel declares in 'Esquisse d'une phénoménologie et d'une métaphysique de l'espérance': 'La vérité est [...] que j'ai conscience en espérant de renforcer [...] un certain lien [...]. Ce lien, de toute évidence, est d'essence religieuse' (*HV*: 65).

In order for human time to be truly meaningful, then — to be able (freely) to value, and to express one's values through hope for the future, or through faith in others — a metaphysics must, Marcel seems to be suggesting, be grounded in God's eternity; for as (finite and imperfect) temporal beings, our freedom to do so is not intelligible unless dependent upon God's will and grace. Marcel's analyses of human experience thus appear to be founded on a Christian — and in particular, Augustinian — understanding of time, which, this section has argued, equally enables us to account for the 'curious blend between metaphysics and morality' that Adams has commented on, and which has complicated this book's attempts to place Marcel's philosophical position. Does this mean that Sartre was right to define Marcel as a Christian philosopher of existence? Not necessarily, the next section will now argue, for there are other aspects of Marcel's philosophy that (at the very least) render this Augustinian reading problematic.

Challenging Augustinian Time

For all the parallels that have been suggested between Marcellian and Augustinian time, other aspects of Marcel's texts reveal his understanding of time to be much less Augustinian, calling into question whether such a reading of Marcel — although possible — is a legitimate way of interpreting Marcel's philosophical project as a whole. Whereas in Augustine, for example, communion with the eternal is deferred to the after-life, to 'that day when, purified and molten by the fire of your love, I flow together to merge into you' (1998: 244–45, XI. xxix.39), the 'autre monde' that Marcel speaks of is not absolutely separate from this one.[23] Rather, Marcel postulates a gradation between (temporal) *existence* and (eternal) *être*, which is more akin to Plotinus's (205–270 CE) understanding of time and eternity before his philosophy was Christianized (notably by Augustine).[24] As Philip Turetzky tells us, 'Plotinus and Neoplatonism had a profound and continuing influence on Augustine, even after his conversion to Christianity'; but 'despite this influence, the Christian outlook and doctrines break [...] with Neoplatonism' (1998: 57).[25] Particularly significant here is the break with Plotinus's doctrine of emanation, which attempts to account for time's relation to eternity[26] by means of a descending hierarchy of three 'hypostases' (the One, the Intellectual-Principle/*Nous*, the (World) Soul):[27] foundational metaphysical forms that express themselves by overflowing their own bounds, generating a subordinate substance of lesser reality and in so doing creating a gradation between the One's (original) eternity and worldly, spatio-temporal matter.[28] The break can be traced back to the first Christian council of bishops at Nicaea (325 CE), convened by Constantine I to address, among other matters, Arius of Alexandria's suggestion that Christ's finitude rendered him metaphysically inferior to God. Arius's position was overruled by the council, which decided instead to declare God the Father, Son, and Holy Spirit of one ontological substance (*homoousios*). Thus, as John Rist writes:

> after Nicaea, forms of Platonism which might look like (or be claimed to look like) the subordinationism of Arius were increasingly impossible for orthodox Christians. One traditional part of Platonism was thus excluded, and Christians who read Plotinus after Nicaea [...] now found it necessary to telescope the Plotinian hypostases of the One and *Nous*, making the Forms God's thoughts [..., or else] run[ning] the risk of being damned as Arians.[29] (1996: 395)

In this way, the Nicene Creed led to the decisive separation of God from time and the world;[30] and consequently, 'Augustine is less ready than Plotinus to talk of being *united* with God', for, Richard Sorabji notes, 'there was always a danger of blasphemy in the suggestion that a human might *become* God' (1983: 172). Marcel, on the other hand, refuses an absolute divide between God and the world which would prevent humans from sharing in his eternity:[31] 'Ma conviction la plus intime, la plus inébranlable — et si elle est hérétique tant pis pour l'orthodoxie — c'est [...] que Dieu ne veut nullement être aimé par nous *contre* le créé, mais glorifié à travers le créé et en partant de lui' (*EA*: I, 169), he writes in his diary on 4 March 1933.[32] As he declares as early as 27 January 1914: 'Je crois qu'il faut reconnaître ce qui est absolument vrai chez Plotin; l'idée que Dieu n'est véritablement pour nous qu'en tant que nous participons à lui. Seulement, il importe de transposer dans l'ordre de l'esprit, dans l'ordre subjectif, tout ce qui subsiste chez Plotin d'émanatisme objectif' (*JM*: 36).

This (subjective) emanationism avoided the difficulties Marcel confronted in his earliest philosophical writings (collected, for the most part, in the first half of the *Journal*), and can be seen directly to prefigure Marcel's (phenomenological) concrete philosophy of existence. Marcel's early philosophy sought, first and foremost, to render religious faith intelligible. He explains in his 'Testament philosophique' (1968): 'Le problème central qui se posait à moi [...] était de déterminer, peut-être vaudrait-il mieux dire, d'apprécier le type de réalité qui pouvait [...] être légitimement attribué, faut-il dire aux essences auxquelles se suspendait l'affirmation religieuse' (1969: 256). At this point in time, Marcel understands 'existence' in terms of the spatially and temporally manifest: 'il n'y a d'existence que de ce qui est objectif', he writes on 17 January 1914. This leads Marcel to dismiss all proofs of, or positive affimations concerning, the existence of God as extraneous: 'Dieu ne peut pas être traité comme un objet empirique et [...] par conséquent l'existence ne peut lui convenir' (*JM*: 33; 27 January 1914). However, testifying to the intelligibility of religious faith becomes highly problematic if God's existence cannot be affirmed on some level. 'L'affirmation portant sur Dieu [doit se définir]', Marcel therefore asserts, 'par un certain mode qui est incompatible avec la notion d'un existant, etc., autrement dit *la foi*' (*JM*: 39; 28 January 1914). At this stage, then, Marcel's philosophy remains strictly dualistic, owing to this separation of reason and faith;[33] conscious intellectualizing reflection can only falsify or negate the transcendent reality that Marcel wishes to affirm.[34] But as Charles Widmer asks: 'si la foi est à ce point transcendante, que devient sa relation à l'histoire [...]? Ne perd-elle pas son contenu concret [...]?' (1971: 53). Similar questions might also be asked of the relation between faith and the concrete experience of the individual who believes: 'Qu'est-ce que cet "individuel" sur lequel porte l'argumentation?', enquires Du Bos (1931: 141). Is not the significance of the here and now utterly dismissed by such absolutist subordination to the transcendent?

Yes, is the answer; and this is precisely what Marcel realizes during the First World War. As he writes in his diary on 17 March 1920: 'Il me semble bien aujourd'hui que je n'ai pas fait autrefois la part assez grande au réalisme dans mon effort pour déterminer le contenu réel de la vie religieuse' (*JM*: 231–32). Marcel thus announces his new 'métaphysique sensualiste' (*JM*: 305; 24 May 1923); and personal and interpersonal experience become increasingly important as his philosophy develops further. From the second part of the *Journal* onward — especially following the 1925 article 'Existence et objectivité' — faith and the empirical no longer mutually exclude one another. The meaning of 'existence' instead changes, so that it does not entail objectifying, spatio-temporal judgement. Now referred to as an 'indubitable' (*JM*: 312), 'existence', like God, is equally betrayed by intellectualizing reflection — with the crucial implication that it now does become possible to attribute existence to God, because existence is not one and the same as objectivity.[35] 'Grâce à la découverte d'un nouvel immédiat, celui de la sensation, du corps propre, de l'existence, Marcel [... s'oriente] vers une philosophie réaliste', reports Widmer. 'La relation à Dieu est [maintenant] conçue dans le prolongement de la communion humaine. Dieu est le Toi absolu' (1971: 13–14).

In Marcel's more established concrete philosophy, then, the distinction between 'existence' and 'être' is no longer so stark: there is a continuum between the two. What Marcel has come to learn from his deeper, phenomenological engagement with experience is, as he tells us in 'La Fidélité créatrice' (1940), that *'croire*, au sens fort[, ...] c'est toujours croire en un *toi*, c'est-à-dire en une réalité personnelle ou supra-personnelle susceptible d'être invoquée et comme placée au-delà de tout jugement portant sur une donnée objective quelconque' (*RI*: 220). For this reason, in Marcel's view, 'les attributs de Dieu tels que les définit la théologie rationnelle: la simplicité, l'inaltérabilité, etc., ne prennent [...] une valeur que si nous parvenons à y reconnaître les caractères d'un Toi' (*RI*: 53).[36] It is thus in an interpersonal sense, Marcel decides, that faith in God can be considered intelligible;[37] and if the human relation to God cannot be grasped in objectively determinable terms, this is only analogous to the difficulty of pinpointing what it is that founds a (genuine) relation of love between two human beings.[38] '[Il faut] faire appel ici à des expériences très humbles, très immédiates, que la philosophie a en général le grand tort soit de dédaigner parce qu'elles lui paraissent triviales' (*RI*: 14), insists Marcel in the introduction to *Du refus à l'invocation* (1940); 's'il est fait systématiquement abstraction de ces soubassements concrets, comme c'est le cas pour toute philosophie qui entend se constituer autour de la pensée en général, c'est-à-dire d'un esprit dépersonnalisé — du même coup la réalité religieuse devient inintelligible' (*RI*: 15).[39]

Thus, Marcel now clearly rejects the traditional Augustinian approach to God, which presents divine eternity as absolutely separate from temporal human existence. The human relation to God is instead personal, in continuum with interpersonal relations. What Marcel's philosophy draws attention to, it therefore seems, are the limitations of the traditional Christian model of time. For Marcel, the ontology of (unChristianized) Neoplatonism seems a more concrete and (therefore) philosophically satisfactory account — because the gradation it postulates allows human time to be related more substantially to eternity, and because this view does not completely denigrate the here and now, by making any and all Being dependent on the (eternal) reality of a separate — and thus abstract and impersonal — God.[40] Arguably, the core of Marcel's philosophy of time should be related not to Augustine, but rather to an earlier Plotinian model for which God is not important.

Hence, the extent to which a specifically Christian context is important for understanding Marcel's work — in particular, his discussions of time and eternity — is entirely debatable, as Marcel gives us good reason to question whether it is necessary to read his philosophy with God at all. Marcel continually insists that he has no interest in proclaiming the truth of established religious dogma, proving the existence of God, or articulating divine attributes. He detests theodicy and dogmatic theological discourse, asserting in one of his earliest notebooks (1913-14) that 'le problème de l'existence de Dieu [est un] problème absolument dépourvu de signification métaphysique, [qui] n'a pu se poser que pour un intellectualisme grossier' (*FP*: 94).[41] In fact, in the preface to *Le Seuil invisible* (1914), Marcel defines religion as 'la foi dans la valeur absolue de la vie, non pas la divinisation d'un phénomène naturel' (1914: 8); and this outlook does not seem to change, even after his conversion. Unlike Augustine, who does not question his faith in God but merely finds it difficult to understand and justify, Marcel's writings do not presuppose a commitment to Christianity: 'Il est certain qu'un adepte de la philosophie concrète, telle que je l'entends, n'est pas nécessairement un chrétien; on ne peut même pas dire à la rigueur qu'il soit engagé sur une voie qui, logiquement, devrait le conduire au christianisme' (*RI*: 108), he writes in 'Ébauche d'une philosophie concrète' (1940). Although, therefore, 'il est très possible que l'existence des données chrétiennes fondamentales soit requise *en fait* pour permettre à l'esprit de concevoir certaines des notions dont j'ai esquissé l'analyse [...], elles ne la supposent pas' (*HP*: 241), Marcel contends in the 1933 lecture 'Position et approches'.

Furthermore, faith in God is questioned by Marcel. On a personal level he remains somewhat uncertain about the authenticity of his own commitment to Catholicism, for he finds religious faith difficult, and this leads him occasionally to identify with non-believers more than with believers.[42] Marcel does not seem afraid to question God on a philosophical level either: 'Puis-je définir Dieu comme présence absolue?' (*EA*: I, 90), he asks, for example, in a diary entry dated 11 March 1931; and more radically, in response to the question (raised in the 1950 Gifford Lecture 'La Mort et l'Espérance') as to whether 'l'on peut dissocier radicalement la foi en un Dieu conçu dans sa sainteté, de toute affirmation portant sur la destinée de l'unité intersubjective formée par des êtres qui s'aiment' (*ME II*: 156), Marcel goes so far as to concede that 'il n'y a peut-être aucun sens à assigner un caractère *littéralement* supra-terrestre à cet invisible où la destinée inter-subjective est appelée à se poursuivre et à s'accomplir' (*ME II*: 158).

In this respect, then, and contrary to the case of Augustine's *Confessions*, a Ricœurian reading of time and eternity in Marcel appears entirely apt, for Marcel seems to want to keep his philosophical affirmations separate from theology:[43] 'je me propose de parler [...] en philosophe et non en théologien' (*PST*: 193), he states categorically in the lecture 'La Rencontre avec le mal' (published 1968); 'il n'est pas question [...] de confondre les mystères enveloppés dans l'expérience humaine en tant que telle — la connaissance, l'amour, par exemple — et les mystères révélés, comme l'Incarnation ou la Rédemption' (*HP*: 243), he asserts in 'Position et approches'. With respect to his philosophical project, therefore, one might conclude — as Marcel himself comments on 28 January 1914 — that 'le problème de Dieu fait place au problème de la foi qui est le problème véritable' (*JM*: 39). Experiences with faith-like structures such as love, hope, and fidelity, are clearly at the heart of his concerns, for it is to their reality that Marcel wishes to testify, in spite of their (logical) unverifiability.

In order for these forms of faith to be intelligible, Marcel argues that a notion of eternity is needed, in addition to an engagement with time. However, this section has called into question the suggestion that such eternity be identified with God; and yet, as Gillman notes, Marcel's writings still present us with two vocabularies: that of Being and that of God. 'What, then, is the relation between the two?' he asks. Why do we need both? (1980: 216)

Philosophy, Ethics, and Theology: The Meanings of Eternity in Marcel

Central to this final section is the question as to whether Marcel's two vocabularies, which — as Gillman (1980: 216) observes — appear to be referring to the same reality, must necessarily be understood as contradictory. Neither the first nor second sections' interpretations seem to be applicable to Marcel's philosophy as a whole; but does this require us to side with one over the other, or can the two perspectives in fact be read together? I will argue for the latter, and suggest that Marcel can therefore be seen as a forerunner of the 'new phenomenology' that has marked recent French thought with its (controversial) 'theological turn'. After first outlining why I believe this connection can be made and what holding such a philosophical position entails, I will compare Marcel with Lévinas in order to illustrate this reading in more detail, and further emphasize its implications for interpreting time and eternity. I will then draw attention to the inconsistency of Marcel's philosophy, before finally drawing conclusions about how, in light of this chapter's discussion, time and eternity might be understood in his work.

This new French philosophical trajectory, stimulated by Lévinas and Henry (1922-2002) and continued by thinkers such as Lacoste (1929-), Marion (1946-) and Chrétien (1952-),[44] 'is premised', Jeffrey Kosky observes, 'on the idea that philosophical discourse can admit and describe religious phenomena through the discovery of a new phenomenological principle or a new mode of phenomenality[: ...] a phenomenology that rejects the anterior condition of a horizon and that challenges the primacy of the I' (2000: 113).[45] In fact, such intentions have been Marcel's from the beginning. As explained in the second section, from his earliest work onward, Marcel has tried to understand religious faith on its own terms — that is, existentially, and independent of religious dogma: 'il s'agit pour moi d'arriver à déterminer la place et le sens du *je crois* dans l'économie métaphysique et spirituelle; et [...] je ne suis nullement tenu dans cette tentative de faire abstraction des données que me fournissent les religions existantes', he writes in 'La Fidélité créatrice'. What is the key to avoiding such abstraction? A phenomenological approach, Marcel decides. 'C'est à partir du *je crois* que se définit une fidélité, quelle qu'elle soit' (*RI*: 219), he insists.

Initially, of course, Marcel had no religious faith of his own to analyse (recall that his conversion was not until 23 March 1929); but for Marcel this was no reason to dismiss the faith of others. On the contrary, Marcel felt it was his duty to take faith seriously. As a philosopher genuinely concerned with investigating the reality of existence, he aspired to be able to place all aspects of human experience — faith included. 'La foi était une réalité: c'était mon devoir en tant que philosophe de m'appliquer à la penser, d'en découvrir les conditions d'intelligibilité' (1969: 256), he explained in his 'Testament philosophique' (1968).[46] And just as Marcel, here, was open to something beyond the horizon of his own 'I', so in general, he argues, must we reject the primacy of the self and its intentionality if we are to understand anything of Being — for Being transcends us, is a mystery; it cannot objectively be grasped like the answer to a definable problem. The implication of this, as Marcel acknowledges, is that nothing can be affirmed about the truth of faith or religion absolutely; but equally, this means that religious faith cannot simply be dismissed from philosophy on the grounds that it cannot be comprehended completely. In 'Position et approches' Marcel declares:

> la reconnaissance du mystère ontologique, où j'aperçois comme le réduit central de la métaphysique, [...] n'entraîne [...] aucunement l'adhésion à une religion déterminée, mais [...] permet cependant à celui qui s'est élevé jusqu'à elle d'entrevoir la possibilité d'une révélation tout autrement que ne pourrait le faire celui qui n'ayant pas dépassé les bornes du problématisable, reste en deçà du point où le mystère de l'être peut être aperçu et proclamé. (*HP*: 243–44)

Long before phenomenologists such as Henry, Marion, or Chrétien, therefore, Marcel was already seeking a way, through phenomenology, to admit the transcendent — and in particular the religious — into philosophical discussion.

Such a venture does not conform to 'traditional' phenomenology, however, and thus the recent, more widespread shift in French philosophy has roused strong opposition — especially from Dominique Janicaud. A brief consideration of Janicaud's contention underlines what is at stake for philosophy here. Responding first to Lévinas,[47] and then to Henry, Marion, and Chrétien, Janicaud complains that the bounds of phenomenology have been transgressed, because God has been introduced into a discipline where God does not belong. 'Phénoménologie et théologie font deux' (2009: 144, 149), Janicaud declares adamantly; but in the case of this new movement, he feels, 'la phénoménologie a été prise en otage [*sic*] par une théologie qui ne veut pas dire son nom' (2009: 74). Janicaud's critique stems from a Husserlian understanding of phenomenology, where phenomenology is conceived as a 'rigorous science',[48] and where God is thus excluded from the phenomenological reduction.[49] A question is therefore raised concerning what the phenomenological method is, and what it can hope to achieve. Janicaud objects to new phenomenology's lack of neutrality and its blurring of boundaries between phenomenology and theology, which he believes confounds the empirical with the ideal by blending immanence — the (allegedly) proper object of phenomenology — with transcendence. 'La phénoménologie n'est pas toute la philosophie. Elle n'a rien à gagner à une parade de ses mérites ni à une surévaluation de ses possibilités' (2009: 63), he maintains; 'il s'agit de savoir si l'on peut manipuler l'expérience ou s'il faut, au contraire, la décrire patiemment pour la connaître' (2009: 75). Another way of wording the question with which this section began, therefore, would be to ask whether immanence (which might be related to Marcel's vocabulary of Being, insofar as Being is temporal) and transcendence (identifiable with Marcel's references to eternity and to God) are necessarily mutually exclusive. Certainly, it has become traditional to see them as such,[50] and it is in defence of this tradition that Janicaud rejects new phenomenology. As Kosky advises, however, perhaps 'one should not let the historical construction of the philosophy of religion as a discipline lead one to assume that transcendence is absolutely foreign to it' (2000: 115).[51]

In the late lecture 'Philosophie, théologie négative, athéisme' (published 1968), Marcel in fact questions himself rhetorically about his philosophy's relation to theology: 'N'êtes-vous pas sorti du domaine de la philosophie pour pénétrer dans celui de la religion?' (*PST*: 269), he asks, before explaining in his defence: 'Cette recherche vise à déterminer les conditions dans lesquelles l'affirmation de Dieu peut intervenir, sans qu'ait été franchi le seuil de la conversion proprement dite, c'est-à-dire en deçà de l'adhésion à un crédo déterminé dans un certain contexte ecclésial spécifique' (*PST*: 270). Similar to the new phenomenologists, then, phenomenology's task is, for Marcel, to describe the full range of possible experience; and this entails challenging the assumption that phenomenological analysis coincides with intentional objectivity. As Lévinas states in 'Dieu et la philosophie' (1975), '[il faut] se demander [...] si le sens qui en philosophie est sens [la manifestation objective; la présence], n'est pas déjà une restriction du sens' (*DVI*: 96). For Marcel and the new phenomenologists, intelligibility means something more. All, in one way or another, invoke Marion's 'third reduction',[52] which brackets the horizon of objectivity or of Being that the phenomenologies of Husserl and Heidegger fail to question.[53] One must not define phenomenology in such a way as to preclude the possibility of alterity preceding intentionality, they contend.[54] Transcendence must be testified to, and this is why Marcel insists on introducing eternity into his phenomenological analyses in addition to time; why he emphasizes the 'mystery' of being, while nevertheless asserting that this can be approached, through his (phenomenological) *approches concrètes*.

With respect to religion, therefore, Marcel and the new phenomenologists demand that we ask what kind of god should be bracketed in the phenomenological reduction, instead of making absolutist decisions concerning any and all gods (or forms of transcendence) and the nature of their relation to the human from the start.[55] After all, (actual) human experience testifies to the possibility of religious faith as a mode of being, and Marcel and the new phenomenologists want to explore the potentialities of its existential significance.[56] 'Nous aurons à nous interroger sur les conditions dans lesquelles il nous est possible, en deçà d'une révélation proprement dite, d'énoncer quelque affirmation sur ce qu'est Dieu ou du moins sur ce qu'il n'est pas ou ne peut pas être' (*ME II*: 8),[57] states Marcel in his first 1950 Gifford Lecture, 'Qu'est-ce que l'être?'; and for such a task, he asserts on 26 March 1943, 'il faudra faire intervenir la notion d'un indubitable existentiel par opposition à l'indubitable objectif' (*PI*: 124).[58] The god all agree should be excluded, therefore, is the god targeted by Heidegger in his critique of 'onto-theology':[59] a god who is appropriated solely for philosophical purposes, used to explain away difficult problems and ground eternally the (unquestioned) unity and transparency of metaphysical accounts — and indeed theological accounts, to the extent that theology draws upon philosophy for its own justification.[60] 'Dès que Dieu est traité comme un Cela métaphysique sur lequel on fait porter des jugements qui sont censés devoir concorder les uns avec les autres, c'est le chaos', admits Marcel in his *Journal* on 30 November 1920. 'Mais faut-il que Dieu soit Cela ou ne soit rien?' (*JM*: 254).[61] Such an either/or decision rests on the assumption that immanence and transcendence — and by extension reason and faith, time and eternity — are diametrically opposed.[62] However, if human subjectivity is taken to be defined by transcendence (signified by eternity) as well as (temporal) immanence, the possibility of an intersection between reason and faith, and therefore between phenomenology and theology, is opened up.[63] It may in fact be possible for Marcel's vocabularies of Being and God to coexist without antagonism.[64]

For such a redefinition of the subject to be philosophically productive, however, the irreducibility of the transcendent must be preserved — on an interdisciplinary level, where a distinction must be maintained between the two fields in order to prevent phenomenology from subsuming theology, as well as within phenomenological analysis itself, where the absolute nature of religious (and, in general, transcendent) phenomena must not be compromised by the self and its intentionality. With respect to the former, as J. Aaron Simmons observes, the new phenomenologists tend to '[insist] that what they are doing is philosophy and not theology' (2008: 916), because they are raising 'the "religious" [...] more as a possibility for thought than as an actuality to be affirmed (even though some of [the philosophers] *personally* do)' (2008: 917). This applies equally to Marcel, who, in a 1949 Gifford Lecture ('Réflexion primaire et réflexion seconde') when discussing the 'repère existentiel' that is so central to his philosophy, declares:

cette recherche est phénoménologique et non pas ontologique. Je veux dire que je ne me demande pas du tout ici si dans l'ordre de l'être il y a un existant absolu qui ne pourrait être que Dieu et qui conférerait lui-même l'existence [...]. Je pars de moi-même en tant que je prononce des jugements d'existence, et je me demande s'il y a un sens par rapport auxquels ces jugements se disposent ou s'organisent. (*ME I*: 103)

In general, therefore, Marcel's conclusions are often qualified with phrases such as 'en se plaçant bien entendu en deçà de la religion proprement dite' (*RI*: 14), 'en deçà de toute spécification théologique' (*ME II*: 171), or 'sans que celle-ci soit forcément référée à un absolu de caractère métaphysique ou religieux' (1954a: 92),[65] for Marcel explicitly insists that it is not for him to use his philosophy to defend his own personal beliefs.[66] In the first of his 1950 Gifford Lectures ('Qu'est-ce que l'être?') he states: 'nous n'avons sûrement pas à poser en principe et dès l'abord que l'être en tant qu'être [...] se confonde nécessairement avec ce que la conscience croyante ou l'âme croyante désigne sous le nom de Dieu';[67] 'c'est le témoignage de la conscience croyante qui peut seul décider ce qui peut ou non être regardé comme Dieu. Je poserai en principe [...] qu'il n'est au pouvoir d'aucune philosophie [...] de procéder à un coup d'état instaurant comme Dieu quelque chose que la conscience croyante refuse de reconnaître comme tel' (*ME II*: 8). Thus, in the same Gifford Lecture, Marcel asserts that 'l'exigence de Dieu n'est autre que l'exigence de transcendance découvrant son visage authentique' (*ME II*: 7)[68] — transcendence which, as Marcel maintains in *Homo viator* (1944), is to be interpreted 'au sens précis et séculaire de ce mot' (*HV*: 7); he will not cross over into theology and draw on religious sources of authority in order to 'prove' his arguments.[69]

In this way, then, it is true to say (as the second section suggested) that God is not essential to the transcendent eternity Marcel is arguing for.[70] And yet at the same time, God cannot be taken out of Marcel's philosophy — for the very same reason in fact, because to exclude God completely would simply be to affirm the other pole of the binary that Marcel wants to oppose, a binary which equates the immanent and temporal with the rational, and separates these absolutely from transcendence, eternity, and faith.[71] Rather, I believe, God and religion in Marcel's philosophy should be interpreted in a way parallel to Merold Westphal's understanding of the religious in Lévinas — namely in Kierkegaardian terms, as 'teleologically suspended' by the ethical. This is not a reduction, as Westphal stresses. Ethics does not subsume religion, and render 'God' simply another word for the Other's ungraspable transcendence. Rather, religion 'is negated in its claim to autonomy, to self-sufficiency and completeness'; but it is nevertheless 'affirmed in relation to [...] that which draws it into a larger whole of which it is not the first principle' (1995: 153) — that is, ethics.

Indeed, as was explained in Chapter 4, philosophy, as Lévinas conceives it, is not ontology in the first instance but ethics. This is because Lévinas wants, primarily, to '[rechercher] la signification de l'au-delà, de la transcendance' (*DVI*: 114, note 15) — but, he maintains, transcendence (including that of God) cannot signify so long as ontology's presentifying discourse dominates.[72] Like Marcel, therefore, Lévinas is not interested in proofs for the existence or non-existence of God, which attribute an onto-theological sense of eternity to His being and thereby commit, as Westphal phrases it, 'the fallacy of misplaced concreteness' (2001: 17). 'Ce qui est recherché ici', writes Lévinas in the foreword to *De Dieu qui vient à l'idée* (1982), 'c'est la *concrétude phénoménologique* dans laquelle cette signification pourrait signifier ou signifie, même si elle tranche sur toute phénoménalité' (*DVI*: 7) — the idea of *trancher sur toute phénoménalité* being key, for the metaphysics of presence can only be broken if philosophy allows phenomena to be given without being entirely present, that is, for the possible disruption of my time by an Other.[73] 'L'éthique n'est pas un moment de l'être — il est autrement et mieux que l'être, la possibilité même de l'au-delà' (*DVI*: 114), declares Lévinas in 'Dieu et la philosophie'.

Because of this desire to ground philosophy in concrete phenomenal experience, however, if, as Lévinas states in *Autrement qu'être*, 'le problème de la transcendance et de Dieu et le problème de la subjectivité irréductible à l'essence — irréductible à l'immanence essentielle — vont ensemble' (*AE*: 33),[74] God is not 'le "premier autrui" ou l'"absolument autrui"'. Rather, he explains in 'Dieu et la philosophie', God is 'autre d'altérité préalable à l'altérité d'autrui' (*DVI*: 115) — which Lévinas terms 'illéité', and defines in *Autrement qu'être* as 'la trace laissée par l'Infini [...] à partir du visage [de l'Autre]' (*AE*: 27). In other words, God is always (first) approached via the ethical relation to the Other.[75] 'La métaphysique se joue [...] dans nos rapports avec les hommes', he insists in *Totalité et infini*;

> il ne peut y avoir, séparée de la relation avec les hommes, aucune 'connaissance' de Dieu. Autrui est le lieu même de la vérité métaphysique et indispensable à mon rapport avec Dieu. [...] Autrui n'est pas l'incarnation de Dieu, mais précisément par son visage, [...] la manifestation de la hauteur où Dieu se révèle.[76] (*TI*: 77)

Thus, as Westphal observes, the ethical relation is the 'horizon within which the true God is truly revealed' (1995: 156); the ethical relation teleologically suspends the religious.

In Marcel too, as the second section of this chapter observed, the human relation to God's transcendence is conceived as an extension of interpersonal relations — that is, as a relation which can only ever be approached via the ethical relation to the other.[77] In fact, this is commented on by Lévinas in *Hors sujet* (1987), who seems to identify with Marcel's project in this respect: 'M. Gabriel Marcel a cité sur la relation entre Dieu et l'homme les textes les plus favorables, les plus profonds. Mais j'ai tout de même l'impression qu'il s'agit là d'une transposition de la relation avec le prochain' (HS: 30). Furthermore, as Chapter 4 suggested, the transcendence that eternity represents in Marcel seems, in the first instance, to be ethical (and not religious). As Gallagher (1995) rightly observes, Marcel's references to human 'creaturehood' — that is, to human Being in specific relation to God — need to be appreciated in the context of Marcel's socio-political thought, as responding to the devaluation of life that Marcel senses around him (le monde cassé), rather than as something which compromises Marcel's usual position as mediator between believers and non-believers. Indeed, quite frequently, Marcel himself suggests that if he employs religious terms, these are 'emprunté[s] au langage religieux' (HV: 29) and are only illustrative,[78] and thus he often restates his (invariably ethical) point using more secular vocabulary, to make it clear that specific testimony to God's eternity is not his intention. In Les Hommes contre l'Humain, for example, Marcel concedes that '[mon] usage du terme péché dans un registre philosophique et non théologique pourrait bien éveiller certaines objections'. After all: 'Le péché n'est-il pas essentiellement la rébellion de la créature contre son Créateur [...]?' 'Mais [...] on devra reconnaître', Marcel continues, 'que les incroyants eux-mêmes, en présence des abus, des horreurs systématiques que nous avons vu se généraliser depuis trente ans, ont pris de plus en plus conscience de l'indice de péché qui affecte de semblables monstruosités' (HH: 61). It is this 'émoi universel' that Marcel is referring to, 'l'affleurement d'un sentiment de piété devant la vie' (HH: 62). Similarly, the 'désacralisation' of which Marcel speaks in the late lecture 'La Vie et le Sacré' (published 1968) is primarily a reaction against the devaluation of life, as opposed to an affirmation of anything specifically religious (PST: 151–74).[79] As Marcel laments in his penultimate 1950 Gifford Lecture ('La Mort et l'Espérance'): 'La vie terrestre est apparue de plus en plus généralement comme une espèce de phénomène sans valeur, sans justification intrinsèque' (ME II: 150). Hence, 'il s'agit aujourd'hui pour l'homme de retrouver le sens de l'éternel' (ME II: 166); and what is most important in this respect, declares Marcel in his concluding lecture, 'consiste [...] à affirmer philosophiquement, c'est-à-dire en deçà de toute spécification théologique, l'indissolubilité de l'espérance, de la foi et de la charité' (ME II: 171).

However, the indissolubility of hope and faith, as well as fidelity and love, in Marcel's philosophy stems from their eschatological temporal structure,[80] that is, their unconditional investment in the promise of an unverifiable future with and for others.[81] It is here that a possible, but not inevitable,[82] intersection with the theological opens out, where, for some, God can become existentially significant[83] — though Marcel does not wish to dictate the theological content of such an experience, because his task is only to think this possibility. In the same way as with the Lévinassian ethical relation, which, Newman explains, 'is not simply the approach of another person, but the approach of what makes the

Other an *Other* in the sense of transcendence [*illéité*]' (2000: 121), a notion of the eternal or infinite is necessary in Marcel in order to maintain this ethical openness, and prevent the self from de-temporalizing the relation through objectifying primary reflection. As Lévinas writes in *Totalité et infini*: 'La transcendance est temps et va vers Autrui. Mais Autrui n'est pas terme' (*TI*: 302); but consequently, as Marcel recognizes in a 1970 interview with Boutang, 'la démarcation entre l'éthique et le religieux n'est pas toujours précisable' (*GM*: 93). Phenomenological experience is rendered ambiguous and undecidable.[84]

This uncertainty is, theoretically, the aim. 'Une philosophie de la vie est [...] par son essence même vouée à l'ambiguïté' (*EA*: II, 39), acknowledges Marcel in the 1933 lecture 'Remarques sur l'irréligion contemporaine'; 'cet indubitable [existentiel] est sujet à éclipses [...] justement parce que [cette forme de certitude] n'est pas possédée' (*PI*: 125), he re-affirms on 26 March 1943.[85] And indeed, indeterminacy is generally the message the reader receives as concerns the place of God and religion in Marcel's work: Marcel, like Lévinas, employs a deeply religious vocabulary while refusing to describe God as a distinct reality; what is meant by these terms is always in question, frustrating the reader's desire for interpretative closure. However, Marcel seems ultimately unable to tolerate such ambiguity, one of the most prominent manifestations of this discomfort being his unease concerning the hazy relation between *existence* and *être*. 'Cette question du rapport de l'être et de l'existence [...] m'a toujours occupé et je puis dire toujours troublé' (*ME II*: 25), he confesses in his second 1950 Gifford Lecture ('Existence et être'), despite recognizing, in his reply to a 1968 article by Keen, that 'to claim that existence is different from being [...] would be to objectify and even to materialize concepts, something I have always opposed' (Schilpp and Hahn 1984: 122). The truth is that, in spite of himself, Marcel still wants to be able to speak of something objective, because otherwise his philosophy has nothing definitive to say. He worries 'que je me fasse l'effet d'aller à la dérive', for as he confesses on 27 March 1943: 'je voudrais m'assurer d'abord que cette recherche a une utilité pour mon travail sur la certitude métaphysique' (*PI*: 126). Marcel wants to discover something certain, and he wants to be able to communicate this to others:[86]

> en aucun cas je ne pourrai me satisfaire si je n'ai pas conscience [...] de satisfaire autrui, c'est-à-dire de donner. Chercher, ce ne peut être que chercher pour donner. Mais donner quoi? C'est ici qu'il faut reprendre la notion de certitude.
>
> Il me semble que cette certitude, je cherche d'abord à me la donner à moi-même; et que je tendrai, du même coup, à la communiquer à autrui.[87]
> (*PI*: 127)

The date of this (late-war) diary entry hints at a reason for Marcel's frustration. As is evident from all of his writings, Marcel clearly believes that unconditional love and faith are possible; but with a distinct lack of either around him, he seems to need reassurance that the social situation can be changed. He thus wants his philosophy to be able to say something that can help to encourage a more hopeful outlook — 'une volonté d'inconditionalité' (*HV*: 185) as he refers to in the 1942 article 'Obéissance et fidélité'. 'À l'heure présente, il n'y a probablement pas pour un philosophe conscient de ses responsabilités en même temps que des dangers qui menacent notre planète, de tâche plus imprescriptible que celle qui consiste à retrouver ces assurances existentielles fondamentales, constitutives de l'être humain véritable en tant qu'*Image de Dieu*' (*PST*: 75), he declared in the lecture 'L'Humanisme authentique et ses présupposés existentiels' (published 1968). So just as Lévinas attempts, in *Totalité et infini*, to provide an absolute basis for ethical life through a definitive account of the encounter with the Other's face (*visage*), Marcel seeks to establish an absolute metaphysical grounding for the real possibility of unyielding love and faith.[88] 'En dernière analyse, c'est par rapport à cette constitution d'un organisme spirituel sans doute, mais charnellement enraciné dans l'éternité de Dieu, et par rapport à elle seule, que peut se définir le vœu créateur en tant qu'y prend corps une fidélité elle-même créatrice' (*HV*: 169), he asserts in the 1943 lecture 'Le Vœu créateur comme essence de la paternité'. However, slipping into this mode betrays the theoretical principles underlying Marcel's concrete philosophy, hijacking his phenomenology by lapsing into the onto-theological dogmatism that Janicaud targets in his protests.[89] Marcel may not want to argue for a commitment to religion through his references to something eternal, but in practice his philosophical use of God does sometimes suggest this — hence the possibility of the first section's Augustinian reading of Marcel and conclusions such as Keen's, which assert that 'behind every approach to being and every ethical value [in Marcel] lies the hidden presence of God' (1984: 115). This more rigidifying use of (God's) eternity belittles time in Marcel's work. If Marcel can be linked to the recent theological turn in French thought, he is not quite a 'new' phenomenologist.

In places, one also senses that Marcel might be speaking more as himself than as a philosopher. Granted, one of Marcel's major contentions is that to philosophize about the human is always personal. The existential philosopher must take his or her individual experience seriously, he argues, because life is never lived from a third-person perspective. However, such an approach is difficult, because it then becomes hard to decide how personal philosophical engagement can justifiably be; and arguably, Marcel sometimes oversteps the mark and speaks on behalf of himself personally — referring perhaps (?) to himself, as Christian, when he writes phrases such as 'moi créature, jetée en ce tumulte' (*HV*: 100) — rather than as a philosopher who wants to take personal experience into account. Of course, the 'I' (or 'me') in Marcel's philosophy can, in many cases, be read as conforming to the conventional philosophical 'I', representative not of Marcel in particular but of any other,[90] in whose position the reader is encouraged to imagine themselves. Yet in order to lay the foundations for such empathy, this 'I' must, Marcel believes, be grounded in a living subject if it is to appear as a subject at all. Given Marcel's frequent appeals to autobiography, and the winding, almost associationist character of his philosophy, one therefore feels that this hypothetical philosophical 'I' is also (to varying extents of course) elided with Marcel's 'I', making it difficult to determine who he is speaking as.

Since Marcel tells us, on a number of occasions, that he is personally convinced that there is '[un] accord entre l'exigence philosophique [...] et l'affirmation chrétienne' (*HH*: 92),[91] the rhetoric used in Marcel's essays and lectures — especially the tone on which they end — therefore invites scepticism concerning the rigour of his arguments, for Marcel often makes extensive use of religious metaphors and similes which, in effect, give religion the last word. To cite but a few examples, Marcel's 1943 lecture, 'Le Vœu créateur comme essence de la paternité', ends by describing the essence of paternity as 'l'anticipation frémissante d'une plénitude, d'un plérôme au sein duquel la vie, cessant de s'improviser comme inépuisable et décevante variation sur quelques thèmes donnés, se ressaisit, se concentre, se rassemble autour de la Personne absolue qui, seule, peut lui apposer le sceau infrangible de l'unité' (*HV*: 170). Marcel also speaks frequently of saintliness, which he seems to take for granted as part of a natural progression; but since this is a specifically religious notion, this, again, effectively fails to question God.[92] And in the final lecture of his Gifford series (1950), Marcel concludes with the following remarks:

> à partir du moment où nous rendons nous-mêmes perméables à ces infiltrations de l'invisible, nous qui n'étions peut-être au départ que des solistes [...] tendons à devenir peu à peu les membres fraternels et émerveillés d'un orchestre où ceux que nous appelons indécemment les morts sont sans doute bien plus près que nous de Celui dont il ne faut peut-être pas dire qu'il conduit la symphonie mais qu'il *est* la symphonie dans son unité profonde et intelligible, une unité à laquelle nous ne pouvons espérer accéder qu'insensiblement à travers des épreuves individuelles dont l'ensemble, imprévisible pour chacun de nous, est pourtant inséparable de sa vocation propre. (*ME II*: 188)

Here, Marcel does then maintain that 'tout ceci [...] demeure en-deçà de la révélation proprement dite et du dogme', and that this is (only) 'une voie d'approche'. However, these comments feel rather token, for his closing lines do not give any indication as to what other valid approaches might be, but continue to argue that

> c'est en cheminant, tels des pèlerins, sur ce chemin difficile et semé d'obstacles que nous avons l'espoir de voir briller un jour cette lumière

éternelle dont un reflet n'a cessé de nous éclairer depuis que nous sommes au monde, cette lumière sans laquelle nous pouvons être assurés que jamais nous ne nous serions mis en route. (*ME II*: 188)

The *exigence ontologique* in Marcel, then, although intended to be interpreted as the immanence-transcendence now advocated by the new phenomenologists, is in places more than just structurally comparable to Augustine's yearning. Time in Marcel's philosophy is at times very Augustinian; God is not always as indefinite as Marcel likes to declare.

This over-personalizing may also explain Marcel's excessive response to Sartre's atheism, as well as Marcel's manipulative re-writing of Bergson when he reviews Bergson's work. Marcel often defines Bergson's philosophy as failing to affirm a spiritual dimension to existence, praising it only when the possibility of identifying Bergson's arguments with a more religious position is not ruled out. He is particularly suspicious of Bergson's (reductive?) engagement with biology. In the 1929 article 'Note sur les limites du spiritualisme bergsonien' he writes:

> Je crois à présent que la structure même de sa philosophie exclut jusqu'à la possibilité d'une solution quant aux questions globales que je me permettais de poser. Cette philosophie s'est en effet développée comme une série d'enquêtes, liées entre elles assurément, mais portant sur des objets strictement déterminés. Mais des investigations concernant l'essence de la réalité spirituelle en elle-même et le sens de notre destinée pouvaient-elles s'engager sur un pareil terrain, dans le prolongement de ces recherches? Je suis persuadé du contraire.[93] (1929b: 267)

And in another 1929 article, 'Carence de spiritualité', Marcel makes an open appeal to Bergson to draw the spiritual conclusions Marcel feels are befitting of his attempt to engage with the concrete. Despairing at the contemporary philosophical situation in France, where he sees only presumptuousness and arrogance resulting from a widespread obsession with logic, Marcel identifies *le bergsonisme* as an exception, stating:

> le concret et le spirituel ne pourront être récupérés qu'ensemble, M. Bergson l'a toujours su, l'a toujours affirmé, mais il faut bien hélas! le constater, son enseignement est demeuré jusqu'ici chez nous sans fécondité véritable, et peut-être ce fait tient-il pour une part à ce que sur le problème religieux il ne s'est jamais expliqué. (1929a: 378)

Marcel thus regards the publication of *Les Deux Sources* as 'un événement philosophique très important', for, he asserts, 'le livre vient de compléter et en quelque façon boucler une doctrine qui jusqu'à présent demeurait inachevée' (1932a: 558), because '[Bergson] donne pour la première fois une adhésion explicite non point au christianisme [...] mais à certaines affirmations centrales des grands mystiques chrétiens' (1932b: 415). 'Il ne s'agit certes pas, pour nous catholiques, d'annexer le bergsonisme: ce serait une sottise et une indiscrétion', maintains Marcel in his 1939 article 'Qu'est-ce que le bergsonisme?'. Yet this is precisely what he does appear to do: 'M. Bergson est sans nul doute aux lisières de la foi chrétienne, et peut-être même est-ce trop peu dire', he writes several lines later, before subsequently equating Bergson's notion of open morality (in *Les Deux Sources*) with 'la morale de l'Évangile' (1939: 5). As some of the journals Marcel is writing for are Catholic, however ('Note sur les limites du spiritualisme bergsonien' was published in *La Vie intellectuelle* and 'Qu'est-ce que le bergsonisme?' appeared in *Temps présent*), it is possible that Marcel is not speaking as a philosopher in these instances, but more as a Catholic intellectual.

Marcel does sometimes acknowledge that he can slip out of a philosophical mode, criticizing himself retrospectively in the late lecture 'Passion et sagesse dans le contexte de la philosophie existentielle' (published 1968), for example, for having offered, in his 1930 lecture 'Remarques sur l'irréligion contemporaine', 'une perspective qui me paraît aujourd'hui trop strictement religieuse', and explaining that 'ceci s'explique par le fait que je venais d'adhérer au catholicisme' (*PST*: 282). As he is willing to admit that his personal circumstances and adherences can exert an unwarranted influence on his philosophical discourse, we can at least, therefore, conclude that Marcel does not intend this to happen. His philosophy, though at times difficult to delineate, is evidently in transition, shifting away from onto-theology toward the postmodernism of new phenomenology.

Eternity thus has a number of different meanings in Marcel's philosophy, and these have varying implications for his presentation of time. It is sometimes onto-theological, for, despite himself, Marcel seems to need objective certainty. Here Marcel's philosophy conforms to an idealist metaphysics of presence, which marginalizes the significance of time's movement and change. When Marcel's philosophy is coloured by his personal faith in God or his Catholic allegiances, however, eternity takes on a theological guise, recasting time as Augustinian. Again, the reality of time is subordinated to a more authentic immutable presence. In general, though, time and eternity are not presented as so hierarchically distinct. Marcel's discussion of time with eternity stems primarily from the desire to recognize transcendence itself, the significance of which is recovered through temporal phenomenological analysis, not beyond it; for, as linear organized time is rejected as a model for experience, human time becomes less distinguishable from eternity — an eternity that is not atemporal but is rather bound up with time, as the transcendent otherness that shatters the self-sufficient identity of immanence. This figuration of subjectivity therefore opens the subject up to the other from the start, rendering it immediately ethical. Thus, in the same way as Kosky argues with respect to Lévinas, subjectivity in Marcel becomes construed as responsibility rather than as the pure consciousness one finds in Husserl; and as a consequence, a broader range of phenomena is allowed to appear — including religious phenomena — since, through its interplay with an ethically charged notion of eternity, responsible time is structured eschatologically. It is here that it is possible for God to become existentially significant. It is here that phenomenological eternity becomes possibly theological.

Notes to Chapter 5

1. And this, in itself, has been questioned — by Wetzel (1995) for example, who feels that it is misguided to look for a philosophical treatment of time in the *Confessions*, since this is not Augustine's preoccupation. Rather, Wetzel claims, Augustine's text confesses, to God, his inability to escape sin, while concurrently lamenting his failure to recollect sinful designs on his part.
2. 'Augustine's views on time [...] are motivated in the context of his meditations on the doctrine of creation [...]. [...] if Augustine is not always as clear as we could wish in propounding his philosophy of time, we should remember that this is not his primary interest' (DeWeese 2004: 111).
3. *TR I*: 58, note 2.
4. *TR I*: 54.
5. Boethius (480–524), Anselm (1033–1109), and Aquinas (1225–1275) were among the medieval philosophical theologians who continued this Augustinian tradition. For contemporary defences of God's atemporality, see for example Stump and Kretzmann (1981), Helm (1988), and Leftow (1991).
6. 'It is not in time that you precede times. Otherwise you would not precede all times. [...] Your "years" neither go nor come. [...] You created all times and you exist before all times' (Augustine 1998: 230, XI. xiii.16).
7. 'Who can lay hold on the heart and give it fixity, so that for some little moment it may be stable [...]? [...] In the eternal, nothing is transient, but the whole is present. But no time is wholly present' (Augustine 1998: 228, XI. xi.13).
8. In Augustine's philosophy — as for Aquinas, Cary informs us — 'seeing the essence of God belongs to the created intellect by grace and not by nature' (2000: 70; Cary is citing *Summa Theologica*, I, 12.4).
9. 'Pour autant que nous visons l'existence de la chose en tant que chose, cette existence qui est déjà comme tout obscurcie par la menace du *n'exister plus*, il ne faut certainement pas dire que cette existence-là est du non-être, ce qui à mon sens ne veut rien dire du tout, mais plutôt qu'elle est *à peine* de l'être' (*ME II*: 29–30).
10. *EA*: I, 220; *FP*: 70–71; *HH*: 195; *HV*: 29, 63, 100; *JM*: 86; *RI*: 54; *ME I*: 75; *ME II*: 20.
11. Gallagher (1995: 54) and Keen (1984: 107–08) refer to Marcel's thought as Augustinian for the same reason.
12. For discussion of the 'inward turn' advocated by Augustine, see Cary (2000).
13. For this reason, Harper, who compares the experience of human time to our response when reading a familiar tale such as *The Sleeping Beauty*, identifies especially with Marcel's philosophy of existence. 'We respond delightedly to such a tale as that of *The Sleeping Beauty* because it is in some way familiar to us. Not only is the longing familiar, the fulfilment is also', he writes. 'We cannot long for something we do not know; we know only what is in some way already experienced. [...] Fulfilment is in some sense a

return' (1955: 14), Harper continues — and thus, he explains, 'there is no more suitable approach to this story than by way of the metaphysical analysis of Gabriel Marcel' (1955: 29). Time, in Augustine, is structured in exactly the same way; and this is exemplified in the temporality of Augustine's writings themselves, for Augustine's confessions are of things that God is presumed already to know. In Book X, for example, he writes: 'you hear nothing true from my lips which you have not first told me' (1998: 179, X. ii.2); and in Book XI: 'Lord, eternity is yours, so you cannot be ignorant of what I tell you' (1998: 221, XI. i.1).

14. 'I intend to remind myself of my past foulness and carnal corruptions, not because I love them but so that I may love you, my God. [...] The recalling of my wicked ways is bitter in my memory, but I do it so that you may be sweet to me [...]. You gathered me together from the state of disintegration in which I had been fruitlessly divided. I turned from unity in you to be lost in multiplicity' (Augustine 1998: 24, II. i.1).

15. The principal scriptural reference to the Fall is found in the writings of Saint Paul. See especially Romans 5. 12, 18–19. Marcel's understanding of Original Sin might actually be more closely identified with Paul than with Augustine. Paul does not believe that we are born individually condemned, but rather that the world itself has inherited a structural fault. Similarly, Marcel argues that 'la structure même de notre monde nous la [la trahison de l'être] recommande' (*EA*: I, 148; 23 December 1932), and speaks of the collective nature of sin rather than of something which can be attributed to anyone in particular (*PI*: 115–17, 120–21). Augustine, on the other hand, emphasizes personal guilt (including his own, in the *Confessions*). In *Of True Religion* (*De vera religione*; 390 CE) he declares: 'sin is so much a voluntary evil that it is not sin at all unless it is voluntary. [...] if it is not by the exercise of the will that we do wrong, no one at all is to be censured or warned' (1953: 238, xiv.27).

16. Cf. the above reference Marcel makes to our human 'mission'.

17. Cf. Augustine: 'Let these transient things be the ground on which my soul praises you (Ps. 145: 2), "God, creator of all". But let it not become stuck in them and glued to them with love through the physical senses. For these things pass along the path of things that move towards non-existence. They rend the soul with pestilential desires [...]. But in these things there is no point of rest: they lack permanence' (1998: 62, IV. x.15).

18. '[S]i l'on peut dire que la mort de Dieu au sens nietzschéen a précédé et rendu possible l'agonie de l'homme à laquelle nous assistons — il reste légitime en un certain sens d'affirmer que c'est des cendres de l'homme que Dieu peut et doit ressusciter' (*HV*: 219).

19. Sartre does not, in fact; but if he is considered to be *le chef de l'existentialisme*, this raises questions concerning the compatibility of 'Christian' varieties with Sartre's atheistic position.

20. In *Les Hommes contre l'Humain*, Marcel asserts that 'un homme ne peut être ou rester libre que dans la mesure où il demeure relié au transcendant' (*HH*: 24). 'Peut-être au sens le plus fort n'y a-t-il pas de personne humaine et ne peut-il pas y en avoir; ce ne serait alors qu'en Dieu que la personne deviendrait réalité' (*RI*: 155), he writes in his 1940 essay 'Remarques sur l'acte et la personne'.

21. As already suggested in Chapter 2 (note 38), the depth of Marcel's engagement with Sartre's philosophy is questionable. In this case, there is no acknowledgement of Sartre's (2004) argument that freedom *pour-soi* is at once freedom *pour-autrui*.

22. 'L'homme non religieux, c'est-à-dire non relié, devient alors l'homme du refus' (*HH*: 196).

23. 'Il est sans doute infiniment plus raisonnable d'admettre que si le mot *au delà* a un sens, comme il faut sans doute l'affirmer, ce mot ne saurait désigner un autre lieu où l'on pénétrerait en sortant de ce lieu-ci' (*ME II*: 158).

24. Camus explores the Christianization of Neoplatonism in the thesis he wrote for his *diplôme d'études supérieures* ('Métaphysique chrétienne et néoplatonisme', Université d'Alger, 1936), focusing the last of his four chapters on Augustine. As Ronald Srigley summarizes in the introduction to his translation, Camus explains how 'Plotinus made Greek reason more amenable to faith through his notion of participation. Augustine could [then] use this "softened" version of reason to make [...] Christian teachings [...] seem more plausible to the minds of Greeks and Romans alike' (Camus 2007: 6). Plotinus thus, in Chadwick's words, 'provided Augustine with a model and a vocabulary' (Augustine 1998: xxi) with which to articulate his quest for union with God. See Augustine (1998: Book VII). See also Grandgeorge (1967).
25. '[L]'influence du néo-platonisme, très considérable à l'origine, alla peu à peu en s'effaçant une fois que saint Augustin fut converti au christianisme. [...] saint Augustin reprend les arguments de Plotin, mais il les fait siens; Plotin n'est plus qu'un auxiliaire' (Grandgeorge 1967: 150).
26. 'For Plotinus [...] one important and central element of [his conception of time and eternity] is the linking of eternity with the unchanging and transcendent intelligible world and time with the physical world of becoming' (Smith 1996: 196). As Turetzky reports, this concern is a response to Plato, who 'was never entirely clear about the nature of eternity and its difference from its image, time' (1998: 48) — thus causing Plotinus to question how it is possible for humans to have a share in eternity, if they themselves are in time.
27. See Plotinus (1962).
28. 'One of the effects of Plotinus's attempt to integrate matter into a single universe caused in its totality by the One is a comparative upgrading of the world of physical nature [...] and a softening of the soul–body dualism of the *Phaedo*' (Rist 1996: 391).
29. Rist is right to limit his discussion to the importance of Nicaea for the reception of Platonism within Christianity, for at the time there was no real orthodoxy to speak of; rather, there were many Christianities, amongst which were debated many ideas and beliefs. Constantine's main aim was to unify the Church, and as Chadwick explains, the ambiguity of the Nicene formula enabled it to be particularly inclusive: 'it was not clear whether it meant specific or generic identity, and that enabled bishops of differing standpoints to agree to it' (2001: 198). But the Nicene Creed did not immediately become a badge of orthodoxy (Studer 1993: 107). It was not affirmed as an exclusive statement of faith until the Council of Constantinople (381 CE) — but even following this, the creed's significance for orthodoxy remains a debated question. Behr (2006) is a good starting point for further reference.
30. 'Christians opened up a veritable chasm between divine substance and created substance, as two quite separate, distinct entities, wholly unlike, with different powers and properties. The divine, they believed, is life, truth, goodness, and beauty; created being, on the other hand, *receives* life, truth, goodness, and beauty from the divine, but is, in itself, nothing' (Harrison 2006: 78).
31. The early Augustine, who believed in a special closeness between God and the soul, might have endorsed something similar. Cary informs us: 'A striking feature of the prehistory of the inward turn [...] is that there is so little that distinguishes turning to the soul from turning to God. [...] Later in his career Augustine becomes aware that this pattern of talk is not as Catholic as it should be, and in the *Retractions* he corrects it' (2000: 105). Thus, 'while Plotinus can identify the human mind with the mind of God, the mature Augustine cannot' (2000: 107).
32. This position can be related to Tillich's (1886–1965) notion of 'correlational' theology, which establishes a relation between human beings and God: we are always for-God,

and God is always for-us. 'Non-correlational' theology, on the other hand, emphasizes God's absolute distance and Otherness (1951: 59–66). However, whereas Tillich undertakes an explicitly theological appropriation of existential phenomenology, the status of Marcel's work is more ambiguous.

33. More generally in fact, as Du Bos observes, 'dans la première partie du *Journal métaphysique, réalité* n'implique pas *existence*, [...] elle l'exclut, car les deux ordres sont séparés, et là où l'on impute *réalité* il faut résolument transcender *existence*' (1931: 136).

34. On 31 January 1914, for example, he writes: 'à la rigueur, la réflexion sur la religion poussée au dernier terme de son exercice se nie elle-même', for 'la réflexion est en effet inséparable d'un formalisme quelconque. [...] La foi apparaît en ce sens comme la réflexion qui s'est réfléchie elle-même et par suite s'est niée' (*JM*: 50). This aversion to conceptualization is visible throughout Marcel's thought. In 'Regard en arrière' (1947), for instance, he asserts: 'penser, formuler, juger, au fond, c'est toujours trahir' (*RA*: 304).

35. 'Il faudrait [...] qu'il fût possible, sans attribuer au Toi absolu une objectivité qui ruinerait son essence même, de sauver son existence. Et c'est ici que mes tentatives pour dissocier l'existence et l'objectivité prennent toute leur signification. [...] L'*existence*... il ne saurait en effet être question de voir en elle une qualité qui entrerait en conflit avec les qualités constitutives de l'objet' (*JM*: 304; 24 May 1923).

36. 'Au fond, ce que j'ai voulu dire, c'est que lorsque je cherche comment je peux parler de Dieu, m'exprimer sur Dieu, je suis amené à constater que je peux plutôt lui parler, que parler de lui. Et, si je lui parle, c'est, en effet, comme à un Toi, qui n'est pas empirique, un Toi dont je peux donc dire que c'est un Toi absolu' (*GM*: 70). See also *JM*: 304.

37. 'If you want to understand religious faith, [Marcel] suggests, look first at the ordinary human acts of putting-trust-in or being-faithful-to, of making a promise and bearing witness. You may not understand what it means to believe in God, but you may have some sense of what it means to trust a friend or to be faithful to your wife' (Cain 1963: 89).

38. 'Sans doute il serait absurde de dire que l'amour porte sur de l'inconnaissable; ce qui est mystérieux, ce n'est pas l'objet de l'amour comme tel, mais plutôt le type de rapport que l'amour enveloppe. C'est en vain que l'amant dénombre les caractères, les mérites de l'être aimé; il est certain *a priori* que cet inventaire ne lui rendra pas son amour transparent pour lui-même' (*JM*: 226; 23 February 1920).

39. '[L]e Toi absolu, tout en échappant à la vérification objective, n'en est pas moins réel' (Widmer 1971: 114).

40. 'La valeur comme enjeu de l'existant, ou plutôt comme sceau. [...] La valeur est-elle conférée du dehors? Sûrement non' (*PI*: 214; 5 August 1943). See also *ME II*: 75.

41. See also *DH*: 181; *EA*: I, 37, 169; *HH*: 103; *JM*: x, 36, 65–66, 158–59, 255, 264; *ME II*: 74–75, 80, 133, 141, 173–74, 175; *PST*: 259, 263; *RI*: 53–54, 169, 238.

42. See his correspondence with Du Bos (Marcel 1974a: 33, 43). See also *EA*: I, 26, and *RI*: 158.

43. As does Ricœur. In an interview with François Azouvi and Marc de Launay (1994–95) he explains, regarding the two poles he identifies in his work ('un pôle biblique et un pôle rationnel et critique'): 'J'ai eu le souci — vivant une sorte de double allégeance — de ne pas confondre les deux sphères' (1995: 16).

44. Other philosophers associated with this movement include Derrida, Ricœur, and Courtine (1944–).

45. On the theological turn in recent French philosophy see also Goodchild (2002); Bloechl (2003); Simmons (2008).

46. See also *DH*: 43–44, 46; *EC*: 71; *EPR*: 77; *RA*: 309–10.

47. 'Totalité et infini d'Emmanuel Lévinas est la première œuvre majeure de la philosophie française où ce tournant théologique, à l'intérieur d'une inspiration phénoménologique, soit non seulement discernable, mais explicitement assumé' (Janicaud 2009: 65).
48. See especially Husserl (1965).
49. See Husserl (1931: 173-74, §58).
50. 'If one had to pinpoint a time when the meaning of transcendence was thrown out of philosophy, it would have to be in the nineteenth-century philosophy of religion starting from Kant and running through Hegel to Nietzsche. This short history was worked out within a set of boundaries: either philosophy operates within the limits of phenomenal immanence and a transcendent God is wholly confined to faith (Kant) or else philosophy steps beyond these limits to include God within the field of a now extended immanence (Hegel), with the eventual result that God is no longer God (Nietzsche)' (Kosky 2000: 115).
51. Indeed, Marcel rejects the opposition between immanence and transcendence in *ME I*: 82.
52. The first reduction is the Husserlian reduction, which excludes all that is transcendental, all that cannot be reduced to an object of consciousness — with the consequence that the 'I' is only ever a constituting 'I'. The second reduction is Heidegger's reduction to *Dasein* — that is, to an (existential) intentionality broadened to Being-in-time-and-the-world — but where phenomena are only recognized, only appear to the 'I' if they can be said to 'be'. 'Les formes de la phénoménologie déployées respectivement par Husserl et Heidegger fixent, par avance, des conditions (donc des bornes) à la visibilité de leurs phénomènes: nul n'apparaîtra, sinon comme objet ou comme étant', writes Marion in 'Réponses à quelques questions' (1991). 'La dernière réduction en effet met entre parenthèses toute condition objective ou bien ontique fixée à ce qui se donne, pour l'admettre purement et simplement en tant et autant qu'il se donne' (1991: 70–71).
53. Marcel argues for 'une certaine hiérarchie des modes de vie', but — he states — 'sans que celle-ci soit forcément référée à un absolu de caractère métaphysique ou religieux' (1954a: 92).
54. Thus Marion argues that every phenomenon surges forth as a gift (*don*), and Lévinas puts great emphasis on the subject's passivity. Marcel, too, states that 'le transcendant ne peut en aucune manière être assimilé à un point de vue où nous pourrions nous placer en imagination' (*RI*: 8), and, like Lévinas, refers to such transcendence as 'un appel' (*RI*: 188).
55. In his (two-part) 1966 lecture 'Religion, athéisme, foi', Ricœur asks a similar question with respect to Nietzsche's critique of religion: 'Tout le monde connaît le mot fameux de l'insensé dans *le Gai Savoir*: "Dieu est mort"; mais la question est de savoir d'abord quel dieu est mort; ensuite qui l'a tué [...]; et enfin quelle sorte d'autorité appartient à la parole qui proclame cette mort' (1969: 435).
56. 'The effort to re-think that dimension of our experience which we designate religious cannot begin apart from a critical consideration of what we mean by knowledge and certainty. What will count as an answer to the question of whether God is real [...]?' (Pax 1972: vii).
57. Cf. Ricœur: 'le philosophe est loin d'être en état de désigner une parole qui mériterait véritablement le nom de parole de Dieu; mais il peut désigner le mode d'être qui rend existentiellement possible quelque chose comme une parole de Dieu' (1969: 440–41). See also *PI*: 261–62.
58. Here, an existential thread extends back to Kierkegaard, for whom subjectivity (rather than imposed, objectively stated dogma) was the core of religion (see in particular Kierkegaard 1946); indeed, immediately after this remark Marcel comments: 'Ce

chemin mène à Kierkegaard' (*PI*: 124). Though Kierkegaard had no influence on the development of Marcel's thought, when Marcel later learns of his work he identifies with Kierkegaard's project retrospectively, appreciating especially his use of dramatic expression to explore existential paradoxes (akin, Marcel feels, to his own dramatic investigations of existence in his plays), and his preference for indirect discourse (Marcel 1966). Of course, philosophy and religion can be thought together in different ways. Thus Marcel's belief that 'la religion ne peut se fonder que subjectivement' (*FP*: 17) divided his position from contemporaries of his such as Maritain, who saw himself as working in continuity with the thought of Aquinas, and advocated the Thomistic method '[qui] va par l'intelligence même à l'existence même' (1947: 23). 'Il y a deux manières fondamentalement différentes d'entendre le mot existentialisme', Maritain writes in his *Court traité de l'existence et de l'existant* (1947). 'Dans un cas on affirme la primauté de l'existence, mais comme impliquant et sauvant les essences ou natures, et comme manifestant une suprême victoire de l'intelligence et de l'intelligibilité, — c'est ce que je considère comme l'existentialisme authentique. Dans l'autre cas', however — and here he would place Marcel — 'on affirme la primauté de l'existence, mais comme détruisant ou supprimant les essences ou natures, et comme manifestant une suprême défaite de l'intelligence et de l'intelligibilité, — c'est ce que je considère comme l'existentialisme apocryphe: celui même d'aujourd'hui, et qui "ne signifie plus rien du tout"' (1947: 13).

59. The term 'onto-theology' itself dates back to Kant, who used it in his *Critique of Pure Reason* to refer to the kind of transcendental theology that thinkers such as Anselm (in his ontological argument for the existence of God) carried out, which derived knowledge of God from concepts alone.

60. Heidegger explains: 'When metaphysics thinks of beings with respect to the ground that is common to all beings as such, then it is logic as onto-logic. When metaphysics thinks of beings as such as a whole, that is, with respect to the highest being which accounts for everything, then it is logic as theo-logic' (2002: 74). 'Western metaphysics, however, since its beginning with the Greeks has eminently been both ontology and theology', Heidegger asserts; and this 'onto-theological character [...] has become questionable for thinking not because of any kind of atheism, but from the experience of a thinking which has discerned in onto-theo-logy the still *unthought* unity of the essential nature of metaphysics' (2002: 68). Furthermore, 'man can neither pray nor sacrifice to this god' (2002: 74), for this god is subordinate to the (higher) order of philosophical reason, the absolute coherence of which is worshipped above all else.

61. At this time, Marcel had not yet developed his phenomenological approach, so he finds the subjectivity that he is left with after refusing an objective God equally frustrating: 'Lorsque j'écrivais autrefois "il n'y a pas de vérité possible de Dieu", je voulais dire au fond qu'il ne peut être pensé comme Cela. Mais alors ne sommes-nous pas refoulés dans la pure subjectivité?' (*JM*: 254).

62. 'Il semble que nous nous trouvions en présence d'un dilemme', writes Marcel on 18 May 1943: 'ou bien la certitude se confond avec le salut, au sens religieux de ce mot, et dans ce cas la métaphysique perd toute autonomie et en fin de compte toute réalité. Ou bien c'est au contraire le salut qui se réduit à la certitude — et on verse dans un intellectualisme' (*PI*: 185).

63. 'Je crois profondément qu'il doit y avoir une convergence secrète de la philosophie et de la religion, mais je pense aussi que l'instrument est tout à fait différent dans les deux cas. La religion ne peut en effet s'appuyer que sur la Foi. Je pense au contraire que l'instrument de la philosophie est la réflexion [...]. Mais j'ai tenté de montrer que la réflexion peut se présenter sous deux formes différentes et complémentaires, l'une

d'elles étant purement analytique et réductrice: c'est la réflexion primaire, l'autre étant au contraire récupératrice, ou si l'on veut synthétique, c'est justement elle qui s'appuie sur l'être, non sur une intuition, mais sur un assurance qui se confond avec ce que nous appelons notre âme' (*PST*: 33).
64. '[S]i le dualisme en question [entre l'exigence philosophique et l'affirmation chrétienne] est incompatible avec une certaine idée de la philosophie comme système, cette idée elle-même ne peut certainement plus être admise sans examen, comme elle le fut si longtemps' (*HH*: 92).
65. See also *PI*: 255–56.
66. In *Les Hommes contre l'Humain*, for example, Marcel affirms that no philosopher has 'le droit de poser à priori l'accord entre l'exigence philosophique considérée en elle-même et l'affirmation chrétienne en tant que telle'; 'même si, comme je le crois personnellement, cet accord existe', he adds, 'il ne peut certainement pas être postulé' (*HH*: 92).
67. See also *ME I*: 83.
68. See also *PST*: 234.
69. '[C]e qui est évoqué ici, c'est un horizon théologique et rien de plus' (Marcel 1976: 16).
70. Marcel's conversion to Catholicism cannot therefore be interpreted as the culmination of his philosophical reflections, as Troisfontaines (1954) and Cain (1963: 50) suggest, nor can Tsukada assert, in such a straightforward manner, that Marcel's analyses of the immediate have led him to 'la Vérité qui est Amour, c'est-à-dire Dieu' (1995: 245).
71. In an interview with Ricœur, Marcel therefore describes himself as 'un *philosophe du seuil*, un philosophe qui se tenait, d'une manière assez inconfortable d'ailleurs, sur une ligne médiane, entre les croyants et les non-croyants' (*EPR*: 82).
72. 'L'intelligibilité de la transcendance n'est pas ontologique' (*DVI*: 125).
73. In fact, Lévinas asserts, such a phenomenological possibility is already testified to (formally) in Descartes's meditations, for 'après la certitude du cogito, présent à lui-même dans la deuxième Méditation', he notes in 'Dieu et la philosophie', 'la troisième méditation annonce que "j'ai en quelque façon premièrement en moi la notion de l'infini [...]"' (*DVI*: 106; see also *TI*: 40–41, 85). Despite having an idea of infinity therefore, Descartes finds that its content exceeds what he himself can ever think. 'Ce n'est pas dans la finalité d'une visée intentionnelle que je pense l'infini', Lévinas concludes; 'ma pensée de l'infini plus ancienne que la pensée du fini, est la diachronie même du temps, la non-coïncidence, le dessaisissement même' (*DVI*: 12). Lévinas's major substantive (that is, phenomenological) claim then becomes, as Critchley explains, his assertion 'that the ethical relation of the self to the other corresponds to this picture, concretely fulfilling this [formal] model' (2002: 14–15).
74. See also *AE*: 151.
75. '[A]utrui, par sa signification, antérieure à mon initiative, ressemble à Dieu' (*TI*: 326).
76. For this reason, 'religion' is defined, in Lévinas, as 'le lien qui s'établit entre le Même et l'Autre, sans constituer une totalité' (*TI*: 30). See also *EN*: 19, and *TI*: 79.
77. Despite Marcel's assertion, in a 1973 interview with Lhoste, that his philosophical preoccupation shifted from an initial preoccupation with 'le mystère de Dieu', to centre on 'le mystère de l'homme' (Lhoste 1999: I, interview 3), I would argue that Marcel always starts with the human. For in order, in the first part of the *Journal*, to defend God's unverifiability against the charge that it is recourse to the arbitrary, Marcel actually uses the *cogito* to show that it is 'possible d'établir un lien entre l'invérifiable qu'est le cogito et l'invérifiable qui est Dieu' (*JM*: 37; 28 January 1914). It is only once he has demonstrated, with respect to this more concrete human situation, that his notion of *l'invérifiable* can be considered intelligible, that Marcel feels he can approach the question of God's nature.

78. Although they also accept that Western civilization has been shaped by Christianity: 'nous ne pouvons pas penser comme s'il n'y avait pas eu des siècles de chrétienté' (*HP*: 242), states Marcel in 'Position et approches'.
79. See also *HV*: 132.
80. Or 'messianic' temporal structure (*le messianisme sans messie; le messianique sans messianisme*) in Derridean terms, where 'cette dimension messianique ne dépend d'aucun messianisme, elle ne suit aucune révélation déterminée, elle n'appartient à aucune religion abrahamique' (Derrida 1996: 28). See also Derrida (1993).
81. Despite Marcel's (rather clumsy) tendency to describe intersubjectivity in terms of presence, which is suggestive of a timeless and more definitive unity, when he attempts to define this presence further it is clear that he does not mean to suggest any fusion or closed relation, where otherness can only be lost. Instead he insists upon '[le] caractère non objectif de la présence[, qui] ne revient aucunement à dire qu'elle n'est que subjective. [...] L'intersubjectivité est essentiellement ouverture' (*PI*: 255), he affirms in the 1951 essay 'Présence et immortalité', having suggested, in a diary entry on 14 May 1943, that there is 'une intime corrélation' (*PI*: 187) between presence and transcendance.
82. 'Le philosophe qui est parvenu à l'exigence de transcendance dans sa plénitude, c'est-à-dire qui ne peut se satisfaire ni de ce qui est dans le monde, ni même du monde lui-même considéré en sa totalité — totalité d'ailleurs toujours fictive — peut très bien rester néanmoins en deçà de toute conversion à une religion historique donnée. Il n'y a là aucun passage nécessaire' (*ME II*: 134).
83. Accordingly, Marcel speaks of the 'infra-chrétien' (*HV*: 136), of 'zones péri-chrétiennes' (*HP*: 242), and of 'un arrière-plan théologique' (*ME I*: 83; see also *PST*: 288). In this way, Marcel's philosophy intersects with the tradition of negative theology. As he says to Boutang in 1970: 'chaque fois que nous cherchons à cerner la pensée que nous avons eue en disant qu'il en est, nous risquons ne nous tromper gravement. [...] j'accorde de plus en plus d'importance à la théologie négative' (*GM*: 59). Indeed, on 26 March 1943 he remarks: 'Le seul Dieu en lequel je puisse croire est un Dieu qui accepte, dans un certain sens qui *veut*, qu'il soit possible de douter de lui' (*PI*: 124, note 1). Such argumentation, in Hart's words, 'deconstructs the metaphysics in theology to the extent that it shows all positive theological images of God rely on a ground that opens out onto an abyss' (2000: xxvi). It is, as Hart argues, a necessary tool for the responsible questioning of any metaphysics that is open to the theological. For this reason, therefore, Bradley argues that 'there is a sense in which negative theology is central to modern French philosophy — not because of some more or less hidden theological or nihilistic agenda [...] but because it names [or: is structured to recognize] an essential tension to which all thought must respond in order to be responsible' (2004: 218). Marcel's philosophy attempts to recognize precisely this, although admittedly (as will be demonstrated in the latter part of this chapter), his negative theology does not quite achieve the constant movement between affirmation and negation of which Rubenstein speaks, where one pole, in a sense, becomes the other, and causes their polarity to collapse (2003: 396).
84. *Illéité* is thus also described by Lévinas as '[une] intrigue qu'on est tenté d'appeler religieuse, qui ne se dit pas en termes de certitude ou d'incertitude et ne repose sur aucune théologie positive' (*AE*: 230).
85. Cf. Lévinas: 'cette ambiguïté est nécessaire à la transcendance. La transcendance se doit d'interrompre ses propres démonstration et monstration: sa phénoménalité. Il lui faut le clignotement et la dia-chronie de l'énigme qui n'est pas certitude simplement précaire, mais qui rompt l'unité de l'aperception transcendantale où toujours l'immanence triomphe de la transcendance' (*DVI*: 127).

86. Gouhier comments, following Marcel's 1973 conference paper 'De la recherche philosophique': 'Vous dites [...] "recherche sur la nature de la recherche". Je me demande s'il ne s'agirait pas plutôt d'une recherche sur la nature de la découverte. [...] Or, toute votre œuvre montre que ce n'est pas une recherche pour la recherche, mais bien pour trouver quelque chose' (*EAGM*: 25).
87. Two days later Marcel changes his mind and writes that 'il n'y a rien à retenir de ces dernières indications' (*PI*: 127). However, at the end of April he reaffirms his search for certitude (*PI*: 156–58).
88. For Caputo in general, Lévinas 'wants to back [undecidability] up, as it were, by rooting it in a desire for the Good, a desire for God. [...] I would not be so quick to leave phenomenology as Levinas' (1998: email 5).
89. As early as the first part of his *Journal*, this propensity to over-ontologize is apparent. Marcel wants to ensure that faith does not regress into objective affirmation, and so he introduces the notion of divine grace to perform this philosophical function. 'La grâce reste bien le postulat transcendant et inobjectivable de l'acte de foi' (*JM*: 60), he writes on 5 February 1914. That is, *la grâce* acts as the guarantor of faith, so that Marcel can be certain that his philosophy is able fully to account for faith. And although he recognizes that his early work is too idealist, Marcel does not put a halt to this tendency, but continues, in places, to ground the human in the divine: 'Sans doute peut-on conclure [...] qu'aimer réellement un être, c'est l'aimer en Dieu' (*JM*: 158), writes Marcel affirmatively on 18 December 1918; 'Je ne crois pas que l'idée de transcendance à la limite soit laïcisable' (1937: 179), he asserts forcefully at a 1937 meeting of the *Société française de philosophie*.
90. Anderson comments that when pondering 'what "I" am, [Marcel] is not referring to his own *individual* self but to all I's, all selves, all persons *as such*' (1985: 274).
91. See also *HP*: 242.
92. See *RI*: 190–91.
93. '[Q]uelque effort que fasse M. Bergson pour lui conférer une dignité métaphysique suprême, celle d'agent spirituel absolu, et j'irai jusqu'à dire pour le surnaturaliser, il reste prisonnier des prémisses biologiques de son système' (Marcel 1932a: 558).

GENERAL CONCLUSION

Toward what Metaphysics?

[N]ous sommes *à la recherche de*... Ne songeons pas à remplacer ces points de suspension par un substantif: car c'est justement le substantif qui fait défaut. (*RA*: 292)

In order to assess the success or even just the specificity of a philosophical position, one must first have an understanding of what philosophy is, of its status and role. However, this is precisely what Marcel questioned, challenging, in particular, Western philosophy's inbuilt assumption that to philosophize always entails building a system: 'le propre d'une expérience en cours n'est-il pas [plutôt] de présenter une inconsistance fondamentale?' (*RA*: 291), he asks in 'Regard en arrière' (1947). The problem of consistency that Chapter 1 points to, it therefore emerged, subscribed to the very tradition that Marcel was contesting. His metaphysical discussions of time were not necessarily to be understood in the conventional, ontological sense, but were rather ambiguous; and this, coupled with Marcel's emphasis on the need for a plurality of approaches to the metaphysical, led this book to consider his thought from a number of different angles, paralleling Marcel's own exploratory, and deliberately wandering, enterprise. As such, the book chooses a range of different associations for Marcel (Bergson, Ricœur, Lévinas, Augustine), showing his philosophy to present a multiplicity of perspectives on metaphysics and its task, and thereby demonstrating the possibility of a more positive, productive response to its apparent incoherence — incoherence which may not actually be so undesirable.

Yet if Marcel questions the philosophical tradition he inherits, he nevertheless continues to employ its ontological language as he struggles to resist the temptation to seek — and hence privilege — objectifying coherence. Marcel wants his philosophy to say something definitive, and it is this which reinstates a dichotomy between eternity (as constant and stable) and time (as uncertain and impermanent) and assigns primacy to the former. For this reason, Marcel is revealed to be a transition figure in twentieth-century French thought: between the totalizing methods of scientific positivism (e.g. Duhem and Poincaré, in the wake of Comte) and idealism (both rational, e.g. Benda and Brunschvicg, and spiritual, e.g. Blondel and Bergsonian mysticism), and a post-structuralist approach, most commonly associated with thinkers such as Deleuze, Derrida, Foucault, and Lyotard, but equally attributable to Ricœur, Lévinas — and even, Lawlor would argue, Bergson. Thus, not only has this book been concerned with Marcel's 'unplaced' status; it has also explored broader movements within French thought which have been neglected. The importance of Bergson, for instance, was, before Deleuze, overlooked by existentialists and poststructuralists, who turned a blind eye to this French philosophical heritage and chose instead to recognize German thinkers such as Husserl and Heidegger. The readings of Marcel alongside Bergson, Ricœur, and Lévinas, however, recast twentieth-century French philosophy, revealing suppressed continuities in its intellectual tradition — including a French parallel to Heidegger's reaction against Husserl, manifest in Marcel's attempt to distance himself from Bergson, and precursors to both Ricœur and Lévinas within Marcel's thought itself. Indeed, it is Jean Hering's belief that 'if German phenomenology [...] had remained unknown in France, nevertheless a phenomenology would have been constituted there; and this, to a large extent, would be due to the influence of Gabriel Marcel' (1950: 75).

It should also be noted that this book has not, in fact, been simply concerned with time. Rather, guided by this theme, it has examined a host of other philosophical topics, engaging with issues relating to the nature of the subject, the other, ethics, and religion. This reveals the extent to which questions of time have a bearing on the very possibilities of philosophy itself, for not only do philosophical positions affect how time is understood; how one understands time (already) affects one's understanding of the task and scope of philosophy. Indeed, Robert Dostal has claimed that 'the phenomenology of time [...] can serve as a key for understanding not only the relation of Husserl and Heidegger, but the development of Continental thought throughout [the twentieth] century as well' (2006: 121). However, more enlightening with respect to theoretical movements during this period still, I would argue, are Marcel's phenomenological studies of time, for their indeterminacy (as Marcel is pulled in many directions at once) sketches out a range of possible avenues for a metaphysics other than a metaphysics of presence (combining phenomenology with hermeneutics, for example, or with ethics or a new, possibly theological, notion of givenness (*donation*)), while also drawing attention to a number of dead ends (idealism, objectification, a lack of rigour, over-personalization). This enables us to glean a more intimate understanding of what was at stake for philosophy at the time, and of the problematics that instigated its various shifts. Encountering French philosophy at this crossroads then encourages us to reflect on the route that the philosopher might now take, what s/he might hope to achieve, and — as discussions of the various forms of Marcel's philosophy prompt us to consider — how s/he might be permitted to speak. Above all else, to read Marcel is to be forced to question what it actually means to philosophize.

BIBLIOGRAPHY

Works by Marcel

1912. 'Les Conditions dialectiques de la philosophie de l'intuition', *Revue de métaphysique et de morale*, 20: 638–52
1914. *Le Seuil invisible* (Paris: Grasset) [includes a preface (pp. 1–8), *La Grâce* (pp. 9–209; written March–April 1911), and *Le Palais de sable* (pp. 211–398; written August–September 1913)]
1921. 'Réflexions sur le tragique', *L'Essor*, 13: 1–7
1924. 'Tragique et personnalité', *La Nouvelle Revue française*, 11: 37–45
1925. *Le Quatuor en fa dièse* (Paris: Plon)
1926. 'Note sur l'évaluation tragique', *Journal de psychologie normale et pathologique*, 23: 68–76
1929A. 'Carence de spiritualité', *La Nouvelle Revue française*, 16: 375–79
1929B. 'Note sur les limites du spiritualisme bergsonien', *La Vie intellectuelle*, 5: 267–70
1931. *Trois pièces* (Paris: Plon) [includes *Le Regard neuf* (pp. 1–103; first staged 24 May 1922, at the *Nouvel ambigu*), *Le Mort de demain* (pp. 105–85; written December 1919), and *La Chapelle ardente* (pp. 187–266; first published 1925)]
1932A. 'Henri Bergson et le problème de Dieu', *L'Europe nouvelle*, 15: 558–59
1932B. 'Un événement philosophique', *La Nouvelle Revue des jeunes*, 4: 415–22 [re: Bergson's *Les Deux Sources de la morale et de la religion* (1932)]
1934. '"La Pensée et le mouvant" de M. Bergson', *L'Europe nouvelle*, 17: 662–63
1935. *Journal métaphysique*, 4th edn (Paris: Gallimard) [first published 1927]
1936. *Le Fanal* (Paris: Stock)
1937. 'Subjectivité et transcendance', *Bulletin de la Société française de philosophie*, 37: 161–211 [Jean Wahl gives the presentation, but Marcel participates in the discussion]
1938. 'Responsabilités de la France', *Temps présent*, 25 February, 2.17: 8
1939. 'Qu'est-ce que le bergsonisme', *Temps présent*, 30 June, 3.85: 5
1940A. 'Considérations sur l'égalité', *Études carmélitaines*, 25: 161–71
1940B. *Du refus à l'invocation* (Paris: Gallimard)
1940C. 'Note sur la condamnation de soi', *Esprit*, 8: 17–20
1943. 'Grandeur de Bergson', in *Henri Bergson: essais et témoignages recueillis par Albert Béguin et Pierre Thévenaz* (Neuchâtel: La Baconnière), pp. 29–38
1944. *Homo viator: prolégomènes à une métaphysique de l'espérance* (Paris: Aubier-Montaigne)
1945. 'Autour de Heidegger', *Dieu vivant*, 1: 89–100
1946. 'Finalité essentielle de l'œuvre dramatique', *Revue théâtrale*, 3: 285–95
1947A. 'Existentialisme et pensée chrétienne', *Témoignages*, Cahiers de la Pierre-qui-Vire, 8: 157–69

1947B. 'Regard en arrière', in *Existentialisme chrétien* (see Gilson 1947, below), pp. 291–319

1949A. 'Justice pour Charles Maurras!', *Aspects de la France*, 29 December, 72: 5

1949B. *Vers un autre royaume: deux drames des années noires* (Paris: Plon) [includes *L'Émissaire* (pp. 1–110; Marcel began writing this in 1945), *Le Signe de la croix* (pp. 111–229; first act written 1938, second act written 1942, third act written close to date of publication), and a postface (pp. 231–36)]

1950A. *La Chapelle ardente* (Paris: Table ronde) [first published 1925; first staged 1925, at the *Vieux-Colombier*; also includes 'Note de l'auteur' (pp. 137–42), and *Le Sol détruit* (first version of *La Chapelle ardente*; pp. 143–253)]

1950B. 'Théâtre de l'âme en exil', *Recherches et débats*, 10: 7–14

1951A. *Les Hommes contre l'Humain* (Paris: La Colombe)

1951B. *Le Mystère de l'être I: réflexion et mystère* (Paris: Aubier-Montaigne) ['Gifford Lectures' delivered at University of Aberdeen, 1949]

1951C. *Le Mystère de l'être II: foi et réalité* (Paris: Aubier-Montaigne) ['Gifford Lectures' delivered at University of Aberdeen, 1950]

1951D. *Rome n'est plus dans Rome* (Paris: Table ronde)

1952. 'Théâtre et philosophie: leurs rapports dans mon œuvre', in *Le Théâtre contemporain*, ed. by Henri Gouhier and others, *Recherches et débats*, n.s., 2 (Paris: Fayard), pp. 17–42

1953. 'Lettre préface', in *De l'existence à l'être* (see Troisfontaines 1953, below): I, 9–14

1954A. *Le Déclin de la sagesse* (Paris: Plon)

1954B. 'Notes pour une philosophie de l'amour', *Revue de métaphysique et de morale*, 59: 374–79

1955A. *Croissez et multipliez* (Paris: Plon)

1955B. 'Réponse de Gabriel Marcel à l'enquête sur l'idée de Dieu et ses conséquences', *L'Âge nouveau*, 10: 39–44

1957. 'Schelling fut-il un précurseur de la philosophie de l'existence?', *Revue de métaphysique et de morale*, 62: 72–87

1958A. *Théâtre et religion* (Lyon: Vitte)

(ed.). 1958B. *Un changement d'espérance: à la rencontre du Réarmement moral* (Paris: Plon)

1959. *Présence et immortalité: journal métaphysique (1938–1943) et autres textes* (Paris: Flammarion)

1962. *Fragments philosophiques: 1909–1914* (Louvain: Nauwelaerts)

1964. *La Dignité humaine et ses assises existentielles* (Paris: Aubier-Montaigne) ['William James Lectures' delivered at Harvard, October–December 1961]

1965. *Paix sur la terre: deux discours, une tragédie* (Paris: Aubier-Montaigne) [includes a preface (pp. 7–14), 'Laudatio' (pp. 15–39; speech made by Carlo Schmid during the 1964 ceremony at which Marcel received a peace prize), 'La Philosophie et la Paix' (pp. 41–60; speech Marcel made upon receiving this price), and *Un juste* (pp. 61–176; play written by Marcel in 1918)]

1966. 'Kierkegaard en ma pensée', in *Kierkegaard vivant: colloque organisé par l'Unesco à Paris du 21 au 23 avril 1964*, ed. by Unesco (Paris: Gallimard), pp. 64–80

1967A. 'Mon temps et moi (Temps et valeur)', in *Entretiens sur le temps* (see Hersch and Poirier 1967, below), pp. 11–19; related discussion pp. 20–29

1967B. *Le Secret est dans les îles* (Paris: Plon) [includes 'Le Secret est dans les îles' (pp. 7–24), *Le Dard* (pp. 25–153; first published 1936; first staged 1 March 1937, at the *Théâtre des arts*), *L'Émissaire* (pp. 155–270), and *La Fin des temps* (pp. 271–349; radio play, first broadcast 17 June 1950, by *Radiodiffusion française*)]
1968A. *Être et avoir*, 2 vols (Paris: Aubier-Montaigne) [first published 1935]
1968B. *Pour une sagesse tragique et son au-delà* (Paris: Plon)
1969. 'Testament philosophique', *Revue de métaphysique et de morale*, 74: 253–62; also in *Gabriel Marcel et les injustices de ce temps* (see Marcel 1983, below), pp. 127–37 [presentation given to the *Congrès international de philosophie* in Vienna, September 1968]
1971A. *Coleridge et Schelling* (Paris: Aubier-Montaigne) [*mémoire de diplôme d'études supérieures*; written 1909]
1971B. *En chemin, vers quel éveil?* (Paris: Gallimard)
1973A. *Cinq pièces majeures* (Paris: Plon) [includes *Un homme de Dieu* (pp. 5–104; written 1921), *Le Monde cassé* (pp. 105–216; written 1932, first published 1933), *Le Chemin de crête* (pp. 217–355; written 1935), *La Soif ou Les Cœurs avides* (pp. 357–450; written 1937; first published as *La Soif* in 1938; re-edited under the title *Les Cœurs avides* in 1952), and *Le Signe de la croix* (pp. 451–550; revised version with epilogue, written 1953)]
1973B. *Percées vers un ailleur*s (Paris: Fayard) [includes a preface (pp. i–x), *L'Iconoclaste* (pp. 5–168), a commentary on *L'Iconoclaste* by Marcel Belay (pp. 169–97), *L'Horizon* (pp. 199–365; first published 1945), a postface (pp. 367–78; written 1944), a commentary on *L'Horizon*, by Belay (pp. 379–404), and 'De l'audace en métaphysique' (pp. 405–21)]
1974. 'Dialogue entre Gabriel Marcel et Mme Parain-Vial', *Revue de métaphysique et de morale*, 79: 383–91
1976. 'De la recherche philosophique', in *Entretiens autour de Gabriel Marcel* (see Belay and others 1976, below), pp. 9–19; related discussion pp. 20–52
1981. *L'Existence et la liberté humaine chez Jean-Paul Sartre* (Paris: Vrin) [essay written 1946]
1983. *Gabriel Marcel et les injustices de ce temps: la responsabilité du philosophe*, ed. by Joël Bouëssée and Anne Marcel, Présence de Gabriel Marcel, Cahier IV (Paris: Aubier-Montaigne)
1998. *L'Homme problématique: position et approches concrètes du mystère ontologique*, 2nd edn (Paris: Présence de Gabriel Marcel) [first published 1955]
2005. *La Métaphysique de Royce* (Paris: L'Harmattan) [essay written 1917–18; first published 1945]

Other Works

ADAMS, PEDRO. 1966. 'Gabriel Marcel: Metaphysician or Moralist?', *Philosophy Today*, 10: 182–89
ALEXANDER, IAN W. 1948. 'The Philosophy of Gabriel Marcel in its Relations with Contemporary French Thought' (unpublished doctoral thesis, University of Edinburgh)
ALQUIÉ, FERDINAND. 2005. *Qu'est-ce que comprendre un philosophe* (Paris: Table Ronde) [1956 lecture]

ANDERSON, THOMAS C. 1985. 'The Nature of the Human Self according to Gabriel Marcel', *Philosophy Today*, 29: 273–83
—— 1989. 'Gabriel Marcel's Notions of Being', in *Contributions of Gabriel Marcel to Philosophy* (see Cooney 1989, below), pp. 47–78 [first published in *Philosophy Today*, 19 (1975)]
—— 2006. *A Commentary on Gabriel Marcel's 'The Mystery of Being'* (Milwaukee, WI: Marquette University Press)
ANDREW, J. DUDLEY. 1973. 'The Structuralist Study of Narrative: Its History, Use, and Limits', *Bulletin of the Midwest Modern Language Association*, 6: 45–51
AUGUSTINE OF HIPPO. 1953. *Augustine: Earlier Writings*, selected and translated with introductions by John H. S. Burleigh (London: SCM Press)
—— 1998. *Confessions*, trans. by Henry Chadwick (Oxford: Oxford University Press) [written 397–400 CE]
—— 2003. *The City of God*, trans. by Henry Bettenson (London: Penguin) [written 413–26 CE]
BAKHTIN, MIKHAIL MIKHAÏLOVICH. 1981. *The Dialogic Imagination: Four Essays*, ed. by Michael Holquist, trans. by Caryl Emerson and Michael Holquist (Austin: University of Texas Press) [first published 1975]
BARNES, JONATHAN (ed.). 1984. *The Complete Works of Aristotle: The Revised Oxford Translation*, 2 vols (Princeton, NJ: Princeton University Press)
BARONI, RAPHAËL. 2009. *L'Œuvre du temps: poétique de la discordance narrative* (Paris: Seuil)
BARRETT, WILLIAM. 1990. *Irrational Man: A Study in Existential Philosophy*, 2nd Anchor edn (New York: Doubleday) [first published 1958]
BARS, HENRY. 1991. 'Gabriel Marcel et Jacques Maritain', in *Jacques Maritain et ses contemporains*, ed. by Bernard Hubert and Yves Floucat (Paris: Desclée de Brouwer), pp. 231–54
BEAUVOIR, SIMONE DE. 1963. *La Force des choses* (Paris: Gallimard)
BEHR, JOHN. 2006. 'The Question of Nicene Orthodoxy', in *Byzantine Orthodoxies: Papers from the Thirty-Sixth Spring Symposium of Byzantine Studies, University of Durham, 23–25 March 2002*, ed. by Andrew Louth and Augustine Casiday (Aldershot: Ashgate), pp. 15–26
BEISER, FREDERICK C. 2002. *German Idealism: The Struggle against Subjectivism, 1781–1801* (Cambridge, MA: Harvard University Press)
BELAY, MARCEL, and OTHERS (eds). 1976. *Entretiens autour de Gabriel Marcel: Centre culturel international de Cerisy-la-Salle, 24–31 août 1973*, Présence de Gabriel Marcel (Neuchâtel: La Baconnière)
BENEFIELD, J. J. 1973. 'The Place of God in the Thought of Gabriel Marcel' (unpublished doctoral thesis, University of Canterbury, Christchurch, New Zealand)
BERGSON, HENRI. 1959. *Œuvres* (Paris: Presses universitaires de France)
BERGSON, HENRI, KEITH ANSELL-PEARSON, and JOHN MULLARKEY. 2002. *Henri Bergson: Key Writings* (London: Continuum)
BERNARD, MICHEL. 1952. *La Philosophie religieuse de Gabriel Marcel: étude critique* (Paris: Les Cahiers du Nouvel Humanisme)
BERNASCONI, ROBERT, and DAVID WOOD (eds). 1988. *The Provocation of Levinas: Rethinking the Other* (London: Routledge)

BLACKHAM, HAROLD JOHN. 1951. *Six Existentialist Thinkers* (London: Routledge & Kegan Paul)
BLOECHL, JEFFREY (ed.). 2000. *The Face of the Other and the Trace of God: Essays on the Philosophy of Emmanuel Levinas* (New York: Fordham University Press)
—— (ed.). 2003. *Religious Experience and the End of Metaphysics* (Bloomington: Indiana University Press)
BLUNDELL, BOYD. 2003. 'Creative Fidelity: Gabriel Marcel's Influence on Paul Ricœur', in *Between Suspicion and Sympathy* (see Wierciński 2003, below), pp. 89–102
BOUËSSÉE, JOËL (ed.). 1977. *Gabriel Marcel interrogé par Pierre Boutang*, Présence de Gabriel Marcel (Paris: Jean-Michel Place) [interviews held June 1970]
BOURGEOIS, PATRICK L. 1995. 'Ricœur and Marcel: An Alternative to Postmodern Deconstruction', *Bulletin de la Société américaine de philosophie de langue française*, 7: 164–75
—— 2002. 'Ricœur and Levinas: Solicitude in Reciprocity and Solitude in Existence', in *Ricœur as Another* (see Cohen and Marsh 2002, below), pp. 109–26
—— 2003. 'Catholic Author, Musician, Philosopher: Gabriel Marcel in Postmodern Dialogue', *Renascence*, 55: 193–209
BOUTANG, PIERRE. 1989. 'Gabriel Marcel politique', in *Gabriel Marcel* (see Sacquin 1989, below), pp. 81–89
BRADLEY, ARTHUR. 2004. *Negative Theology and Modern French Philosophy* (London: Routledge)
BROCKELMAN, PAUL. 1985. *Time and Self: Phenomenological Explorations* (New York: Crossroad)
BUSCH, THOMAS W. 1987. *The Participant Perspective*: A *Gabriel Marcel Reader* (Lanham, MD: University Press of America)
—— 1995. 'Secondary Reflection as Interpretation', *Bulletin de la Société américaine de philosophie de langue française*, 7: 176–83
BUTLER, JUDITH. 2005. *Giving an Account of Oneself* (New York: Fordham University Press)
CAIN, SEYMOUR. 1963. *Gabriel Marcel* (New York: Hillary House)
CAMUS, ALBERT. 1951. *L'Homme révolté* (Paris: Gallimard)
—— 2007. *Christian Metaphysics and Neoplatonism*, trans. by Ronald D. Srigley (Columbia: University of Missouri Press) [for the original French — 'Métaphysique chrétienne et néoplatonisme', written 1936 — see Albert Camus, *Essais*, ed. by Roger Quilliot and Louis Faucon (Paris: Gallimard, 1965), pp. 1224–1313]
CAPEK, MILIC. 1950. 'Stream of Consciousness and "Durée Réelle"', *Philosophy and Phenomenological Research*, 10: 331–53
CAPUTO, JOHN D., and EDITH WYSCHOGROD. 1998. 'Postmodernism and the Desire for God: An E-mail Exchange', *Cross Currents*, 48.3: 293–310
CARR, DAVID. 1991. 'Épistémologie et ontologie du récit', in *Paul Ricœur: les métamorphoses de la raison herméneutique*, ed. by Jean Greisch and Richard Kearney (Paris: Cerf), pp. 205–14
—— 1999. *The Paradox of Subjectivity: The Self in the Transcendental Tradition* (Oxford: Oxford University Press)
CARY, PHILLIP. 2000. *Augustine's Invention of the Inner Self: The Legacy of a Christian Platonist* (Oxford: Oxford University Press)

CHADWICK, HENRY. 2001. *The Church in Ancient Society: From Galilee to Gregory the Great* (Oxford: Oxford University Press)

CHENU, JOSEPH. 1948. *Le Théâtre de Gabriel Marcel et sa signification métaphysique* (Paris: Aubier-Montaigne)

CHEVALIER, JACQUES. 1959. *Entretiens avec Bergson* (Paris: Plon)

COHEN, RICHARD A. 2001. *Ethics, Exegesis and Philosophy: Interpretation after Levinas* (Cambridge: Cambridge University Press)

—— 2002. 'Moral Selfhood: A Levinasian Response to Ricœur on Levinas', in *Ricœur as Another* (see Cohen and Marsh 2002, below), pp. 127–60

COHEN, RICHARD A., and JAMES L. MARSH (eds). 2002. *Ricœur as Another: The Ethics of Subjectivity* (Albany: State University of New York Press)

COLLINS, JAMES D. 1968. *The Existentialists*, 7th Gateway edn (Chicago, IL: Regnery) [first published 1952]

COONEY, WILLIAM (ed.). 1989. *Contributions of Gabriel Marcel to Philosophy: A Collection of Essays* (Lewiston, NY: Edwin Mellen)

COOPER, DAVID E. 1999. *Existentialism: A Reconstruction*, 2nd edn (Oxford: Blackwell) [first published 1990]

CRITCHLEY, SIMON. 2002. 'Introduction', in *The Cambridge Companion to Levinas* (see Critchley and Bernasconi 2002, below), pp. 1–32

CRITCHLEY, SIMON, and ROBERT BERNASCONI (eds). 2002. *The Cambridge Companion to Levinas* (Cambridge: Cambridge University Press)

CROWELL, STEVEN. 2001. *Husserl, Heidegger and the Space of Meaning* (Evanston, IL: Northwestern University Press)

CROWLEY, PATRICK. 2003. 'Paul Ricœur: The Concept of Narrative Identity, the Trace of Autobiography', *Paragraph*, 26.3: 1–12

DAIGLE, CHRISTINE (ed.). 2006. *Existentialist Thinkers and Ethics* (Montreal and Kingston: McGill-Queen's University Press)

DELEUZE, GILLES. 2004. *Le Bergsonisme*, 3rd edn (Paris: Presses universitaires de France) [first published 1966]

DELEUZE, GILLES, and FÉLIX GUATTARI. 1991. *Qu'est-ce que la philosophie?* (Paris: Minuit)

DENNETT, DANIEL C. 1992. 'The Self as the Center of Narrative Gravity', in *Self and Consciousness: Multiple Perspectives*, ed. by Frank S. Kessel, Pamela M. Cole, and Dale L. Johnson (Hillsdale, NJ: Erlbaum), pp. 103–15

DERRIDA, JACQUES. 1967A. *La Voix et le Phénomène: introduction au problème du signe dans la phénoménologie de Husserl* (Paris: Presses universitaires de France)

—— 1967B. 'Violence et métaphysique: essai sur la pensée d'Emmanuel Levinas', in *L'Écriture et la Différence* (Paris: Seuil), pp. 117–228 [first published 1964]

—— 1972. *Positions* (Paris: Minuit)

—— 1975. *Marges de la philosophie* (Paris: Minuit) [first published 1972]

—— 1993. *Spectres de Marx: l'état de la dette, le travail du deuil et la nouvelle internationale* (Paris: Galilée)

—— 1996. 'Foi et savoir: les deux sources de la "religion" aux limites de la simple raison', in *La Religion*, ed. by Jacques Derrida and Gianni Vattimo (Paris: Seuil), pp. 9–86

DESCARTES, RENÉ. 1951. *Discours de la méthode suivi des Méditations* (Paris: Union

générale d'éditions) [*Discours* first published 1637; *Méditations* first published 1641 (Latin) and 1647 (French)]
DESMOND, WILLIAM. 1994. 'Philosophies of Religion: Marcel, Jaspers, Levinas', in *Twentieth-century Continental Philosophy*, ed. by Richard Kearney (London: Routledge), pp. 108–43
DEVAUX, ANDRÉ A. 1974. 'Charles Du Bos, Jacques Maritain et Gabriel Marcel, ou Peut-on aller de Bergson à Saint Thomas d'Aquin?', *Cahiers Charles Du Bos*, 18: 87–103
DEWEESE, GARRETT J. 2004. *God and the Nature of Time* (Aldershot: Ashgate)
DOSSE, FRANÇOIS. 1997. *Paul Ricœur: les sens d'une vie* (Paris: La Découverte)
DOSTAL, ROBERT J. 2006. 'Time and Phenomenology in Husserl and Heidegger', in *The Cambridge Companion to Heidegger*, ed. by Charles B. Guignon, 2nd edn (Cambridge: Cambridge University Press), pp. 120–48
DU BOS, CHARLES. 1931. 'Gabriel Marcel', in *Le Roseau d'Or: essais et poèmes*, ed. by Paul Claudel and others (Paris: Plon), pp. 87–165
DUDIAK, JEFFREY. 2001. *The Intrigue of Ethics: A Reading of the Idea of Discourse in the Thought of Emmanuel Levinas* (New York: Fordham University Press)
EAKIN, PAUL JOHN. 1999. *How our Lives Become Stories: Making Selves* (Ithaca, NY: Cornell University Press)
ENGEL, SUSAN. 1999. *Context is Everything: The Nature of Memory* (New York: Freeman)
FOUILLOUX, ÉTIENNE. 1989. 'Un philosophe devient catholique en 1929', in *Gabriel Marcel* (see Sacquin 1989, below), pp. 93–114; related discussion pp. 115–20
FRANCIS, JACQUES. 1989. 'Langage et conflit dans le théâtre de Gabriel Marcel', in *Gabriel Marcel* (see Sacquin 1989, below), pp. 231–48
FREEMAN, MARK. 2003. 'Rethinking the Fictive, Reclaiming the Real: Autobiography, Narrative Time, and the Burden of Truth', in *Narrative and Consciousness: Literature, Psychology, and the Brain*, ed. by Gary Fireman, Ted McVay, and Owen Flanagan (Oxford: Oxford University Press), pp. 115–28
FUCHS, WOLFGANG WALTER. 1976. *Phenomenology and Metaphysics of Presence: An Essay in the Philosophy of Edmund Husserl* (The Hague: Nijhoff)
GALLAGHER, KENNETH T. 1962. *The Philosophy of Gabriel Marcel* (New York: Fordham University Press)
—— 1995. 'Humanity and Creaturehood', *Bulletin de la Société américaine de philosophie de langue française*, 7: 49–58
GILLMAN, NEIL. 1980. *Gabriel Marcel on Religious Knowledge* (Washington, DC: University Press of America)
GILSON, ÉTIENNE (ed.). 1947. *Existentialisme chrétien: Gabriel Marcel* (Paris: Plon)
GOODCHILD, PHILIP (ed.). 2002. *Rethinking Philosophy of Religion: Approaches from Continental Philosophy* (New York: Fordham University Press)
GRANDGEORGE, L. 1967. *Saint Augustin et le néo-platonisme* (Frankfurt: Minerva) [first published Paris, 1896]
GUGELOT, FRÉDÉRIC. 1998. *La Conversion des intellectuels au catholicisme en France (1885–1935)* (Paris: CNRS)
GUTTING, GARY. 2001. *French Philosophy in the Twentieth Century* (Cambridge: Cambridge University Press)
HANLEY, KATHARINE ROSE. 1987. *Dramatic Approaches to Fidelity: A Study in the*

Theater and Philosophy of Gabriel Marcel (1889–1973) (Lanham, MD: University Press of America)

HARPER, RALPH. 1955. *The Sleeping Beauty* (London: Harvill Press)

HARRISON, CAROL. 2006. *Rethinking Augustine's Early Theology: An Argument for Continuity* (Oxford: Oxford University Press)

HART, KEVIN. 2000. *The Trespass of the Sign: Deconstruction, Theology, and Philosophy*, 2nd edn (New York: Fordham University Press) [first published 1989]

HEGEL, GEORG WILHELM FRIEDRICH. 1975. *Hegel's Logic: Being Part One of the 'Encyclopedia of the Philosophical Sciences', 1830*, trans. by William Wallace, 3rd edn (Oxford: Clarendon Press)

—— 1998. *Phenomenology of Spirit*, trans. by A. V. Miller, analysis and foreword by J. N. Findlay (Delhi: Motilal Banarsidass) [first published 1807]

HEIDEGGER, MARTIN. 1962. *Being and Time*, trans. by John Macquarrie and Edward Robinson (Oxford: Blackwell) [first published 1927]

—— 2002. 'The Onto-Theo-Logical Constitution of Metaphysics', trans. by Joan Stambaugh, in *The Religious*, ed. by John D. Caputo (Malden, MA: Blackwell), pp. 67–75 [1957 lecture]

HELD, KLAUS. 2007. 'Phenomenology of "Authentic Time" in Husserl and Heidegger', *International Journal of Philosophical Studies*, 15: 327–47

HELLMAN, JOHN. 1981. *Emmanuel Mounier and the New Catholic Left, 1930–1950* (Toronto: University of Toronto Press)

HELM, PAUL. 1988. *Eternal God: A Study of God without Time* (Oxford: Oxford University Press)

HERING, JEAN. 1950. 'Phenomenology in France', in *Philosophic Thought in France and the United States: Essays Representing Major Trends in Contemporary French and American Philosophy*, ed. by Marvin Farber (Buffalo, NY: University of Buffalo Publications), pp. 67–85

HERMAN, DAVID. 2002. *Story Logic: Problems and Possibilities of Narrative* (Lincoln: University of Nebraska Press)

—— (ed.). 2007. *The Cambridge Companion to Narrative* (Cambridge: Cambridge University Press)

HERSCH, JEANNE, and RENÉ POIRIER (eds). 1967. *Entretiens sur le temps* (Paris: Mouton)

HOCKING, WILLIAM ERNEST. 1954. 'Marcel and the Ground Issues of Metaphysics', *Philosophy and Phenomenological Research*, 14: 439–69

HODGE, JOANNA. 2007. *Derrida on Time* (London: Routledge)

HUGHES, HENRY STUART. 2002. *The Obstructed Path: French Social Thought in the Years of Desperation, 1930–1960* (New Brunswick, NJ: Transaction Publishers) [first published 1968]

HUSSERL, EDMUND. 1931. *Ideas: General Introduction to a Pure Phenomenology*, trans. by W. R. Boyce Gibson (London: Allen & Unwin) [first published 1913; known as *Ideas I*]

—— 1964. *The Phenomenology of Internal Time-Consciousness*, ed. by Martin Heidegger, trans. by James S. Churchill (Bloomington: Indiana University Press) [lectures from 1905 with addenda; first published 1928]

—— 1965. 'Philosophy as Rigorous Science', in *Phenomenology and the Crisis of*

Philosophy, trans. by Quentin Lauer (New York: Harper & Row), pp. 71–147 [1911 article]
—— 1997. *Edmund Husserl: Psychological and Transcendental Phenomenology and the Confrontation with Heidegger (1927–1931)*, trans. and ed. by Thomas Sheehan and Richard E. Palmer (Dordrecht: Kluwer Academic)
HUTTO, DANIEL D. (ed.). 2007. *Narrative and Understanding Persons*, Royal Institute of Philosophy Supplement, LX (Cambridge: Cambridge University Press)
HYPPOLITE, JEAN. 1971. *Figures de la pensée philosophique*, 2 vols (Paris: Presses universitaires de France)
IHDE, DON. 1971. *Hermeneutic Phenomenology: The Philosophy of Paul Ricœur* (Evanston, IL: Northwestern University Press)
JANICAUD, DOMINIQUE. 2009. *La Phénoménologie dans tous ses états: Le Tournant théologique de la phénoménologie française suivi de La Phénoménologie éclatée* (Paris: Gallimard) [*Le Tournant* first published 1990; *La Phénoménologie* first published 1998]
JEFFERSON, ANN. 2005. 'Biography and the Question of Literature in Sartre', in *Sartre Today: A Centenary Celebration*, ed. by Adrian Van den Hoven and Andrew Leak (New York: Berghahn), pp. 179–94
JUDOVITZ, DALIA. 1988. *Subjectivity and Representation in Descartes: The Origins of Modernity* (Cambridge: Cambridge University Press)
KANT, IMMANUEL. 1998. *Critique of Pure Reason*, ed. and trans. by Paul Guyer and Allen W. Wood (Cambridge: Cambridge University Press) [first published 1781, with a second edition 1787]
KEARNEY, RICHARD. 2003. *Strangers, Gods and Monsters: Interpreting Otherness* (London: Routledge)
—— 2005. 'Time, Evil, and Narrative: Ricœur on Augustine', in *Augustine and Postmodernism: Confessions and Circumfession*, ed. by John D. Caputo and Michael J. Scanlon (Bloomington: Indiana University Press), pp. 144–58
KEEN, SAM. 1984. 'The Development of the Idea of Being in Marcel's Thought', in *The Philosophy of Gabriel Marcel* (see Schilpp and Hahn 1984, below), pp. 99–120, with a reply from Marcel pp. 121–22 [article written 1968]
KIERKEGAARD, SØREN. 1946. *Attack upon 'Christendom' 1854–1855*, trans. by Walter Lowrie (London: Oxford University Press)
KIRK, GEOFFREY STEPHEN, JOHN EARLE RAVEN, and MALCOLM SCHOFIELD. 1983. *The Presocratic Philosophers: A Critical History with Selected Texts*, 2nd edn (Cambridge: Cambridge University Press) [first published 1957]
KOSKY, JEFFREY L. 2000. 'Translator's Preface: The Phenomenology of Religion: New Possibilities for Philosophy and for Religion', in *Phenomenology and the 'Theological Turn': The French Debate*, ed. by Dominique Janicaud and others (New York: Fordham University Press), pp. 107–20
—— 2001. *Levinas and the Philosophy of Religion* (Bloomington: Indiana University Press)
LAWLOR, LEONARD. 2003. *The Challenge of Bergsonism: Phenomenology, Ontology, Ethics* (London: Continuum)
LAZARON, HILDA. 1978. *Gabriel Marcel the Dramatist* (Gerrards Cross: Smythe)
LÊ, THÀNH TRI. 1961. *L'Idée de la participation chez Gabriel Marcel: superphénoménologie d'une intersubjectivité existentielle* (Saigon: Nguyen Dinh Vuong)

LEFTOW, BRIAN. 1991. *Time and Eternity* (Ithaca, NY: Cornell University Press)
LESCOE, FRANCIS J. 1974. *Existentialism with or without God* (New York: Alba House)
LÉVINAS, EMMANUEL. 1983. *Le Temps et l'Autre* (Paris: Presses universitaires de France) [four lectures from 1946–47; first published 1979]
—— 1987. *Hors sujet* (Paris: Fata Morgana)
—— 1988. 'The Paradox of Morality: An Interview with Emmanuel Levinas', with Tamra Wright, Peter Hughes, and Alison Ainley, trans. by Andrew Benjamin and Tamra Wright, in *The Provocation of Levinas* (see Bernasconi and Wood 1988, above), pp. 168–80 [interview held in 1986]
—— 1990A. *Autrement qu'être: ou au-delà de l'essence* (Paris: Kluwer Academic) [first published 1974]
—— 1990B. *Totalité et infini: essai sur l'extériorité* (Paris: Kluwer Academic) [first published 1961]
—— 1991. *Entre nous: essais sur le penser-à-l'autre* (Paris: Grasset)
—— 1992. *De Dieu qui vient à l'idée* (Paris: Vrin) [first published 1982]
—— 1993. *Dieu, la mort et le temps* (Paris: Grasset) [two lecture courses presented at the Sorbonne, 1975–76]
LÉVINAS, EMMANUEL, XAVIER TILLIETTE, and PAUL RICŒUR (eds). 1976. *Jean Wahl et Gabriel Marcel* (Paris: Beauchesne)
LHOSTE, PIERRE. 1999. *Entretiens avec Gabriel Marcel*, 3 vols ([Paris]: Institut national de l'audiovisuel) [12 interviews in total, from 1973; interviews unnamed]
LOCKE, JOHN. 1997. *An Essay Concerning Human Understanding*, ed. by Roger Woolhouse (London: Penguin) [first published 1689]
LOUBET DEL BAYLE, JEAN-LOUIS. 1969. *Les Non-conformistes des années 30: une tentative de renouvellement de la pensée politique française* (Paris: Seuil)
MACINTYRE, ALASDAIR. 1981. *After Virtue: A Study in Moral Theory* (London: Duckworth)
MACQUARRIE, JOHN. 1972. *Existentialism* (London: Hutchinson)
MARION, JEAN-LUC. 1991. 'Réponses à quelques questions', *Revue de métaphysique et de morale*, 96: 65–76
MARITAIN, JACQUES. 1947. *Court traité de l'existence et de l'existant* (Paris: Hartmann)
MARSH, JAMES L., JOHN D. CAPUTO, and MEROLD WESTPHAL (eds). 1992. *Modernity and its Discontents* (New York: Fordham University Press)
MARTIN, WALLACE. 1986. *Recent Theories of Narrative* (Ithaca, NY: Cornell University Press)
MARY, ANNE. 2008. 'Drame et pensée: le cas de Gabriel Marcel' (unpublished doctoral thesis (3e cycle), Université Paris IV — Sorbonne)
McQUILLAN, MARTIN (ed.). 2000. *The Narrative Reader* (London: Routledge)
MICHAUD, THOMAS A. 1990. 'Secondary Reflection and Marcelian Anthropology', *Philosophy Today*, 34: 222–28
MONESTIER, MARIANNE. 1999. *Entretiens avec Gabriel Marcel*, 2 vols ([Paris]: Institut national de l'audiovisuel) [6 interviews in total, from 1970: I: (1) 'La Philosophie et la liberté'; (2) 'La Philosophie est-elle un art de vivre?'; (3) 'De la philosophie à la parapsychologie'; II: (4) 'Le Théâtre de l'inquiétude métaphysique'; (5) 'De la communication aux techniques d'avilissement'; (6) 'La Philosophie de l'espérance']

MOORE, F. C. T. 1996. *Bergson: Thinking Backwards* (Cambridge: Cambridge University Press)
MORAN, DERMOT. 2000. *Introduction to Phenomenology* (London: Routledge)
MOUNIER, EMMANUEL. 1962. *Introduction aux existentialismes* (Paris: Gallimard)
MULDOON, MARK S. 2005. 'Ricœur's Ethical Poetics: Genesis and Elements', *International Philosophical Quarterly*, 45: 61–86
—— 2006. *Tricks of Time: Bergson, Merleau-Ponty and Ricœur in Search of Time, Self and Meaning* (Pittsburgh, PA: Duquesne University Press)
MULLARKEY, JOHN (ed.). 1999. *The New Bergson* (Manchester: Manchester University Press)
NEWMAN, MICHAEL. 2000. 'Sensibility, Trauma, and the Trace: Levinas from Phenomenology to the Immemorial', in *The Face of the Other and the Trace of God* (see Bloechl 2000, above), pp. 90–129
NEWTON, ISAAC. 1729. *The Mathematical Principles of Natural Philosophy*, trans. by Andrew Motte, 2 vols (London: printed for Benjamin Motte) [first published 1687]
NIETZSCHE, FRIEDRICH. 1913. *The Will to Power: An Attempted Transvaluation of all Values*, trans. by Anthony M. Ludovici, in *The Complete Works of Friedrich Nietzsche*, ed. by Oscar Levy, 2nd edn, 18 vols (Edinburgh: Foulis), XV [first published posthumously, in 1901]
O'MALLEY, JOHN B. 1966. *The Fellowship of Being: An Essay on the Concept of Person in the Philosophy of Gabriel Marcel* (The Hague: Nijhoff)
ONEGA, SUSANA, and JOSÉ ANGEL GARCÍA LANDA (eds). 1996. *Narratology: An Introduction* (London: Longman)
PARAIN-VIAL, JEANNE. 1966. *Gabriel Marcel et les niveaux de l'expérience* (Paris: Seghers)
—— 1976. 'L'Être et le temps chez Gabriel Marcel', in *Entretiens autour de Gabriel Marcel* (see Belay and others 1976, above), pp. 187–201; related discussion pp. 202–10 [1973 conference paper]
—— 1985A. 'Existence', in *Vocabulaire philosophique de Gabriel Marcel* (see Plourde and others 1985, below), pp. 233–47
—— 1985B. 'Temps-éternité', in *Vocabulaire philosophique de Gabriel Marcel* (see Plourde and others 1985, below), pp. 480–513
—— 1989. *Gabriel Marcel: un veilleur et un éveilleur* (Lausanne: L'Âge d'Homme)
—— (ed.). 1992. *L'Esthétique musicale de Gabriel Marcel*, Présence de Gabriel Marcel, Cahiers II–III (Paris: Aubier-Montaigne) [first published 1980]
PAX, CLYDE. 1972. *An Existential Approach to God: A Study of Gabriel Marcel* (The Hague: Nijhoff)
PEPERZAK, ADRIAAN T. (ed.). 1995. *Ethics as First Philosophy: The Significance of Emmanuel Levinas for Philosophy, Literature and Religion* (New York: Routledge)
PERPICH, DIANE. 2008. *The Ethics of Emmanuel Levinas* (Stanford, CA: Stanford University Press)
PIERCEY, ROBERT. 2005. 'The Role of Greek Tragedy in the Philosophy of Paul Ricœur', *Philosophy Today*, 49: 3–13
PLATO. 1971. *Timaeus and Critias*, trans. and with an introduction and appendix on Atlantis by Desmond Lee (Harmondsworth: Penguin) [written 360 BCE]

PLOTINUS. 1962. *The Enneads*, trans. by Stephen MacKenna, 3rd edn, rev. by Bertram Samuel Page (London: Faber & Faber) [written 250 CE]

PLOURDE, SIMONNE, and OTHERS. 1985. *Vocabulaire philosophique de Gabriel Marcel* (Montreal: Bellarmin; Paris: Cerf)

POIRIER, RENÉ. 1989. 'Témoignage', in *Gabriel Marcel* (see Sacquin 1989, below), pp. 38–45

PROTEVI, JOHN. 1999. '*Inventio* and the Unsurpassable Metaphor: Ricœur's Treatment of Augustine's Time Meditation', *Philosophy Today*, 43: 86–94

QUINN, JOHN M., OSA. 1999. 'Time', in *Augustine through the Ages: An Encyclopedia*, ed. by Allan D. Fitzgerald and others (Grand Rapids, MI: Eerdmans), pp. 832–38

RAMUSSEN, DAVID. 2002. 'Rethinking Subjectivity: Narrative Identity and the Self', in *Ricœur as Another* (see Cohen and Marsh 2002, above), pp. 57–69

RANDALL, ALBERT B. 1995. 'Personhood and the Mystery of Being: Marcel's Ontology of Communion, Presence, and Availability', *Bulletin de la Société américaine de philosophie de langue française*, 7: 93–103

RAOUL, VALERIE. 1983. 'The Diary Novel: Model and Meaning in *La Nausée*', *The French Review*, 56: 703–10

RÉE, JONATHAN. 1987. *Philosophical Tales: An Essay on Philosophy and Literature* (London: Methuen)

—— 1991. 'Narrative and Philosophical Experience', in *On Paul Ricœur* (see Wood 1989, below), pp. 74–83

RÉMOND, RENÉ. 1989. 'Gabriel Marcel, témoin de son temps', in *Gabriel Marcel* (see Sacquin 1989, below), pp. 31–37

RICHARDSON, BRIAN. 2007. 'Drama and Narrative', in *The Cambridge Companion to Narrative* (see Herman 2007, above), pp. 142–55

RICKMAN, HANS PETER (ed.). 1976. *W. Dilthey: Selected Writings*, trans. by Hans Peter Rickman (Cambridge: Cambridge University Press)

RICŒUR, PAUL. 1948. *Gabriel Marcel et Karl Jaspers: philosophie du mystère et philosophie du paradoxe* (Paris: Temps présent)

—— 1968. *Entretiens Paul Ricœur Gabriel Marcel* (Paris: Aubier-Montaigne)

—— 1969. *Le Conflit des interprétations: essais d'herméneutique* (Paris: Seuil)

—— 1976A. 'Entre Gabriel Marcel et Jean Wahl', in *Jean Wahl et Gabriel Marcel* (see Lévinas, Tilliette, and Ricœur 1976, above), pp. 57–87 [1975 conference paper]

—— 1976B. 'Gabriel Marcel et la phénoménologie', in *Entretiens autour de Gabriel Marcel* (see Belay and others 1976, above), pp. 53–74; related discussion pp. 75–94 [1973 conference paper]

—— 1980. 'Narrative Time', *Critical Inquiry*, 7: 169–90

—— 1983. *Temps et récit I: l'intrigue et le récit historique* (Paris: Seuil)

—— 1984A. 'Le Temps raconté', *Revue de métaphysique et de morale*, 89: 436–52

—— 1984B. *Temps et récit II: la configuration du temps dans le récit de fiction* (Paris: Seuil)

—— 1985. *Temps et récit III: le temps raconté* (Paris: Seuil)

—— 1986. *Du texte à l'action: essais d'herméneutique II* (Paris: Seuil)

—— 1988. 'L'Identité narrative', *Esprit*, 7/8: 295–304

—— 1989. 'Entre éthique et ontologie: la disponibilité', in *Gabriel Marcel: colloque* (see Sacquin 1989, below), pp. 157–65 [1988 conference paper]

—— 1990. *Soi-même comme un autre* (Paris: Seuil)
—— 1991. 'Life in Quest of Narrative', in *On Paul Ricœur* (see Wood 1989, below), pp. 20–33
—— 1995. *La Critique et la Conviction: entretien avec François Azouvi et Marc de Launay* (Paris: Calmann-Lévy)
RIST, JOHN. 1996. 'Plotinus and Christian Philosophy', in *The Cambridge Companion to Plotinus*, ed. by Lloyd P. Gerson (Cambridge: Cambridge University Press), pp. 386–413
RODOWICK, DAVID NORMAN. 1997. *Gilles Deleuze's Time Machine* (Durham, NC: Duke University Press)
RUBENSTEIN, MARY-JANE. 2003. 'Unknow Thyself: Apophaticism, Deconstruction, and Theology after Ontotheology', *Modern Theology*, 19: 387–417
RUSSELL, BERTRAND. 1925. *Mysticism and Logic and Other Essays* (London: Longmans, Green)
—— 1991. *History of Western Philosophy: And its Connection with Political and Social Circumstances from the Earliest Times to the Present Day*, 2nd edn (London: Routledge) [first published 1946]
RYAN, MARIE-LAURE. 2007. 'Toward a Definition of Narrative', in *The Cambridge Companion to Narrative* (see Herman 2002, above), pp. 22–35
SACQUIN, MICHÈLE (ed.). 1989. *Gabriel Marcel: colloque organisé par la Bibliothèque nationale et l'association 'Présence de Gabriel Marcel', 28–30 septembre 1988* (Paris: Bibliothèque nationale)
SARTRE, JEAN-PAUL. 1938. *La Nausée* (Paris: Gallimard)
—— 1977. *Sartre*, film produced by Alexandre Astruc and Michel Contat, with the participation of Simone de Beauvoir and others (Paris: Gallimard)
—— 1996. *L'Existentialisme est un humanisme* (Paris: Gallimard) [lecture presented October 1945; first published 1946]
—— 2004. *L'Être et le Néant*, ed. by Arlette Elkaïm-Sartre (Paris: Gallimard) [first published 1943]
SAYRE, PATRICIA A. 1997. 'Personalism', in *A Companion to Philosophy of Religion*, ed. by Philip L. Quinn and Charles Taliaferro (Malden, MA, and Oxford: Blackwell), pp. 129–35
SCHACTER, DANIEL L. 1996. *Searching for Memory: The Brain, the Mind, and the Past* (New York: BasicBooks)
SCHECHTMAN, MARYA. 1996. *The Constitution of Selves* (Ithaca, NY: Cornell University Press)
—— 2007. 'Stories, Lives and Basic Survival', in *Narrative and Understanding Persons* (see Hutto 2007, above), pp. 155–78
SCHILPP, PAUL ARTHUR, and LEWIS EDWIN HAHN (eds). 1984. *The Philosophy of Gabriel Marcel* (La Salle, IL: Open Court)
SEVERSON, ERIC. 2009. 'Proximity and Diachrony: Levinas and the Philosophy of Time', unpublished conference paper, *Society for Phenomenology and Existential Philosophy*, 48th Annual Meeting, 31 October, Key Bridge Marriott Hotel, Arlington, VA
SIMMONS, J. AARON. 2008. 'God in Recent French Phenomenology', *Philosophy Compass*, 3: 910–32
SMITH, ANDREW. 1996. 'Eternity and Time', in *The Cambridge Companion to*

Plotinus, ed. by Lloyd P. Gerson (Cambridge: Cambridge University Press), pp. 196–216

SORABJI, RICHARD. 1983. *Time, Creation and the Continuum: Theories in Antiquity and the Early Middle Ages* (London: Duckworth)

SPIEGELBERG, HERBERT. 1994. *The Phenomenological Movement: A Historical Introduction*, with the collaboration of Karl Schuhmann, 3rd edn (Dordrecht: Kluwer) [first published 1960]

STRATTON-LAKE, PHILIP. 1998. 'Marcel', in *A Companion to Continental Philosophy*, ed. by Simon Critchley and William R. Schroeder (Oxford: Blackwell), pp. 340–46

STRAUS, ERWIN W., and MICHAEL A. MACHADO. 1984. 'Marcel's Notion of Incarnate Being', in *The Philosophy of Gabriel Marcel* (see Schilpp and Hahn 1984, above), pp. 123–55 [article written 1970]

STRAWSON, GALEN (ed.). 2005. *The Self?* (Oxford: Blackwell)

STUDER, BASIL. 1993. *Trinity and Incarnation: The Faith of the Early Church*, trans. by Matthias Westerhoff, ed. by Andrew Louth (Edinburgh: T&T Clark)

STUMP, ELEONORE, and NORMAN KRETZMANN. 1981. 'Eternity', *Journal of Philosophy*, 78: 429–58

SWEETMAN, BRENDAN. 2002. 'Gabriel Marcel: Ethics within a Christian Existentialism', in *Phenomenological Approaches to Moral Philosophy: A Handbook*, ed. by John J. Drummond and Lester Embree (Dordrecht: Kluwer), pp. 269–88

—— 2008. *The Vision of Gabriel Marcel: Epistemology, Human Person, the Transcendent* (Amsterdam: Rodopi)

TATTAM, HELEN. 2007. 'Existentialist Ethical Thought in the Theatre of Gabriel Marcel, Albert Camus, and Jean-Paul Sartre' (unpublished master's thesis, University of Nottingham)

—— 2010A. 'Storytelling as Philosophy: The Case of Gabriel Marcel', *Romance Studies*, 28: 223–34

—— 2010B. 'Philosophers' Stories: Gabriel Marcel and Narrative', *American Catholic Philosophical Quarterly*, 84: 711–27

TAYLOR, CHARLES. 1989. *Sources of the Self: The Making of the Modern Identity* (Cambridge: Cambridge University Press)

TILLICH, PAUL. 1951. *Systematic Theology, Volume 1* (Chicago, IL: University of Chicago Press) [vols 2 and 3 published 1957 and 1963]

TRAINOR, PAUL. 1988. 'Autobiography as Philosophical Argument: Socrates, Descartes, and Collingwood', *Thought*, 63: 378–96

TREANOR, BRIAN. 2006A. *Aspects of Alterity: Lévinas, Marcel, and the Contemporary Debate* (New York: Fordham University Press)

—— 2006B. 'Constellations: Gabriel Marcel's Philosophy of Relative Otherness', *American Catholic Philosophical Quarterly*, 80: 369–92

TROISFONTAINES, ROGER. 1953. *De l'existence à l'être: la philosophie de Gabriel Marcel*, 2 vols (Louvain: Nauwelaerts-Vrin)

—— 1954. *Un philosophe de la foi: Gabriel Marcel*, Collection Convertis du XXe siècle (Brussels: Foyer Notre Dame)

TSUKADA, SUMIYO. 1995. *L'Immédiat chez H. Bergson et G. Marcel* (Louvain: Peeters)

TURETZKY, PHILIP. 1998. *Time* (London: Routledge)
VAN DEN HENGEL, JOHN. 2002. 'Can there be a Science of Action?', in *Ricœur as Another* (see Cohen and Marsh 2002, above), pp. 71–92
VANHOOZER, KEVIN J. 1991. 'Antecedents to *Time and Narrative*', in *On Paul Ricœur* (see Wood 1989, below), pp. 34–54
VANNI, MICHEL. 2004. *L'Impatience des réponses: l'éthique d'Emmanuel Lévinas au risque de son inscription pratique* (Paris: CNRS)
VIGORITO, JOHN V. 1984. 'On Time in the Philosophy of Marcel', in *The Philosophy of Gabriel Marcel* (see Schilpp and Hahn 1984, above), pp. 391–417, with a reply from Marcel pp. 418–19 [article written 1969]
VOLKOFF, VLADIMIR. 1989. 'Gabriel Marcel et la monarchie', in *Gabriel Marcel* (see Sacquin 1989, above), pp. 72–80
WAHL, JEAN. 1954. *Les Philosophies de l'existence* (Paris: Colin)
WALDENFELS, BERNHARD. 1995. 'Response and Responsibility in Levinas', in *Ethics as First Philosophy* (see Peperzak 1995, above), pp. 39–52
WESTPHAL, MEROLD. 1995. 'Levinas's Teleological Suspension of the Religious', in *Ethics as First Philosophy* (see Peperzak 1995, above), pp. 151–60
—— 2001. *Overcoming Onto-Theology: Toward a Postmodern Christian Faith* (New York: Fordham University Press)
WETZEL, JAMES. 1995. 'Time after Augustine', *Religious Studies*, 31: 341–57
WIERCIŃSKI, ANDRZEJ (ed.). 2003. *Between Suspicion and Sympathy: Paul Ricœur's Unstable Equilibrium* (Toronto: Hermeneutic Press)
WIDMER, CHARLES. 1971. *Gabriel Marcel et le théisme existentiel* (Paris: Cerf)
WOLFF, ERNST. 2010. 'The Quest for a Post-Metaphysical Access to the Human: From Marcel to Heidegger', *Journal of the British Society for Phenomenology*, 41: 132–49
WOOD, DAVID C. 1989. *The Deconstruction of Time* (Atlantic Highlands, NJ: Humanities Press International)
—— (ed.). 1991. *On Paul Ricœur: Narrative and Interpretation* (London: Routledge)
ZAHAVI, DAN. 2003. 'Phenomenology and Metaphysics', in *Metaphysics, Facticity, Interpretation: Phenomenology in the Nordic countries*, ed. by Dan Zahavi, Sara Heinämaa, and Hans Ruin (Dordrecht and London: Kluwer Academic), pp. 3–22
—— 2006. *Subjectivity and Selfhood: Investigating the First-Person Perspective* (Cambridge, MA: MIT Press)
—— 2007. 'Self and Other: The Limits of Narrative Understanding', in *Narrative and Understanding Persons* (see Hutto 2007, above), pp. 179–20

INDEX

absolute knowledge, *see* certainty, objectivity, proof
abstraction 8, 15–17, 27, 30, 33, 34, 35–36, 40 nn. 12 & 17, 42 n. 39, 47, 54, 55, 65, 71 n. 1, 72 nn. 8 & 17, 97, 106, 115 n. 56, 116 n. 74, 170, 172
Adams, Pedro 88, 167
Alexander, Ian W. 10 n. 15, 39 n. 1, 75 n. 59
Alquié, Ferdinand 108
alterity, *see* other/otherness
ambiguity 6, 9, 21, 63, 68, 94, 136, 137, 138, 142, 144, 154, 179, 186 n. 29, 187 n. 32, 191 n. 85, 193, 194
analytic, *see* abstraction, logic
Anderson, Thomas 114 n. 52, 192 n. 90
anti-dogmatism 3, 7, 8, 54, 105, 107–08, 115 n. 64, 123, 146 n. 16, 154, 156, 170, 172, 179, 180–81, 188 n. 58
appearance 21, 23–24, 30, 31–32, 37, 41 n. 21, 40 nn. 8 & 14, 48–50, 51, 63, 71 n. 1, 77–78 n. 98, 78 n. 99, 96, 102, 139, 181, 184, 188 n. 52; *see also* phenomenology
appropriation 3, 23, 119, 120–21, 125, 128, 132, 133, 142, 144, 175, 187 n. 32; *see also* being and having distinction
Aquinas, Thomas, *see* Thomism
Aristotle 87, 111 n. 5
atemporal 5, 17, 18, 20, 49, 140, 162, 182, 184 n. 5
atheism 165, 166, 182, 185 n. 19, 188 n. 55, 189 n. 60
Augustine of Hippo 34, 45 n. 77, 86–92, 93, 100, 111 nn. 3 & 7 & 8 & 9 & 11, 111–12 n. 15, 160–63, 164–65, 167, 168, 170, 171, 180, 182, 183, 184 nn. 1 & 2 & 5 & 6 & 7 & 8 & 11 & 12, 185 nn. 13 & 14 & 15 & 17, 186 nn. 24 & 25 & 31
authenticity 2, 9, 14, 15, 18, 19, 24, 25–26, 31, 34, 38, 45 n. 75, 47, 61, 62, 63, 66, 67–68, 69, 72 n. 8, 74 n. 46, 77 n. 90, 79, 80, 86, 96, 103, 113 nn. 34 & 42, 115 n. 59, 119, 127, 144, 162, 163
autobiography 9, 103, 114 n. 55, 114–15 n. 56, 115 n. 60, 116 n. 71, 164, 181; *see also* diary form, narrative form, personal
availability, *see* disponibilité
avoir, see being and having distinction

bad faith 74 n. 38, 117 n. 84, 130
Bakhtin, Mikhail Mikhaïlovich 104–05, 115 n. 63
Baroni, Raphaël 109, 110, 157 n. 7
Barrett, William 40 n. 17
Bars, Henry 10 n. 15, 44 n. 64
being 10 n. 20, 14, 17, 22, 23–24, 25, 26, 29–30, 31, 34, 38, 40 n. 8, 42–43 n. 50, 43 n. 55, 47–48, 52, 53–54, 57–60, 61–62, 63–68, 72 nn. 9 & 10 & 14 & 15, 74 n. 37, 75 n. 57, 76 n. 82, 77 nn. 90 & 93 & 95, 78 n. 102, 79–83, 84 nn. 2 & 8, 86, 89, 91, 96, 98, 103–04, 106, 107, 109, 115 n. 58, 119, 126–27, 145 n. 2, 164–65, 170, 172, 173, 174–75, 178, 179, 180, 189 n. 60
  and existence distinction (*être* vs. *existence*) 9, 62, 63–64, 65, 68, 77 n. 94, 80, 92, 162–63, 167, 169, 179, 184 n. 9
  and having distinction (*être* vs. *avoir*) 21–23, 41 n. 36, 45 n. 73
being-toward-death 18, 127–28, 129–31, 137, 141
Belay, Marcel 130
Benefield, J. J. 1, 10 nn. 15 & 18
Bergson, Henri viii, 9, 11, n. 32, 14–20, 21, 22, 23–30, 31, 32–46, 47–48, 52, 54, 56, 57, 61, 65, 70, 73 n. 24, 81–83, 100, 101, 105, 113 nn. 38 & 42 & 43, 143, 182–83, 192 n. 93, 193, 194
Bernard, Michel 77 n. 90
binary 63–64, 80, 84 n. 3, 134, 176; *see also* dualism
Blundell, Boyd 78 n. 109, 114 n. 50
body 11 n. 34, 21, 35, 54–55, 186 n. 28
Bourgeois, Patrick 114 n. 50, 115 n. 66
Bradley, Arthur 191 n. 83
Bradley, Francis Herbert 7, 11 n. 30
Brockelman, Paul 113 n. 36
Buber, Martin 40 n. 16, 119–20
Busch, Thomas 112–13 n. 30, 114 n. 50, 114–15 n. 56
Butler, Judith 149 n. 48, 151 n. 69, 152 n. 82

Cain, Seymour 1, 5, 6, 39 n. 1, 41 n. 19, 75 n. 57, 102, 187 n. 37, 190 n. 70
Camus, Albert 3, 10 n. 19, 11 n. 28, 12 n. 37, 40 n. 16, 76 nn. 77 & 79, 153 n. 101, 186 n. 24
Capek, Milic 40 n. 17, 41 n. 29

INDEX

Caputo, John 117 n. 87, 192 n. 88
Carr, David 84 n. 2, 95, 103–04
Cary, Phillip 184 nn. 8 & 12, 186 n. 31
causation 5, 20, 51, 77 n. 97, 88, 89, 149 n. 41, 165, 186 n. 28
certainty 16, 31, 54, 55, 68, 95, 112 n. 22, 115 n. 60, 121, 127, 132, 179, 183, 188 n. 56, 189 n. 62, 192 n. 89
Chadwick, Henry 45 n. 77, 186 nn. 24 & 29
change 19, 24–26, 32–33, 37–38, 66, 68, 76 n. 65, 82, 91, 94–95, 98, 109, 110, 130, 131, 156, 162, 166, 183
Chenu, Joseph 122, 146 n. 17, 148 n. 34
Christianity, *see* creation, Marcel and Christianity/religion, philosophy and Christianity
chronology, *see* time and succession
Claudel, Paul 3, 140
cogito 16, 35, 45 n. 74, 54, 56, 73 nn. 24 & 27 & 28 & 29 & 30 & 31, 74 n. 38, 92, 94, 95, 100, 112 n. 22; *see also* Descartes, dualism
Cohen, Richard 150 n. 59, 152 n. 72, 154
Coleridge, Samuel Taylor 3
Collins, James 6, 10 n. 1, 30
communication 2, 15, 19, 24, 25, 26, 43 n. 54, 44 n. 71, 101, 104, 106–08, 116 n. 74, 121, 122, 123, 137–38, 147 n. 25, 151 n. 63, 152 n. 81, 156, 166, 179–81, 186 n. 24, 187 n. 36, 189 n. 58; *see also* language
communion, *see* intersubjectivity, unity
community, *see* intersubjectivity
conceptual thought, *see* abstraction, objectification, objectivity, representation
consistency 1, 6, 14, 31, 39, 40 n. 16, 48, 52, 63, 79, 81, 83, 123, 145, 161, 172, 193; *see also* unity of Marcel's œuvre
contingency 34, 37, 38, 46 n. 92, 47, 51, 62, 63, 72, 99–100, 101, 114 n. 45
conversion viii, 3-4, 24, 55, 70, 156, 162–63, 164, 167, 171, 173, 174, 186 n. 25, 190 n. 70, 191 n. 82
Cooney, William 5, 20
Cooper, David 2–3, 10 nn. 1 & 6 & 19
creation, Christian doctrine of 162, 163, 164–65, 166, 178, 180, 184 nn. 2 & 6 & 8, 185 n. 15, 186 n. 30
creativity 23, 26, 30, 42 n. 45, 45 n. 84, 87, 93, 97, 98, 100–01, 110, 111 n. 14, 113 nn. 34 & 38 & 41, 114 n. 44, 116 n. 79, 117 n. 84, 128, 140, 180
Critchley, Simon 190 n. 73
Crowley, Patrick 109, 110, 118 n. 90, 156

Dasein 51, 72 nn. 10 & 14 & 15, 77 nn. 94 & 95, 188 n. 52; *see also* Heidegger
death 58, 75 n. 57, 135–36, 152 n. 73
  life after death 43 n. 51, 167
  see also being-toward-death, finitude, immortality
deduction, *see* logic
Deleuze, Gilles 1, 5, 83, 194
Derrida, Jacques 7, 79–81, 84 nn. 2 & 4 & 5, 109, 111, 150 n. 58, 187 n. 44, 191 n. 80, 194; *see also* metaphysics of presence
Descartes, René 115 n. 60, 190 n. 73; *see also* cogito
Desmond, William 43 n. 55, 165
Devaux, André 10 n. 15, 44 n. 64, 70
DeWeese, Garrett 86, 111 nn. 3 & 8, 162, 184 n. 2
dialectic 14, 27–28, 29, 30, 36, 37, 38, 39, 44 n. 63, 45 n. 89, 47, 49–50, 65, 66, 71 n. 4, 76 n. 82, 81, 86, 95, 96, 97, 98, 101, 112 n. 23, 114 n. 45
diary form 6, 9, 108, 115 n. 60, 117 n. 84; *see also* autobiography, narrative form, personal
Dilthey, Wilhelm 39 n. 5
disponibilité 56, 60, 74 nn. 36 & 37, 76 n. 67, 78 n. 105, 108, 130, 131, 142, 144, 166
dogmatism, *see* anti-dogmatism
Dostal, Robert 84 n. 2, 194
dualism 31, 32, 45 n. 74, 76 n. 81, 186 n. 28, 190 n. 64; *see also* binary, cogito
Du Bos, Charles viii, 3, 70, 73 n. 26, 78 n. 108, 169, 187 nn. 33 & 42
Dudiak, Jeffrey 129, 134, 151 n. 66
durée 17, 18–20, 22, 23, 24–25, 26, 28–29, 30, 32–33, 34, 35–36, 38, 43 n. 50, 44 nn. 66 & 71, 45 n. 82, 47, 100, 101, 126

Eakin, Paul John 114 n. 55
empiricism 17, 31, 34, 35, 51, 55, 95, 113 n. 34, 169, 174, 187 n. 36
engagement 3–4, 17, 104, 108, 123–24, 136, 147 nn. 24 & 25 & 29 & 30, 148 n. 31, 165, 178, 179–80, 182
Engel, Susan 45 n. 84
epistemology 16, 66, 68, 74 n. 39, 79, 80, 81, 103, 116 n. 66, 162; *see also* certainty, uncertainty
essence, *see* being
eternal:
  return 119, 150 n. 56
  presence/present 32–33, 34, 39, 47, 66, 80, 86, 90, 91, 130, 134, 140, 144, 162, 163, 171, 180, 184 n. 7
eternity 9, 14, 26–27, 31–33, 34, 35–36, 37–39, 46 n. 93, 47, 48, 59–62, 63, 64–65, 66, 68, 70, 71 n. 6, 75 nn. 57 & 59, 76 nn. 69 & 71, 80,

Index

86, 88, 89, 90, 91, 92, 94, 95, 96, 98, 99–100, 101, 109, 110–11, 114 n. 45, 130, 132, 144, 145, 156–57, 171–72, 178–80, 183–84, 193
   and God 90, 160–62, 166–68, 170, 172, 174–77, 178, 180, 181–82, 183–84, 184 n. 7, 185 n. 13, 186 n. 26
ethics 11 nn. 28 & 31, 56, 68–70, 78 n. 105, 88, 104, 108–09, 120–21, 123–25, 126, 129, 133–34, 136–37, 139, 143–44, 146 n. 8, 147 n. 25, 148 n. 35, 150 n. 55, 152 n. 82, 153 nn. 91 & 98 & 99, 154–57, 164–65, 167, 176–79, 180, 183–84, 190 n. 73, 194; *see also* justice, normativity, responsibility
être, *see* being
existence, *see* being, self, subjectivity, time as experienced/lived
existentialism 1, 2–3, 10 nn. 2 & 7 & 19, 165, 166, 185 n. 19, 189 n. 58
expression, *see* communication, language

faith/fidelity 2, 11 n. 31, 58–60, 62, 75 n. 63, 76 nn. 69 & 81, 97, 124, 127, 137, 139, 153 n. 100, 156, 161, 166–67, 171–72, 176, 178, 180
   religious faith 31–32, 167, 127–28, 168–70, 171, 172–73, 175, 176, 183, 186 n. 24, 187 nn. 34 & 37, 188 n. 50, 189–90 n. 63, 192 n. 89
fiction 12 n. 37, 93, 105, 109, 116 n. 66, 191 n. 82
fidelity, *see* faith/fidelity
finitude 37–38, 47, 49, 52, 63, 64, 75 n. 57, 76 n. 81, 86, 127, 129, 137, 149 n. 41, 155, 162, 164, 166, 167, 168, 190 n. 73; *see also* creation, death
first-person, *see* autobiography, diary form, narrative form, personal, subjectivity
flux, *see* change
forgiveness 131–32, 135, 150 nn. 54 & 55
form, *see* genre, philosophy and form
Francis, Jacques 131, 150 n. 55
freedom 23, 62, 119, 125, 132, 134, 141, 144, 145 n. 4, 149 n. 41, 150 n. 56, 151 n. 67, 151–52 n. 71, 153 n. 101, 165–66, 167, 185 nn. 20 & 21
Freeman, Mark 116 n. 66
Fuchs, Wolfgang 79–80, 84 n. 2
future 60, 75 n. 65, 79, 83, 86, 88–89, 91, 96, 98–99, 100, 111, 112 n. 24, 113 nn. 41 & 42, 127, 129, 130, 131, 132–33, 135, 136, 138, 139, 143, 152 nn. 72 & 80, 162, 167, 178

Gallagher, Kenneth 1, 3, 5, 77 n. 90, 78 nn. 101 & 109, 178, 184 n. 11
genre 6–7, 9, 11 n. 28, 24, 43 n. 54, 111–12 n. 15, 124–25; *see also* diary form, narrative form, philosophy and form, theatre
Gillman, Neil 4–5, 69, 153 n. 97, 172
Gilson, Étienne 6, 10 n. 2
God, *see* eternity and God, Marcel and Christianity/religion, philosophy and Christianity, religion, unity with God
Gouhier, Henri 192 n. 86
Grandgeorge, L. 186 nn. 24 & 25
Gugelot, Frédéric 3–4

Hanley, Katharine Rose 122
Harper, Ralph 184–85 n. 13
Harrison, Carol 186 n. 30
Hart, Kevin 191 n. 83
having, *see* being and having distinction
Hegel, Georg Wilhelm Friedrich 11 n. 29, 40 n. 17, 43 n. 55, 44 n. 63, 49–50, 65–66, 76 n. 78, 77 nn. 97 & 98, 109, 117 n. 87, 145 n. 6, 150 n. 58, 188 n. 50
Heidegger, Martin ix, 3, 17, 18, 24, 40 n. 16, 41 nn. 19 & 21, 42 n. 41, 43 n. 53, 51–52, 59, 65, 67–68, 72 n. 16, 81, 87, 114 n. 48, 126–27, 129, 131, 137, 174, 175, 188 n. 52, 189 n. 60, 194; *see also* being-toward-death, *Dasein*
Hering, Jean 194
hermeneutics 28, 45 n. 89, 51–52, 59, 63, 65, 71 n. 6, 84, 87, 100, 109, 110, 112 n. 21, 155, 194
Hersch, Jeanne 114 n. 53
Hocking, William 8, 11 n. 32
Hodge, Joanna 151 n. 65
hope 60–63, 78 n. 101, 142, 167, 171–72, 178, 180
Husserl, Edmund 17, 18, 41 nn. 19 & 23 & 25, 42 nn. 46 & 49, 43 n. 57, 51, 52, 72 nn. 9 & 12, 79, 80, 84 n. 2, 87, 173, 174, 184, 188 n. 52, 194

idealism 7–8, 9, 11 n. 31, 16, 17, 42 n. 46, 47, 48–50, 65–66, 71 n. 4, 77 n. 97, 82, 83, 86, 102, 143, 157, 183, 192 n. 89, 193, 194
identity, *see* self
imagination, *see* creativity
immanence 35, 51, 174, 175, 176, 177, 182, 183, 188 nn. 50 & 51, 191 n. 85
immediate, *see* present
immortality 58, 60, 64, 75 n. 57, 163
impersonal, *see* abstraction, logic, objectivity
inauthenticity 22, 23, 32, 56, 63, 67–68, 80
indeterminacy, *see* ambiguity, unverifiability
infinite/infinity 41 n. 29, 45 n. 77, 51, 69, 71 n. 6, 75 n. 56, 99, 125, 128, 134, 137, 141, 149 n. 46, 177, 179, 190 n. 73

infinite time 129, 133, 135
intersubjectivity 2, 30, 38, 47–48, 52–54, 56, 60, 62, 63, 64–65, 66, 67–68, 69–70, 73 n. 23, 74 nn. 37 & 38 & 40, 75 nn. 56 & 57, 77 n. 90, 80, 103–04, 106–07, 110, 115 n. 58, 116 n. 75, 116–17 n. 79, 117 n. 88, 119–20, 121–22, 125, 126, 135, 137, 139–40, 143–44, 152 n. 86, 153 nn. 90 & 100, 166, 170, 171, 177, 178, 180, 191 n. 81; *see also* disponibilité, love, other/otherness, presence
intuition 11 n. 29, 18, 24, 25–26, 27–28, 29–30, 32, 38, 42 nn. 49 & 50, 43 n. 54, 44 nn. 64 & 65 & 70, 52, 57, 71 n. 1, 78 n. 98, 88, 100, 190 n. 63

Janicaud, Dominique 173–74, 180, 188 n. 47
Jaspers, Karl 2, 3, 10 n. 4, 40 n. 16, 42 n. 41
Jefferson, Ann 117 n. 84
Judovitz, Dalia 115 n. 60, 116 n. 77
justice 69, 70, 114 n. 54, 121, 123, 138, 142–43, 146 n. 8, 148 nn. 30 & 31; *see also* ethics

Kant, Immanuel 11 n. 29, 20, 41 n. 34, 48, 50, 66, 71 nn. 1 & 4, 72 n. 9, 77 n. 97, 87, 111 n. 7, 161, 188 n. 50, 189 n. 59
Kearney, Richard 91, 155, 157 n. 7
Keen, Sam 63, 166–67, 179, 180
Kierkegaard, Søren 3, 40 n. 16, 176, 188–89 n. 58
Kosky, Jeffrey 172, 174, 183–84, 188 n. 50

language 19, 21, 26, 33, 42 n. 39, 43 n. 51, 44 n. 70, 55, 65, 68, 69, 70, 74 nn. 40 & 42, 76 n. 65, 77 n. 86, 83, 88, 96, 103, 107–08, 109, 111 n. 14, 115 n. 66, 117 n. 88, 117–18 n. 89, 118 n. 90, 126, 134, 137, 140, 141, 143, 144, 146 n. 8, 148 nn. 40 & 41, 150 n. 58, 151 nn. 63 & 66 & 67, 152 nn. 83 & 84 & 85 & 86 & 87, 156, 161, 164, 166, 169, 172, 174, 175, 176, 178, 179, 183, 185 n. 83, 193
Lawlor, Leonard 24, 45 n. 85, 81–83, 194
Lazaron, Hilda 122, 146 n. 14, 152 n. 75
Lê, Thành Tri 39
Lévinas, Emmanuel 9, 111, 119–20, 125–29, 131–39, 141–45, 145 nn. 1 & 5, 145–46 n. 6, 146 n. 8, 148 nn. 39 & 40, 148–49 n. 41, 149 nn. 42 & 43 & 44 & 45 & 46, 150 nn. 56 & 58 & 59 & 60 & 61 & 62, 151 nn. 63 & 64 & 65 & 66 & 67 & 68 & 69 & 70, 151–52 n. 71, 152 nn. 72 & 74 & 79 & 80 & 81 & 82 & 83 & 84 & 86, 153 nn. 91 & 92 & 99, 154–55, 156, 157, 172, 173, 175, 176–77, 178–79, 180, 183–84,

188 nn. 47 & 54, 190 nn. 72 & 73 & 75 & 76, 191 nn. 84 & 85, 192 n. 88, 194
linear, *see* time and succession
logic 7, 16, 20–21, 40 nn. 8 & 16, 41 n. 34, 43 n. 55, 54, 55, 73 n. 27, 80, 84 n. 4, 100, 104, 106, 171, 182, 189 nn. 59 & 60
Loubet del Bayle, Jean-Louis 21, 147 n. 20
love 56–60, 64–65, 69, 75 n. 57, 132, 135, 139, 142–43, 153 nn. 92 & 94 & 97 & 100, 166–67, 170, 171, 178, 180, 185 nn. 14 & 17
loyalty, *see* faith/fidelity

Marcel, Gabriel:
  autobiographical works:
    En chemin, vers quel éveil? 3, 10 n. 17, 11 n. 33, 12 n. 37, 15, 16, 28–29, 40 nn. 12 & 13, 41 n. 21, 43 n. 58, 70, 73 nn. 19 & 23 & 25, 77 n. 89, 78 n. 108, 103, 113 n. 42, 117–18 n. 89, 121, 122, 123, 138, 146–47 n. 20, 147 nn. 24 & 25 & 29, 148 n. 31
    'Regard en arrière' 7, 8, 11 nn. 30 & 33, 16, 17, 40 nn. 12 & 13, 84 n. 8, 113 n. 34, 187 n. 34, 193
  and Christianity/religion 2, 3–4, 31–32, 43 n. 51, 142–43, 157, 161, 164–67, 168–69, 170–71, 172–73, 174–76, 178–79, 180, 181–83, 184, 185 n. 22, 187 nn. 34 & 37, 188 n. 53, 189–90 n. 63, 191 nn. 82 & 83
  fictional prose 12 n. 37
  interviews 7
    with Marianne Monestier 10 n. 17, 40 n. 13, 68, 113 n. 42, 122, 140, 146 n. 12
    with Paul Ricœur 14–15, 143, 146 n. 7, 147 n. 21, 190 n. 71
    with Pierre Boutang 3, 14, 27, 29, 30, 40 n. 13, 53, 83, 84, 122–23, 124, 138, 147 n. 21, 179, 187 n. 36, 191 n. 83
    with Pierre Lhoste 7, 114 n. 54, 123, 138, 146 n. 12, 147 n. 21, 190 n. 77
  introductions, prefaces, postfaces and responses:
    to *Un changement d'espérance* 4
    to Keen's article 63, 179
    to *Paix sur la terre* 70
    to *Rome n'est plus dans Rome* 146 n. 16
    to *Le Seuil invisible* 156, 157 n. 6, 171
    to Troisfontaines's volumes 25, 115 n. 59
    to *Vers un autre royaume* 146 n. 16
  journal articles:
    'Autour de Heidegger' 41 n. 21, 64, 77 n. 94
    'Carence de spiritualité' 182

'Les Conditions dialectiques de la philosophie de l'intuition' 27
'Considérations sur l'égalité' 140
'Un événement philosophique' 182
'Existence et objectivité' 40 n. 9
'Finalité essentielle de l'œuvre dramatique' 122, 138, 152 n. 85
'Grandeur de Bergson' 27
'Henri Bergson et le problème de Dieu' 182, 192 n. 93
'Justice pour Charles Maurras!' 147–48 n. 30
'Kierkegaard en ma pensée' 189 n. 58
'Note sur l'évaluation tragique' 157 n. 6
'Note sur les limites du spiritualisme bergsonien' 28, 182, 183
'Qu'est-ce que le bergsonisme?' 182–83
'Réflexions sur le tragique' 157 n. 6
'Théâtre de l'âme en exil' 139
'Tragique et personnalité' 42 n. 45, 157 n. 6
musical essays:
'Bergsonisme et musique' 25, 30
'Méditation sur la musique' 14
'Musique comprise et musique vécue' 42 n. 48, 43 n. 54
'La Musique dans ma vie et mon œuvre' 42 n. 39
philosophical presentations and publications:
contributions to discussion at 1973 Cerisy-la-Salle conference 12 n. 37, 68, 124, 146 n. 16, 148 n. 34
Déclin de la sagesse, Le 176, 188 n. 53
Dignité humaine et ses assises existentielles, La (William James Lectures) 7, 11 n. 31, 15, 26, 27, 29, 30, 39 n. 5, 53, 58, 61, 65, 73 n. 23, 74 nn. 36 & 40 & 48, 75 nn. 51 & 52 & 53 & 60, 77 n. 89, 100–01, 114 nn. 47 & 51, 122, 124, 140–41, 144, 145, 145–46 n. 6, 146 n. 13, 148 n. 38, 152 n. 87, 164
Existence et la liberté humaine chez Jean-Paul Sartre, L' 165
Être et avoir 9, 21, 25, 26, 28, 34, 41 n. 36, 44 n. 71, 45 n. 88, 46 nn. 90 & 94, 53, 56, 57, 59, 60, 61, 65, 66, 67, 68, 70, 71 n. 5, 72 n. 17, 73 nn. 23 & 32, 74 nn. 33 & 36 & 42, 75 nn. 55 & 58 & 60, 75–76 n. 65, 76 nn. 71 & 72 & 75, 99, 103, 108, 114 n. 51, 153 n. 94, 163, 165, 168, 171, 179, 185 n. 15
Fragments philosophiques 11 n. 29, 170, 189 n. 58
Gabriel Marcel et les injustices de ce temps: la responsabilité du philosophe 147 n. 24
Homme problématique: position et approches concrètes du mystère ontologique, L' 7, 21–22, 24, 26, 42 n. 43, 53, 54, 56, 57, 58, 61, 69, 73 nn. 18 & 30, 74 nn. 34 & 36 & 47, 75 nn. 51 & 53, 76 n. 67, 113 n. 42, 171, 173, 191 nn. 78 & 83
Hommes contre l'Humain, Les 64, 69, 76 n. 78, 103, 164, 166, 178, 181, 185 nn. 20 & 22, 190 nn. 64 & 66
Homo viator 23, 40 n. 12, 45 n. 75, 58, 60, 61, 66–67, 74 n. 36, 76 nn. 70 & 73, 77 nn. 89 & 94, 156, 163, 164, 166, 167, 176, 178, 180, 181, 185 n. 18, 191 n. 83
Journal métaphysique 8, 9, 11 nn. 32 & 34, 16, 20, 21, 26, 27–28, 29, 31–34, 35–36, 37–38, 40 n. 9 & 13, 45 n. 73, 45 n. 90, 46 nn. 91 & 93 & 94 & 96, 47, 48–51, 54, 55, 59, 61, 63, 68–69, 71 n. 6, 72 nn. 8 & 17, 73 nn. 18 & 26 & 31, 74 n. 39, 75 nn. 51 & 56 & 63, 76 nn. 76 & 81 & 82, 77 n. 89, 78 n. 100, 97, 102, 104, 107, 108, 113 nn. 31 & 42, 114 n. 47, 115 n. 62 & 64, 117 n. 81, 120, 143, 163, 168, 169, 171, 175, 187 nn. 33 & 34 & 35 & 36 & 38, 189 n. 61, 190 n. 77, 192 n. 89
Métaphysique de Royce, La 11 nn. 31 & 32, 71 n. 6
'Mon temps et moi' 23, 29, 51, 52, 71, 81
Mystère de l'être, Le (Gifford Lectures) 7, 16, 22, 26, 42 n. 43, 53, 54, 56, 58–59, 62, 63, 73 nn. 21 & 23 & 27, 74 n. 38, 75 n. 51, 96, 97, 98–99, 105–06, 110, 113 nn. 33 & 40 & 42, 116 nn. 68 & 72 & 73 & 74 & 75, 153 n. 90, 163, 171, 175, 175–76, 178, 179, 181–82, 184 n. 9, 185 n. 23, 187 n. 40, 188 n. 51, 191 nn. 82 & 83
Pour une sagesse tragique et son au-delà 12 n. 35, 17, 20, 26, 30, 40 n. 18, 41 n. 21, 44 n. 70, 54, 69, 76 n. 68, 77 n. 85, 108, 113 n. 32, 114 nn. 47 & 54, 116 n. 78, 116–17 n. 79, 117 nn. 80 & 82, 141, 166, 171, 174, 178, 180, 183, 189–90 n. 63, 191 n. 83
Présence et immortalité 8, 9, 22, 23, 53, 61, 65, 72 n. 13, 77 n. 93, 103, 107, 113 nn. 39 & 41, 114 n. 51, 146 n. 16, 164–65, 175, 179, 185 n. 15, 187 n. 40, 188 n. 57, 188–89 n. 58, 189 n. 62, 191 nn. 81 & 83, 192 n. 87

INDEX

'De la recherche philosophique' 74 n. 38, 107, 124, 146 n. 12, 190 n. 69
Du refus à l'invocation 10 n. 4, 17, 18, 20, 26, 55, 58, 66, 71 n. 5, 74 nn. 36 & 43 & 46, 77 n. 91, 114 n. 47, 115 n. 61, 139, 140, 164, 170, 171, 172, 176, 185 n. 20, 188 n. 54
'Subjectivité et transcendance' 70, 192 n. 89
'Testament philosophique' 10 n. 17, 70, 72 n. 17, 74 n. 36, 121, 124, 168–69, 173
Théâtre et religion 56
theatre:
Chapelle ardente, La 130–31, 135, 138, 148 n. 35, 149 nn. 52 & 53
Chemin de crête, Le 155, 157 n. 4
Croissez et multipliez 124
Dard, Le 116 n. 70, 124, 129–30, 135, 144–45, 155
Duchesse de Modène, La 11 n. 24
Émissaire, L' 116 n. 70, 124, 136
Fanal, Le 135–36, 152 nn. 75 & 76
Grâce, La 127–28
Homme de Dieu, Un 116 n. 70, 131–32, 135, 150 n. 54
Horizon, L' 130–31, 149 nn. 49 & 50
Iconoclaste, L' 116 n. 70
Juste, Un 136, 148 n. 35, 155, 157 n. 3
Monde cassé, Le 7, 42 n. 43, 116 n. 70, 132–33, 135, 138, 139, 147 n. 23, 150 n. 56, 155
Mort de demain, La 75 n. 54, 130–31, 136, 148 n. 35, 149 nn. 50 & 51
Palais de sable, Le 116 n. 70
Regard neuf, Le 148 n. 35
Rome n'est plus dans Rome 124, 137, 153 n. 101
Signe de la croix, Le 124
Marion, Jean-Luc 172, 173, 174, 188 nn. 52 & 54
Maritain, Jacques 3, 4, 10 n. 15, 44 n. 64, 70, 147 n. 24, 189 n. 58
Mary, Anne 135, 137, 138, 145, 146 n. 14, 152 n. 73
materialism 16, 17, 35, 82
meaning 6, 17, 18, 23–24, 51, 68, 72 n. 14, 75 n. 57, 84 n. 2, 88, 96, 97, 98, 100, 101, 107, 108, 111, 115 n. 56, 117 n. 80, 125, 127, 129, 137, 141, 145, 156, 161, 167, 170, 175, 177, 178, 183, 184, 187 n. 35, 189 n. 58, 190 n. 75; *see also* language
memory 19, 34–38, 45 nn. 77 & 82 & 84 & 88 & 89, 46 nn. 91 & 92 & 94, 47, 82–83, 89, 91, 97, 98, 113 n. 33, 134–35, 150 n. 62, 151 n. 68, 152 n. 80, 185 n. 14
Merleau-Ponty, Maurice 3, 18, 40 n. 16, 41 n. 19

metaphysics 5, 8, 9, 10–11 n. 20, 11 nn. 29 & 32, 15–16, 17, 19–20, 25–26, 28, 32–33, 39–40 n. 6, 42 n. 48, 43 nn. 54 & 55, 49–50, 55–56, 58, 59–60, 61, 64–69, 73 nn. 23 & 32, 74 n. 38, 75 n. 51, 78 n. 109, 79, 81–83, 84 n. 2, 91, 103, 108–09, 114 n. 51, 117 nn. 81 & 87, 120, 125, 145 n. 3, 146 n. 17, 157, 157 n. 6, 161, 165, 166–67, 168, 169, 170, 172, 173, 175, 176, 177, 179–80, 184–85 n. 13, 188 n. 53, 189 nn. 60 & 62, 191 n. 83, 192 n. 93, 193, 194; *see also* being
of presence 7, 79–83, 84 nn. 2 & 3 & 5, 105, 108, 109, 110–11, 119, 120, 123, 125–26, 138, 139, 142, 144, 148 n. 41, 174, 177, 183, 194; *see also* Derrida
Michaud, Thomas 43–44 n. 62
moral/morality, *see* ethics, justice, normativity, responsibility
Moran, Dermot 17, 40 n. 14, 51
mortality, *see* being-toward-death, death, finitude
Mounier, Emmanuel 4, 147 n. 24
Muldoon, Mark 29, 87, 89, 154
music 24–25, 30, 42 n. 39, 43 nn. 54 & 55 & 56 & 57 & 58, 44 n. 71, 116 n. 78, 141
mystery 28–29, 45 n. 77, 68, 74 n. 47, 75 n. 52, 76 n. 68, 77 n. 95, 78 n. 101, 99, 114 n. 54, 127, 139, 152 nn. 79 & 85, 161, 164, 171, 174, 187 n. 38, 190 n. 77; *see also* problem and mystery distinction, unverifiability

narrative:
definition of 12 n. 36, 157, 157 n. 7
form 9, 12 n. 37, 101–03, 104–09, 110, 111–12 n. 15, 113 nn. 30 & 35 & 36 & 37, 113–14 n. 44, 115 n. 60, 116 n. 66, 117 n. 84, 120, 121, 122, 124–25, 137, 144–45, 146 n. 15, 154, 156–57
identity 93–104, 109–10, 112 nn. 17 & 19 & 23 & 24 & 25, 112–13 n. 30, 113 n. 36, 114 n. 55, 115 nn. 57 & 58, 121, 124, 155, 156
time 87–88, 92–93, 94, 96, 97, 100, 101, 104, 107, 109, 110–11, 111 n. 2, 113 n. 37, 114 nn. 48 & 54, 116 n. 66, 120, 122, 125, 144, 145, 154–55, 156–57
see also diary form, fiction, Ricœur
Natorp, Paul 24, 43 n. 52
necessity 16, 19, 31–32, 36, 37, 42 n. 46, 50, 55, 65, 71 n. 1, 72 n. 8, 73 n. 28, 77–78 n. 98, 86, 98, 99–100, 101, 107, 109, 110, 113 n. 36, 114 nn. 45 & 55, 117 n. 84, 133, 138, 150 nn. 56 & 58, 166, 170, 171, 172, 174, 176, 179, 191 nn. 82 & 83 & 85, 193

Neoplatonism 161, 167–68, 170, 186 nn. 24 & 25
Newman, Michael 133, 150 n. 59, 78–79
Newton, Isaac 111 n. 9
Nieztsche, Friedrich 3, 28, 40 n. 16, 43 n. 55, 55, 73 n. 29, 112 n. 22, 185 n. 18, 188 nn. 50 & 55
non-existence 88, 89, 162, 177, 185 n. 17
normativity 69, 134, 136–37, 144; *see also* ethics, responsibility

objectification 8, 15, 19, 20, 24, 26, 28, 35, 42 n. 39, 43 n. 60, 45 n. 89, 48, 56, 57, 60, 74 n. 40, 82, 84 n. 8, 86, 94, 111 n. 2, 124, 125, 127, 142, 144, 148 n. 39, 179, 193, 194; *see also* representation, totalization
objectivity 7–8, 16–17, 19–22, 24–25, 31–32, 35, 40 n. 16, 44 n. 64, 54–56, 57–58, 59, 60, 61, 63, 71 n. 6, 73 n. 26, 75 nn. 52 & 59, 78 n. 109, 80–81, 82, 86, 90, 94, 95, 98, 101, 105, 106, 108, 113 n. 31, 116 nn. 67 & 71 & 79, 120, 153 n. 90, 163, 164, 169–70, 173, 174–75, 179, 183, 187 nn. 34 & 35 & 39, 188 n. 52, 189 n. 61, 191 n. 81, 192 n. 89
O'Malley, John 115 n. 65
ontology, *see* being, metaphysics
other/otherness 38, 54, 74 n. 37, 75 n. 50, 95, 102, 104, 106–07, 108, 110–11, 115 n. 58, 117 n. 84, 119, 120–21, 122, 125–26, 127–29, 130, 131, 132, 133–35, 136, 137, 138, 139, 140, 141–44, 145 nn. 3 & 4, 145–46 n. 6, 148 n. 39, 148–49 n. 41, 149 nn. 43 & 44 & 46 & 48, 150 nn. 52 & 55 & 58 & 59, 151 nn. 63 & 67 & 69, 151–52 n. 71, 153 nn. 91 & 92, 154–55, 174, 176–77, 178–79, 180, 183–84, 187 n. 32, 190 nn. 73 & 76, 191 n. 81; *see also* ethics, intersubjectivity
and time 110–11, 119, 125–26, 128–29, 131, 132, 133–35, 136, 138, 141–42, 144, 145, 148–49 n. 41, 151 n. 65, 154, 156–57, 160, 177
Oxford Group, *see* Réarmement moral

paradox 58, 74 n. 46, 86, 88, 89, 91, 92, 100, 115 n. 60, 117 n. 84, 119, 123, 131, 133, 150 n. 56, 161, 189 n. 58
Parain-Vial, Jeanne 5, 11 n. 21, 30, 41 n. 24, 56, 64–65, 71 n. 2, 76 n. 69, 77 n. 84
participation 30, 52, 56, 59, 74 n. 34, 104, 114 n. 54, 121, 123, 141, 153 n. 97, 168, 186 n. 24
passivity 27, 37, 44 n. 65, 51, 97, 115 n. 58, 127, 128, 151 nn. 64 & 66 & 67 & 69 & 71, 188 n. 54
past 18, 34–38, 45 n. 85, 46 nn. 90 & 91, 75 n. 63, 79, 82–83, 84 n. 7, 86, 88–89, 91, 93, 96, 97, 98–99, 100, 109, 111, 112 n. 24, 113 nn. 40 & 41, 127, 131, 132, 133–36, 139, 141, 150 n. 62,
151 nn. 64 & 66 & 68, 152 nn. 72 & 80, 162, 185 n. 14; *see also* memory
Pax, Clyde 188 n. 56
perception 8, 16, 17, 24, 35, 36, 37, 40 n. 8, 42 n. 46, 45 n. 82, 82, 84 nn. 5 & 7
Perpich, Diane 134
person, *see* self
personal 7, 9, 16, 17, 40 n. 16, 54, 87, 97, 115 n. 65, 117 n. 84, 121, 140, 170, 171, 175, 176, 180–81, 182–83, 190 n. 66, 194; *see also* autobiography, diary form, narrative form, subjectivity
personalism 4
phenomenology 17, 40 n. 14, 48, 51–53, 55–56, 57–58, 59, 60, 62, 63, 65–69, 70, 72 nn. 7 & 9, 78 nn. 99 & 100 & 101, 79–82, 84, 84 nn. 1 & 2, 86–87, 90–91, 92, 93, 100, 102, 103, 106, 110, 111 nn. 2 & 13, 114 n. 55, 142, 145 n. 1, 160, 168, 169–70, 172–77, 179, 180, 182, 183–84, 187 n. 32, 188 nn. 47 & 50 & 52 & 54, 189 n. 61, 190 n. 733, 191 n. 85, 192 n. 88, 194
philosophy:
and Christianity 164, 167–68, 170, 185 n. 19, 186 nn. 24 & 25 & 29 & 30, 191 n. 78
and form 5, 6–7, 8–9, 10 n. 19, 12 nn. 35 & 37, 24–26, 43 nn. 54 & 55, 101–03, 104–09, 110–11, 111–12 n. 15, 114–15 n. 56, 115 n. 60, 116 nn. 66 & 71 & 77, 117 n. 84, 120, 121, 122, 124–25, 137, 144–45, 154, 155, 156–57, 180–81, 190 n. 64, 194–95
Plato 33, 40 n. 17, 45 n. 77, 83, 168, 186 nn. 26 & 29
Plotinus 33, 161, 167–68, 170, 186 nn. 24 & 25 & 26 & 27 & 28 & 31
Plourde, Simonne 78 n. 102
Poirier, René 123, 147 n. 24
possession, *see* appropriation, being and having distinction
postmodern 81–82, 115 n. 66, 155, 183, 194
presence 29, 54, 57–59, 62, 66, 68–69, 74 nn. 37 & 48, 75 nn. 52 & 53 & 56 & 58, 76 n. 67, 79, 107, 110, 116 n. 75, 125, 138, 140, 143, 152 nn. 74 & 81 & 84, 153 n. 100, 191 n. 81
metaphysics of, *see* metaphysics of presence
present 14, 15, 18, 20, 26, 27, 28–29, 34, 35, 36, 37, 38, 39, 45 nn. 75 & 82 & 84 & 85, 46 nn. 92 & 96, 54, 75–76 n. 65, 76 n. 78, 79, 82–83, 84 nn. 5 & 7, 87, 88–89, 96, 97, 98, 99, 100, 108, 110–11, 125, 126, 127, 129, 130, 133, 134, 135, 136, 138, 139, 149 n. 42, 150 nn. 58 & 62, 151 nn. 64 & 66 & 67 & 68, 152 n. 83, 162, 169, 170
three-fold 89, 90–91, 162
see also eternal present

primary reflection, *see* reflection
problem and mystery distinction 8, 57–59, 65, 70, 74 nn. 42 & 43, 75 nn. 52 & 58, 78 n. 109, 103, 173; *see also* mystery
promise 59–60, 139, 156, 178, 187 n. 37
proof 16, 31, 55, 169, 176, 177; *see also* certainty, logic
Protevi, John 112 n. 15, 161
Proust, Marcel 113 n. 33

Quinn, John 111 n. 8, 164

Ramussen, David 112 n. 24
Raoul, Valerie 117 n. 84
reader 93, 104–07, 115 n. 60, 117 n. 84, 154, 179, 181
Réarmement moral 4, 10 n. 17
Rée, Jonathan 12 n. 37, 115 n. 60
reflection, modes of 23–24, 26, 27, 28, 43–44 n. 62, 50–51, 52, 55, 56–57, 62, 65, 70, 72 n. 8, 73 n. 32, 112–13 n. 30
religion 127, 172, 174–75, 176–77, 188 nn. 50 & 55 & 56 & 57, 188–89 n. 58, 191 nn. 80 & 84
    definitions of 171, 176, 190 n. 76
    see also eternity and God, faith/fidelity, Marcel and Christianity/religion, philosophy and Christianity
Rémond, René 78 n. 108, 123, 147 nn. 23 & 24
representation 17, 18, 20, 24, 25, 41 n. 29, 59, 71 n. 1, 72 n. 8, 82, 100, 114 n. 44, 124, 125, 150 n. 62, 151 nn. 64 & 66 & 68, 155
response (existential) 57, 61, 108, 151 n. 64, 152 nn. 74 & 80, 114 n. 44, 130, 132, 133, 134, 135–36, 137, 148 n. 35, 150 n. 61, 151 nn. 65 & 66, 154–55, 156, 184 n. 13
responsibility 17, 67–68, 108, 128–30, 132–38, 139, 142, 143, 147 n. 24, 150 nn. 56 & 61, 151 nn. 64 & 66 & 70 & 71, 152 nn. 80 & 83, 154–55, 165, 180, 184, 191 n. 83; *see also* engagement, ethics, normativity
rhetoric 70, 104, 106, 115 n. 60, 116 n. 77, 174, 181
Ricœur, Paul 9, 68, 70, 75 n. 61, 78 nn. 105 & 109, 86–88, 90–95, 96–97, 98, 99–101, 102, 103–04, 105, 106–07, 108, 109, 110, 111–12 n. 15, 112 nn. 19 & 22 & 24 & 25, 113 nn. 30 & 35 & 44, 114 nn. 48 & 50, 115 n. 58, 116 nn. 66 & 67, 117 n. 87, 118 n. 90, 119, 121, 143, 145, 150 n. 59, 153 n. 99, 154, 155–57, 160–61, 171, 187 nn. 43 & 44, 188 nn. 55 & 57, 194
Rist, John 168, 186 nn. 28 & 29
Royce, Josiah 7, 11 n. 31
Rubenstein, Mary-Jane 191 n. 83

Russell, Bertrand 71 n. 5, 111 n. 7

Sartre, Jean-Paul 2, 3, 10 n. 7, 11 n. 28, 12 n. 37, 18, 74 n. 38, 165
Schechtman, Marya 112 n. 19
Schelling, Friedrich Wilhelm Joseph von 3, 7, 11 n. 29
science 15, 16, 17, 19, 25, 31, 32, 39 nn. 5 & 6, 40 n. 17, 43 n. 60, 71 n. 5, 72 n. 9, 112 n. 21, 116 nn. 72 & 79, 173, 182, 192 n. 93, 193; *see also* empiricism, materialism
secondary reflection, *see* reflection
self 16, 20–21, 43 n. 51, 45 nn. 75 & 77, 51, 52, 53, 74 n. 33, 76 n. 81, 92, 93, 94, 95, 96, 97, 98–99, 100, 101, 110, 112 nn. 20 & 25, 113 n. 36, 114 n. 55, 119, 121, 124, 131, 134, 145 n. 1, 150 n. 56, 151 nn. 63 & 65, 156, 163, 164, 165; *see also cogito*, narrative identity, subjectivity
sensation 20, 27–28, 35, 54, 58, 60, 69, 73 n. 27, 75 n. 56, 116 n. 75, 117 n. 81, 121, 141, 156, 165, 169, 178
significance, *see* meaning
Simmons, J. Aaron 175
Smith, Andrew 186 n. 26
solitude 108, 117 n. 84, 122, 140–41
Sorabji, Richard 111 nn. 3 & 8, 168
space 8, 10 n. 20, 11 n. 34, 18–20, 34, 35, 38, 41 nn. 25 & 26, 43 n. 56, 45 n. 82, 52, 65, 71 nn. 1 & 5, 91, 94, 104, 105, 115 n. 63, 126, 148 n. 41, 150 n. 58, 168, 169
Spiegelberg, Herbert 41 n. 19, 55, 66–67
story, *see* narrative form
Stratton-Lake, Philip 2
Straus, Erwin and Michael Machado 164
Strawson, Galen 112 n. 19
subject, *see* self, subjectivity
subjectivity 2, 9, 16, 17, 24–25, 35, 49, 56, 71 n. 1, 77 nn. 95 & 97, 79, 82, 89–90, 97, 102, 108, 110, 111 n. 7, 115 n. 60, 116 n. 79, 119, 128–29, 133–34, 137–38, 139, 145 n. 1, 151 n. 63, 151 n. 67, 153 n. 91, 168, 175, 177, 183–84, 188–89 n. 58, 189 n. 61, 191 n. 81; *see also* personal
succession, *see* time and succession
Sweetman, Brendan 1, 10 n. 15, 44 n. 64, 153 n. 98
synthesis, *see* unity
system, *see* abstraction, totalization

technology 21–22, 42 n. 41, 57
teleology 64, 110, 155, 165, 176, 177
theatre 7, 9–10, 11 n. 28, 12 n. 37, 42 n. 43, 56,

220 INDEX

75 n. 54, 105, 116 n. 70, 120, 122–25, 126, 127–28, 129–33, 135–39, 144–45, 146 nn. 7 & 12 & 13 & 14 & 15 & 16 & 17, 147 nn. 21 & 23 & 25, 148 nn. 34 & 35 & 36 & 38, 149 nn. 49 & 50 & 51 & 52 & 53, 150 n. 54, 152 nn. 72 & 73 & 75 & 76 & 85, 153 n. 101, 154, 155, 156, 189 n. 58

Thomism 4, 10 n. 15, 44 n. 64, 189 n. 58

Tillich, Paul 186–87 n. 32

time, philosophical presentation/theory of 5–6, 9, 10, 14, 17–20, 22–23, 25, 26, 29, 31–34, 35, 38–39, 40 n. 17, 41 nn. 23 & 24, 45 nn. 73 & 82, 47–52, 62, 63–64, 65, 66, 68, 70, 71, 71 nn. 1 & 5 & 6, 72 n. 8, 75 n. 59, 77 n. 86, 77–78 n. 98, 80–83, 86–92, 100, 104, 109, 111, 111 nn. 2 & 7 & 8 & 9 & 15, 113 n. 42, 114 n. 45, 119, 120, 121, 125, 126–27, 128–29, 133–35, 138–39, 140, 141–42, 144, 145, 148–49 n. 41, 151 nn. 65 & 66, 154, 156, 157, 160–61, 162–63, 164–65, 166–68, 170, 171–72, 174, 175, 177, 180, 182, 183–84, 184 nn. 1 & 2 & 6 & 7, 185 n. 13, 186 n. 26, 193, 194

as experienced/lived 9, 18–19, 22–23, 26, 28–29, 38–39, 40 n. 17, 43 n. 57, 47, 54, 58, 59, 60–62, 63, 64, 68, 71 n. 6, 72 nn. 7 & 8 & 15, 75 n. 59, 77 n. 84, 81, 86, 87, 88, 94, 98, 125, 129–33, 135–37, 139, 149 n. 50, 151 n. 66, 162–63, 166, 169, 170, 184 n. 13

and narrative, *see* narrative time

and otherness, *see* other/otherness and time

and succession 1, 5, 18, 19–20, 22, 26, 41 nn. 26 & 29, 43 n. 56, 59, 60, 72 n. 8, 75 n. 59, 96, 98, 99, 100, 101, 106, 151 n. 65, 162, 165, 183

totalization 8, 9, 16, 82, 104, 109, 111, 119, 125, 126, 128, 140, 141, 143, 144, 145 n. 2, 149 nn. 41 & 42 & 45, 154, 156, 193; *see also* objectification

tragedy/*le tragique* 28–29, 39, 44 n. 67, 61, 62, 70, 102, 155–56, 157 n. 6

Trainor, Paul 12 n. 37

transcendence 2, 16, 24, 31–32, 34, 38, 45 n. 75, 46 n. 92, 50–51, 55, 56–57, 59, 60, 61, 66, 70, 72 n. 8, 74 n. 37, 75 nn. 57 & 59, 76 n. 65, 78 n. 101, 98, 100, 108, 110, 128, 138, 141, 148 n. 34, 151 n. 63, 152 n. 83, 154, 163–64, 165, 166, 169, 173, 174–75, 176–78, 179, 182, 183, 185 n. 20, 186 n. 26, 187 n. 33, 188 nn. 50 & 51 & 54, 190 n. 72, 191 nn. 81 & 82 & 85, 192 n. 89

transcendental 71 nn. 1 & 4, 84 n. 2, 145 n. 1, 188 n. 52, 189 n. 59, 191 n. 85

Treanor, Brian 69, 142–43, 153 n. 91

Troisfontaines, Roger 5, 25, 72 n. 8, 102, 190 n. 70

Tsukada, Sumiyo 18, 26, 29, 38, 39 n. 1, 44 n. 66, 190 n. 70

Turetzky, Philip 111 n. 9, 167, 186 n. 26

uncertainty 137, 141, 171, 179

unity 16, 24–25, 30, 40 n. 8, 47, 51, 60, 64–65, 66, 67, 68, 76 n. 65, 77 n. 97, 80, 86, 89, 91–92, 95, 98–99, 101, 104, 109, 110, 116 n. 75, 117 n. 87, 135, 139–40, 157, 161, 162–63, 171, 175, 181, 185 n. 14, 189 n. 60, 191 nn. 81 & 85

with God 167, 168, 169, 186 n. 24

of Marcel's *œuvre* 7, 9–10, 102, 139, 144, 145, 146 nn. 7 & 13 & 17, 172, 193, 194; *see also* consistency

universal 12 n. 37, 15–16, 39 n. 5, 44 n. 71, 54, 60, 64, 72 n. 9, 75 n. 57, 95, 103, 105, 108, 115 n. 60, 117 n. 84, 139, 140–41, 178, 192 n. 90

unverifiability 31, 35, 95, 171, 178, 187 n. 39, 190 n. 77; *see also* ambiguity, mystery, uncertainty

value 27, 31, 37–38, 39 n. 6, 46 n. 93, 47, 50, 58, 60, 62, 63, 64, 65, 67, 69, 70, 75 nn. 56 & 57 & 60 & 65, 79, 97, 98–99, 113 n. 36, 116 n. 79, 117 n. 80, 140, 165–67, 170, 171, 178, 180, 187 n. 40

van den Hengel, John 115 n. 58

Vigorito, John 11 n. 21, 77 n. 86, 113 n. 33

violence 84 n. 4, 125, 128, 148 n. 39, 148–49 n. 41, 149 nn. 45 & 46, 150 n. 58, 160

Westphal, Merold 176–77

Wetzel, James 111 n. 8, 184 n. 1

Widmer, Charles 84 n. 1, 169, 187 n. 39

Wood, David 52, 65, 67, 79, 81, 84 n. 2, 144

World War I viii, 8, 11 n. 33, 16, 21, 29, 53, 102, 130, 143, 146–47 n. 20, 148 n. 35, 155, 169

World War II 4, 41 n. 19, 123, 124, 136, 148 n. 35

Zahavi, Dan 43 nn. 52 & 53, 84 n. 2, 103